CRC Handbook
of
Materials Science

Volume II: Metals, Composites, and Refractory Materials

Editor

Charles T. Lynch, Ph.D.

Senior Scientist for Environmental Effects,
Metals Behavior Branch, USAF
Air Force Wright Aeronautical Materials Laboratory
Wright-Patterson Air Force Base, Ohio

CRC Press, Inc.
Boca Raton, Florida

CRC Handbook of Materials Science

Volume I: General Properties

Volume II: Metals, Composites, and Refractory Materials

Volume III: Nonmetallic Materials and Applications

Volume IV: Wood

Library of Congress Cataloging in Publication Data

Lynch, Charles T.
  CRC handbook of materials science.

  Vol. 2 includes bibliography and index.
  CONTENTS:
—v.2. Metals, composites, and refractory materials.
  1. Materials — Handbooks, manuals, etc.
TA403.4.L94   620.1'12   73-90240
IBSN-0-87819-232-8(v.2)

Direct all inquiries to CRC Press, Inc., 2000 Corporate Blvd., N.W., Boca Raton, Florida, 33431.

© 1975 by CRC Press, Inc.
Second Printing, 1980
Third Printing, 1983
Fourth Printing, 1984
Fifth Printing, 1985
Sixth Printing, 1987

International Standard Book Number 0-87819-232-8 (Volume II)
International Standard Book Number 0-87819-234-4 (Complete Set)

Library of Congress Card Number 73-90240
Printed in the United States

PREFACE

It has been the goal of the *CRC Handbook of Materials Science* to provide a current and readily accessible guide to the physical properties of solid state and structural materials. Interdisciplinary in approach and content, it covers the broadest variety of types of materials consistent with a reasonable size for the volumes, including materials of present commercial importance plus new biomedical, composite, and laser materials. This volume, Metals, Composites, and Refractory Materials, is the second of the three-volume *Handbook*; General Properties is the first. Volume III, Nonmetallic Materials and Applications, also contains a section on materials information and referral publications, data banks, and general handbooks.

During the approximately four years that it has taken to formulate and compile this *Handbook*, the importance of materials science has taken on a new dimension. The term "materials limited" has come into new prominence, enlarged from the narrower consideration of technical performance of given materials in given conditions of stress, environment, and so on, to encompass the availability of materials in commerce at a reasonable price. Our highly industrialized society, with its immense per capita consumption of raw materials, today finds itself facing the long-prophesized shortages of materials in many diverse areas of our economy. Those future shortages have become today's problems. As we find ourselves "materials limited" with respect to availability and price, with a growing concern for where our raw materials come from and how supplies may be manipulated to our national disadvantage, increased economic utilization of all our resources, and particularly our materials resources, becomes an American necessity. With this changing background the purpose for this type of compilation has broadened beyond a collection of data on physical properties to one of concern for comparative properties and alternative employment of materials. Therefore, at this time it seems particularly appropriate to enter this new addition to the CRC Handbook Series.

Most of the information presented in this *Handbook* is in tabular format for easy reference and comparability of various properties. In some cases it has seemed more advisable to retain written sections, but these have been kept to a minimum. The importance of having critically evaluated property data available on materials to solve modern problems is well understood. In this *Handbook* we seek to bridge the gap between uncritical data collections carrying all the published information for a single material class and general reference works with only limited property and classification data on materials. On the basis of advice from many and varied sources, numerous limitations and omissions have been necessary to retain a reasonable size. This reference is particularly aimed at the nonexperts, or those who are experts in one field but seek information on materials in another field. The expert normally has his own specific original sources available to guide him in his own area of expertise. He often needs assistance, however, to get started on something new. There is also considerable general information of interest to almost all scientists, engineers, and many administrators in the field of materials and materials applications. Comments and suggestions, and the calling to our attention of typographical errors, will be welcomed and are encouraged.

My sincere thanks is extended to all who have advised on the formulation, content, and coverage of this *Handbook*. I am grateful to many colleagues in industry, academic circles, and government for countless suggestions and specific contributions, and am particularly indebted to the Advisory Board and Contributors who have put so much of their time, effort, and talent into this compilation. Special appreciation is extended to the editorial staff of CRC Press, to Karen A. Gajewski, the Administrative Editor, and to Gerald A. Becker, Director of Editorial Operations.

I want to pay special tribute to my wife, Betty Ann, for her magnificent patience, encouragement, and assistance, and to our children, Karen, Ted Jr., Richard, and Thomas, for giving their Dad some space, quiet, and assistance in the compilation of a considerable amount of data.

Charles T. Lynch
Fairborn, Ohio
December 1974

THE EDITOR

Charles T. Lynch, Ph.D., is Senior Scientist for Environmental Effects in the Metals Behavior Branch of the Air Force Wright Aeronautical Materials Laboratory, Wright-Patterson Air Force Base, Ohio.

Dr. Lynch graduated from The George Washington University in 1955 with a B.S. degree in chemistry. He received his M.S. and Ph.D. degrees in analytical chemistry in 1957 and 1960, respectively, from the University of Illinois, Urbana.

Dr. Lynch served in the Air Force for several years before joining the Air Force Materials Laboratory as a civilian employee in 1962. Prior to his current position, he served as a research engineer, group leader for ceramic research, and Chief of the Advanced Metallurgical Studies Branch.

Dr. Lynch is a member of the American Chemical Society, American Ceramic Society, American Association for the Advancement of Science, Ohio Academy of Science, New York Academy of Science, Sigma Xi-RESA, and the Metallurgical Society of the AIME. He holds 13 patents and has published more than 60 research papers, over 70 national and international presentations, and one book, *Metal Matrix Composites,* written with J. P. Kershaw and published by The Chemical Rubber Company (now CRC Press) in 1972.

CONTRIBUTORS

C. Howard Adams
SPI Research Associate
National Bureau of Standards
Washington, D.C. 20234

Allen M. Alper
Director of Research and Engineering
Chemical and Metallurgical Division
GTE Sylvania, Incorporated
Towanda, Pennsylvania 18848

Ray E. Bolz
Vice President and Dean of the Faculty
Worcester Polytechnic Institute
Worcester, Massachusetts 01609

Allen Brodsky
Radiological Physicist
Mercy Hospital
Pittsburgh, Pennsylvania 15219

D. F. Bunch
Atomics International
Canoga Park, California 91304

Donald E. Campbell
Senior Research Associate – Chemistry
Research and Development Division, Technical
  Staffs Services Laboratories
Corning Glass Works
Sullivan Park
Corning, New York 14830

William B. Cottrell
Director, Nuclear Safety Program
Oak Ridge National Laboratory
Oak Ridge, Tennessee 37830

Joseph E. Davison
Assistant Professor of Materials Engineering
University of Dayton
300 College Park
Dayton, Ohio 45409

Edward B. Fernsler (retired)
114 Willoughby Avenue
Huntington, West Virginia 25705

Francis S. Galasso
Chief, Materials Science
United Aircraft Research Laboratories
East Hartford, Connecticut 06108

Henry E. Hagy
Senior Research Associate – Physics
Research and Development Division, Technical
  Staffs Services Laboratories
Corning Glass Works
Sullivan Park
Corning, New York 14830

C. R. Hammond
Emhart Corporation
P.O. Box 1620
Hartford, Connecticut 06102

Michael Hoch
Professor, Department of Materials Science and
  Metallurgical Engineering
University of Cincinnati
Clifton Avenue
Cincinnati, Ohio 45221

Bernard Jaffe
Vernitron Piezoelectric Division
232 Forbes Road
Bedford, Ohio 44146

Richard N. Kleiner
Section Head, Ceramics Department
Precision Materials Group
Chemical and Metallurgical Division
GTE Sylvania, Incorporated
Towanda, Pennsylvania 18848

George W. Latimer, Jr.
Group Leader, Analytical Methods
Mead Johnson Company
Evansville, Indiana 47721

Robert I. Leininger
Project Director, Biomaterials
Biological, Ecological, and Medical Sciences
  Department
Battelle/Columbus Laboratories
505 King Avenue
Columbus, Ohio 43201

Robert S. Marvin
 Office of Standard Reference Data
 National Bureau of Standards
 U.S. Department of Commerce
 Washington, D.C. 20234

Eugene F. Murphy
 Director, Research Center for Prosthetics
 U.S. Veterans Administration
 252 Seventh Avenue
 New York, New York 10001

A. Pigeaud
 Research Associate
 Department of Metallurgy and Material
  Science
 University of Cincinnati
 Clifton Avenue
 Cincinnati, Ohio 45221

B. W. Roberts
 Director, Superconductive Materials Data
  Center
 General Electric Corporate Research and
  Development
 Box 8
 Schenectady, New York 12301

Gail D. Schmidt
 Chief, Radioactive Materials Branch
 Division of Radioactive Materials and Nuclear
  Medicine
 Bureau of Radiological Health
 U.S. Public Health Service
 Rockville, Maryland 20852

James E. Selle
 Senior Research Specialist
 Mound Laboratory
 Monsanto Research Corporation
 Miamisburg, Ohio 45342

Gertrude B. Sherwood
 Office of Standard Reference Data
 National Bureau of Standards
 U.S. Department of Commerce
 Washington, D.C. 20234

Ward F. Simmons
 Associate Director, Defense Metals Information
  Center
 Battelle/Columbus Laboratories
 505 King Avenue
 Columbus, Ohio 43201

George L. Tuve
 2625 Exeter Road
 Cleveland Heights, Ohio 44118

A. Bennett Wilson, Jr.
 Executive Director, Committee on Prosthetics
  Research and Development
 National Research Council
 Washington, D.C. 20037

TABLE OF CONTENTS
VOLUME II

Section 1

Metals

Note: For more extensive information and data on specific alloys the reader should refer to general references such as:

Metals Handbook Committee, *Metals Handbook,* Vol. 1, 8th ed., American Society for Metals, Metals Park, Oh., 1961.
Hauck, J. E., Ed., *1973 Materials Selector,* Reinhold, Stamford, Conn., 1972.
Smithells, C. J., *Metals Reference Book,* Vol. 1–3, 4th ed., Butterworths, London, 1967.

and the handbooks of commercial producers of metals such as steel, aluminum, etc.

1.1 FERROUS ALLOYS

Table 1.1–1
SAE ALLOY STEEL COMPOSITIONS

SAE No.	C	Mn	P	S	Si	Ni	Cr	Other	Corresponding AISI No.
					Composition[a], %				
1330	0.28–0.33	1.60–1.90	0.035	0.040	0.20–0.35	—	—	—	1330
1335	0.33–0.38	1.60–1.90	0.035	0.040	0.20–0.35	—	—	—	1335
1340	0.38–0.43	1.60–1.90	0.035	0.040	0.20–0.35	—	—	—	1340
1345	0.43–0.48	1.60–1.90	0.035	0.040	0.20–0.35	—	—	Mo	1345
4012	0.09–0.14	0.75–1.00	0.035	0.040	0.20–0.35	—	—	0.15–0.25	4012
4023	0.20–0.25	0.70–0.90	0.035	0.040	0.20–0.35	—	—	0.20–0.30	4023
4024	0.20–0.25	0.70–0.90	0.035	0.035–0.050	0.20–0.35	—	—	0.20–0.30	4024
4027	0.25–0.30	0.70–0.90	0.035	0.040	0.20–0.35	—	—	0.20–0.30	4027
4028	0.25–0.30	0.70–0.90	0.035	0.035–0.050	0.20–0.35	—	—	0.20–0.30	4028
4032	0.30–0.35	0.70–0.90	0.035	0.040	0.20–0.35	—	—	0.20–0.30	—
4037	0.35–0.40	0.70–0.90	0.035	0.040	0.20–0.35	—	—	0.20–0.30	4037
4042	0.40–0.45	0.70–0.90	0.035	0.040	0.20–0.35	—	—	0.20–0.30	—
4047	0.45–0.50	0.70–0.90	0.035	0.040	0.20–0.35	—	—	0.20–0.30	4047
4118	0.18–0.23	0.70–0.90	0.035	0.040	0.20–0.35	—	0.40–0.60	0.08–0.15	4118
4130	0.28–0.33	0.40–0.60	0.035	0.040	0.20–0.35	—	0.80–1.10	0.15–0.25	4130
4135	0.33–0.38	0.70–0.90	0.035	0.040	0.20–0.35	—	0.80–1.10	0.15–0.25	—
4137	0.35–0.40	0.70–0.90	0.035	0.040	0.20–0.35	—	0.80–1.10	0.15–0.25	4137
4140	0.38–0.43	0.75–1.00	0.035	0.040	0.20–0.35	—	0.80–1.10	0.15–0.25	4140
4142	0.40–0.45	0.75–1.00	0.035	0.040	0.20–0.35	—	0.80–1.10	0.15–0.25	4142
4145	0.43–0.48	0.75–1.00	0.035	0.040	0.20–0.35	—	0.80–1.10	0.15–0.25	4145
4147	0.45–0.50	0.75–1.00	0.035	0.040	0.20–0.35	—	0.80–1.10	0.15–0.25	4147
4150	0.48–0.53	0.75–1.00	0.035	0.040	0.20–0.35	—	0.80–1.10	0.15–0.25	4150
4161	0.56–0.64	0.75–1.00	0.035	0.040	0.20–0.35	—	0.70–0.90	0.25–0.35	4161
4320	0.17–0.22	0.45–0.65	0.035	0.040	0.20–0.35	1.65–2.00	0.40–0.60	0.20–0.30	4320
4340	0.38–0.43	0.60–0.80	0.035	0.040	0.20–0.35	1.65–2.00	0.70–0.90	0.20–0.30	4340
4419	0.18–0.23	0.45–0.65	0.035	0.040	0.20–0.35	—	—	0.45–0.60	4419
4422	0.20–0.25	0.70–0.90	0.035	0.040	0.20–0.35	—	—	0.35–0.45	—
4427	0.24–0.29	0.70–0.90	0.035	0.040	0.20–0.35	—	—	0.35–0.45	—

Table 1.1–1 (continued)
SAE ALLOY STEEL COMPOSITIONS

Composition[a], %

SAE No.	C	Mn	P	S	Si	Ni	Cr	Other	Corresponding AISI No.
4615	0.13–0.18	0.45–0.65	0.035	0.040	0.20–0.35	1.65–2.00	—	0.20–0.30	4615
4617	0.15–0.20	0.45–0.65	0.035	0.040	0.20–0.35	1.65–2.00	—	0.20–0.30	—
4620	0.17–0.22	0.45–0.65	0.035	0.040	0.20–0.35	1.65–2.00	—	0.20–0.30	4620
4621	0.18–0.23	0.70–0.90	0.035	0.040	0.20–0.35	1.65–2.00	—	0.20–0.30	4621
4626	0.24–0.29	0.45–0.65	0.035	0.04 max	0.20–0.35	0.70–1.00	—	0.15–0.25	4626
4718	0.16–0.21	0.70–0.90	—	—	—	0.90–1.20	0.35–0.55	0.30–0.40	4718
4720	0.17–0.22	0.50–0.70	0.035	0.040	0.20–0.35	0.90–1.20	0.35–0.55	0.15–0.25	4720
4815	0.13–0.18	0.40–0.60	0.035	0.040	0.20–0.35	3.25–3.75	—	0.20–0.30	4815
4817	0.15–0.20	0.40–0.60	0.035	0.040	0.20–0.35	3.25–3.75	—	0.20–0.30	4817
4820	0.18–0.23	0.50–0.70	0.035	0.040	0.20–0.35	3.25–3.75	—	0.20–0.30	4820
5015	0.12–0.17	0.30–0.50	0.035	0.040	0.20–0.35	—	0.30–0.50	—	5015
5046	0.43–0.48	0.75–1.00	0.035	0.040	0.20–0.35	—	0.20–0.35	—	—
5060	0.56–0.64	0.75–1.00	0.035	0-040	0.20–0.35	—	0.40–0.60	—	—
5115	0.13–0.18	0.70–0.90	0.035	0.040	0.20–0.35	—	0.70–0.90	—	
5120	0.17–0.22	0.70–0.90	0.035	0.040	0.20–0.35	—	0.70–0.90	—	5120
5130	0.28–0.33	0.70–0.90	0.035	0.040	0.20–0.35	—	0.80–1.10	—	5130
5132	0.30–0.35	0.60–0.80	0.035	0.040	0.20–0.35	—	0.75–1.00	—	5132
5135	0.33–0.38	0.60–0.80	0.035	0.040	0.20–0.35	—	0.80–1.05	—	5135
5140	0.38–0.43	0.70–0.90	0.035	0.040	0.20–0.35	—	0.70–0.90	—	5140
5145	0.43–0.48	0.70–0.90	0.035	0.040	0.20–0.35	—	0.70–0.90	—	5145
5147	0.46–0.51	0.70–0.95	0.035	0.040	0.20–0.35	—	0.85–1.15	—	5147
5150	0.48–0.53	0.70–0.90	0.035	0.040	0.20–0.35	—	0.70–0.90	—	5150
5155	0.51–0.59	0.70–0.90	0.035	0.040	0.20–0.35	—	0.70–0.90	—	5155
5160	0.56–0.64	0.75–1.00	0.035	0.040	0.20–0.35	—	0.70–0.90	—	5160
6118	0.16–0.21	0.50–0.70	0.035	0.040	0.20–0.35	—	0.50–0.70	0.10–0.15	6118
6150	0.48–0.53	0.70–0.90	0.035	0.040	0.20–0.35	—	0.80–1.10	0.15 Mo	6150
8115	0.13–0.18	0.70–0.90	0.035	0.040	0.20–0.35	0.20–0.40	0.30–0.50	0.08–0.15	8115
8615	0.13–0.18	0.70–0.90	0.035	0.040	0.20–0.35	0.40–0.70	0.40–0.60	0.15–0.25	8615
8617	0.15–0.20	0.70–0.90	0.035	0.040	0.20–0.35	0.40–0.70	0.40–0.60	0.15–0.25	8617
8620	0.18–0.23	0.70–0.90	0.035	0.040	0.20–0.35	0.40–0.70	0.40–0.60	0.15–0.25	8620

Table 1.1–1 (continued)
SAE ALLOY STEEL COMPOSITIONS

SAE No.	C	Mn	P	S	Si	Ni	Cr	Other	Corresponding AISI No.
8622	0.20–0.25	0.70–0.90	0.035	0.040	0.20–0.35	0.40–0.70	0.40–0.60	0.15–0.25	8622
8625	0.23–0.28	0.70–0.90	0.035	0.040	0.20–0.35	0.40–0.70	0.40–0.60	0.15–0.25	8625
8627	0.25–0.30	0.70–0.90	0.035	0.040	0.20–0.35	0.40–0.70	0.40–0.60	0.15–0.25	8627
8630	0.28–0.33	0.70–0.90	0.035	0.040	0.20–0.35	0.40–0.70	0.40–0.60	0.15–0.25	8630
8637	0.35–0.40	0.75–1.00	0.035	0.040	0.20–0.35	0.40–0.70	0.40–0.60	0.15–0.25	8637
8640	0.38–0.43	0.75–1.00	0.035	0.040	0.20–0.35	0.40–0.70	0.40–0.60	0.15–0.25	8640
8642	0.40–0.45	0.75–1.00	0.035	0.040	0.20–0.35	0.40–0.70	0.40–0.60	0.15–0.25	8642
8645	0.43–0.48	0.75–1.00	0.035	0.040	0.20–0.35	0.40–0.70	0.40–0.60	0.15–0.25	8645
8650	0.48–0.53	0.75–1.00	0.035	0.040	0.20–0.35	0.40–0.70	0.40–0.60	0.15–0.25	–
8655	0.51–0.59	0.75–1.00	0.035	0.040	0.20–0.35	0.40–0.70	0.40–0.60	0.15–0.25	8655
8660	0.56–0.64	0.75–1.00	0.035	0.040	0.20–0.35	0.40–0.70	0.40–0.60	0.15–0.25	–
8720	0.18–0.23	0.70–0.90	0.035	0.040	0.20–0.35	0.40–0.70	0.40–0.60	0.20–0.30	8720
8740	0.38–0.43	0.75–1.00	0.035	0.040	0.20–0.35	0.40–0.70	0.40–0.60	0.20–0.30	8740
8822	0.20–0.25	0.75–1.00	0.035	0.040	0.20–0.35	0.40–0.70	0.40–0.60	0.30–0.40	8822
9254	0.51–0.59	0.50–0.80	0.035	0.040	1.20–1.60	–	0.50–0.80	–	–
9255	0.51–0.59	0.70–0.95	0.035	0.040	1.80–2.20	–	–	–	9255
9260	0.56–0.64	0.75–1.00	0.035	0.040	1.80–2.20	–	–	–	9260
9310	0.08–0.13	0.45–0.65	0.025	0.025	0.20–0.35	3.00–3.50	1.00–1.40	0.08–0.15	–

Composition[a], %

[a]Small quantities of other elements are present which are acceptable to the following amounts: 0.35 Cu, 0.25 Ni, 0.20 Cr, and 0.06 Mo.

Table courtesy of American Iron and Steel Institute.

Table 1.1−2
AISI-SAE STANDARD CARBON STEELS

A. Free-machining Grades

Composition[a], %

AISI No.	C	Mn	P	S	SAE No.
			Resulfurized		
−	0.08−0.13	0.50−0.80	0.040 max	0.08−0.13	1108
1109	0.08−0.13	0.60−0.90	0.040 max	0.08−0.13	1109
1110	0.08−0.13	0.30−0.60	0.040 max	0.08−0.13	1110
1116	0.14−0.20	1.10−1.40	0.040 max	0.16−0.23	1116
1117	0.14−0.20	1.00−1.30	0.040 max	0.08−0.13	1117
1118	0.14−0.20	1.30−1.60	0.040 max	0.08−0.13	1118
1119	0.14−0.20	1.00−1.30	0.040 max	0.24−0.33	1119
1132	0.27−0.34	1.35−1.65	0.040 max	0.08−0.13	1132
1137	0.32−0.39	1.35−1.65	0.040 max	0.08−0.13	1137
1139	0.35−0.43	1.35−1.65	0.040 max	0.13−0.20	1139
1140	0.37−0.44	0.70−1.00	0.040 max	0.08−0.13	1140
1141	0.37−0.45	1.35−1.65	0.040 max	0.08−0.13	1141
1144	0.40−0.48	1.35−1.65	0.040 max	0.24−0.33	1144
1145	0.42−0.49	0.70−1.00	0.040 max	0.04−0.07	1145
1146	0.42−0.49	0.70−1.00	0.040 max	0.08−0.13	1146
1151	0.48−0.55	0.70−1.00	0.040 max	0.08−0.13	1151
		Resulfurized and Rephosphorized			
1211	0.13 max	0.60−0.90	0.07−0.12	0.10−0.15	1211
1212	0.13 max	0.70−1.00	0.07−0.12	0.16−0.23	1212
1213	0.13 max	0.70−1.00	0.07−0.12	0.24−0.33	1213
1215	0.09 max	0.75−1.05	0.04−0.09	0.26−0.35	1215
12L14	0.15 max	0.85−1.15	0.04−0.09	0.26−0.35	12L14

B. Nonresulfurized Grades

Composition[a], %

AISI No.	C	Mn	P max	S max	SAE No.
−	0.06 max	0.35 max	0.040	0.050	1005
1006	0.08 max	0.25−0.40	0.040	0.050	1006
1008	0.10 max	0.30−0.50	0.040	0.050	1008
1010	0.08−0.13	0.30−0.60	0.040	0.050	1010
−	0.08−0.13	0.60−0.90	0.040	0.050	1011
1012	0.10−0.15	0.30−0.60	0.040	0.050	1012
−	0.11−0.16	0.50−0.80	0.040	0.050	1013
1513	0.10−0.16	1.10−1.40	0.040	0.050	1513
1015	0.13−0.18	0.30−0.60	0.040	0.050	1015
1016	0.13−0.18	0.60−0.90	0.040	0.050	1016
1017	0.15−0.20	0.30−0.60	0.040	0.050	1017
1018	0.15−0.20	0.60−0.90	0.040	0.050	1018
1518	0.15−0.21	1.10−1.40	0.040	0.050	1518
1019	0.15−0.20	0.70−1.00	0.040	0.050	1019
1020	0.18−0.23	0.30−0.60	0.040	0.050	1020

Table 1.1–2 (continued)
AISI-SAE STANDARD CARBON STEELS

B. Nonresulfurized Grades (continued)

Composition[a], %

AISI No.	C	Mn	P max	S max	SAE No.
1022	0.18–0.23	0.70–1.00	0.040	0.050	1022
1522	0.18–0.24	1.10–1.40	0.040	0.050	1522
1023	0.20–0.25	0.30–0.60	0.040	0.050	1023
1524	0.19–0.25	1.35–1.65	0.040	0.050	1524
1025	0.22–0.28	0.30–0.60	0.040	0.050	1025
1525	0.23–0.29	0.80–1.10	0.040	0.050	1525
1026	0.22–0.28	0.60–0.90	0.040	0.050	1026
1526	0.22–0.29	1.10–1.40	0.040	0.050	1526
1527	0.22–0.29	1.20–1.50	0.040	0.050	1527
1030	0.28–0.34	0.60–0.90	0.040	0.050	1030
1035	0.32–0.38	0.60–0.90	0.040	0.050	1035
1536	0.30–0.37	1.20–1.50	0.040	0.050	1536
1037	0.32–0.38	0.70–1.00	0.040	0.050	1037
1038	0.35–0.42	0.60–0.90	0.040	0.050	1038
1039	0.37–0.44	0.70–1.00	0.040	0.050	1039
1040	0.37–0.44	0.60–0.90	0.040	0.050	1040
1541	0.36–0.44	1.35–1.65	0.040	0.050	1541
1042	0.40–0.47	0.60–0.90	0.040	0.050	1042
1044	0.43–0.50	0.30–0.60	0.040	0.050	1044
1045	0.43–0.50	0.60–0.90	0.040	0.050	1045
1046	0.43–0.50	0.70–1.00	0.040	0.050	1046
1547	0.45–0.51	1.35–1.65	0.040	0.050	1547
1548	0.44–0.52	1.10–1.40	0.040	0.050	1548
1049	0.46–0.53	0.60–0.90	0.040	0.050	1049
1050	0.48–0.55	0.60–0.90	0.040	0.050	1050
1551	0.45–0.56	0.85–1.15	0.040	0.050	1551
1552	0.47–0.55	1.20–1.50	0.040	0.050	1552
1053	0.48–0.55	0.70–1.00	0.040	0.050	1053
1055	0.50–0.60	0.60–0.90	0.040	0.050	1055
1060	0.55–0.65	0.60–0.90	0.040	0.050	1060
1561	0.55–0.65	0.75–1.05	0.040	0.050	1561
–	0.60–0.70	0.60–0.90	0.040	0.050	1065
1566	0.60–0.71	0.85–1.15	0.040	0.050	1566
–	0.65–0.75	0.40–0.70	0.040	0.050	1069
1070	0.65–0.75	0.60–0.90	0.040	0.050	1070
1572	0.65–0.76	1.00–1.30	0.040	0.050	1572
–	0.70–0.80	0.50–0.80	0.040	0.050	1074
–	0.70–0.80	0.40–0.70	0.040	0.050	1075
1078	0.72–0.85	0.30–0.60	0.040	0.050	1078
1080	0.75–0.88	0.60–0.90	0.040	0.050	1080

Table 1.1–2 (continued)
AISI-SAE STANDARD CARBON STEELS

B. Nonresulfurized Grades (continued)

Composition[a], %

AISI No.	C	Mn	P max	S max	SAE No.
1084	0.80–0.93	0.60–0.90	0.040	0.050	1084
–	0.80–0.93	0.70–1.00	0.040	0.050	1085
–	0.80–0.93	0.30–0.50	0.040	0.050	1086
1090	0.85–0.98	0.60–0.90	0.040	0.050	1090
1095	0.90–1.03	0.30–0.50	0.040	0.050	1095

[a]When silicon is required, it is added from 0.10 to 0.30%. In some steels, a silicon content of 0.30 to 2.20% can be specified. Silicon is not generally added to resulfurized or rephophorized steels.

Table courtesy of American Iron and Steel Institute.

Table 1.1–3
STANDARD TYPES OF STAINLESS AND HEAT-RESISTING STEELS[a]

Chemical Ranges and Limits

Chemical composition, % (maximum unless otherwise shown)

Type number	C	Mn max	P, max	S, max	Si, max	Cr	Ni	Mo	Zr	Se	Ti	Nb-Ta	Ta	Al	N
201[b]	0.15 max	5.50/ 7.50	0.060	0.030	1.00	16.00/ 18.00	3.50/ 5.50								0.25 max
202[b]	0.15 max	7.50/ 10.00	0.060	0.030	1.00	17.00/ 19.00	4.00/ 6.00								0.25 max
301[b]	0.15 max	2.00	0.045	0.030	1.00	16.00/ 18.00	6.00/ 8.00								
302[b]	0.15 max	2.00	0.045	0.030	1.00	17.00/ 19.00	8.00/ 10.00								
302B[b]	0.15 max	2.00	0.045	0.030	2.00/ 3.00	17.00/ 19.00	8.00/ 10.00								
303[b]	0.15 max	2.00	0.20	0.15 min	1.00	17.00/ 19.00	8.00/ 10.00	0.60 max[c]							
303Se[b]	0.15 max	2.00	0.20	0.06	1.00	17.00/ 19.00	8.00/ 10.00		0.60 max[c]	0.15 min					
304[b]	0.08 max	2.00	0.045	0.030	1.00	18.00/ 20.00	8.00/ 12.00								
304L[b]	0.03 max	2.00	0.045	0.030	1.00	18.00/ 20.00	8.00/ 12.00								
305[b]	0.12 max	2.00	0.045	0.030	1.00	17.00/ 19.00	10.00/ 13.00								
308[b]	0.08 max	2.00	0.045	0.030	1.00	19.00/ 21.00	10.00/ 12.00								
309[b]	0.20 max	2.00	0.045	0.030	1.00	22.00/ 24.00	12.00/ 15.00								
309S[b]	0.08 max	2.00	0.045	0.030	1.00	22.00/ 24.00	12.00/ 15.00								
310[b]	0.25 max	2.00	0.045	0.030	1.50	24.00/ 26.00	19.00/ 22.00								
310S[b]	0.08 max	2.00	0.045	0.030	1.50	24.00/ 26.00	19.00/ 22.00								
314[b]	0.25 max	2.00	0.045	0.030	1.50/ 3.00	23.00/ 26.00	19.00/ 22.00								
316[b]	0.08 max	2.00	0.045	0.030	1.00	16.00/ 18.00	10.00/ 14.00	2.00/ 3.00							

Table 1.1–3 (continued)
STANDARD TYPES OF STAINLESS AND HEAT-RESISTING STEELS

Chemical Ranges and Limits (continued)

Chemical composition, % (maximum unless otherwise shown)

Type number	C	Mn, max	P, max	S, max	Si, max	Cr	Ni	Mo	Zr	Se	Ti	Nb-Ta	Ta	Al	N
316L[b]	0.03 max	2.00	0.045	0.030	1.00	16.00/ 18.00	10.00/ 14.00	2.00/ 3.00							
317[b]	0.08 max	2.00	0.045	0.030	1.00	18.00/ 20.00	11.00/ 15.00	3.00/ 4.00							
321[b]	0.08 max	2.00	0.045	0.030	1.00	17.00/ 19.00	9.00/ 12.00				5×C min				
347[b]	0.08 max	2.00	0.045	0.030	1.00	17.00/ 19.00	9.00/ 13.00					10×C min			
348[b]	0.08 max	2.00	0.045	0.030	1.00	17.00/ 19.00	9.00/ 13.00					10×C min	0.10 max		
403[d]	0.15 max	1.00	0.040	0.030	0.50	11.50/ 13.00									
405[e]	0.08 max	1.00	0.040	0.030	1.00	11.50/ 14.50								0.10/ 0.30	
409	0.08 max	1.00	0.045	0.045	1.00	10.50/ 11.75					6×C min 0.75 max				
410[d]	0.15 max	1.00	0.040	0.030	1.00	11.50/ 13.50									
414[d]	0.15 max	1.00	0.040	0.030	1.00	11.50/ 13.50	1.25/ 2.50								
416[d]	0.15 max	1.25	0.06	0.15 min	1.00	12.00/ 14.00		0.60 max[c]							
416 Se[d]	0.15 max	1.25	0.06	0.06	1.00	12.00/ 14.00			0.60 max[c]	0.15 min					
420[d]	Over 0.15 max	1.00	0.040	0.030	1.00	12.00/ 14.00									
430[e]	0.12 max	1.00	0.040	0.030	1.00	14.00/ 18.00									
430F[e]	0.12 max	1.25	0.06	0.15 min	1.00	14.00/ 18.00		0.60 max[c]							
430F Se[e]	0.12 max	1.25	0.06	0.06	1.00	14.00/ 18.00			0.60 max[c]	0.15 min					
431[d]	0.20 max	1.00	0.040	0.030	1.00	15.00/ 17.00	1.25/ 2.50								

Table 1.1–3 (continued)
STANDARD TYPES OF STAINLESS AND HEAT-RESISTING STEELS

Chemical Ranges and Limits (continued)

Chemical composition, % (maximum unless otherwise shown)

Type number	C	Mn, max	P, max	S, max	Si, max	Cr	Ni	Mo	Zr	Se	Ti	Nb-Ta	Ta	Al	N
440A[d]	0.60/ 0.75	1.00	0.040	0.030	1.00	16.00/ 18.00		0.75 max							
440B[d]	0.75/ 0.95	1.00	0.040	0.030	1.00	16.00/ 18.00		0.75 max							
440C[d]	0.95/ 1.20	1.00	0.040	0.030	1.00	16.00/ 18.00		0.75 max							
446[e]	0.20 max	1.50	0.040	0.030	1.00	23.00/ 27.00									0.25 max
501[d]	Over 0.10	1.00	0.040	0.030	1.00	4.00/ 6.00		0.40/ 0.65							
502[d]	0.10 max	1.00	0.040	0.030	1.00	4.00/ 6.00		0.40/ 0.65							

[a]Subject to tolerances for check, product, or verification analyses.
[b]Not heat treatable.
[c]Added at producer's option.
[d]Heat treatable.
[e]Essentially not heat treatable.

Courtesy of American Iron and Steel Institute.

Table 1.1–4
GRAY IRONS – CAST

Class 20

Typical chemical composition, percent by weight: C, 3.5–3.8; Mn, 0.50–0.70; Fe, balance; S, 0.08–0.13; Si, 2.4–2.6; P, 0.2–0.8

Physical constants and thermal properties
Density, lb/in.^{3}: 0.25
Coefficient of thermal expansion, (70–200°F) in./in./°F × 10^{-6}: 6
Modulus of elasticity, psi: tension, 12 × 10^6
Melting point, °F: ~2150
Thermal conductivity, Btu/ft^2/hr/in./°F, 70°F: 336–360
Electrical resistivity, ohms/cmil/ft, 70°F: 300–1200

Short Time Tensile Properties

As Cast

Temperature,°F	T.S., psi	Y.S., psi, 0.2% offset	Elong., in 2 in., %	Hardness, Brinell
70	20,000–25,000	–	–	140–180

Fatigue Strength

As Cast

Test temperature, °F	Stress, psi
70	10,000 endurance limit

Table 1.1–5
GRAY IRONS – CAST

Class 40

Typical chemical composition, percent by weight: C, 3.0–3.2; Mn, 0.45–0.65; Fe, balance; S, 0.07–0.12; Si, 1.90–2.20; P, 0.10–0.25

Physical constants and thermal properties
Density, lb/in.^{3}: 0.26
Coefficient of thermal expansion, (70–200°F) in./in./°F × 10^{-6}: 6
Modulus of elasticity, psi: tension, 17 × 10^6
Melting point, °F: ~2150
Thermal conductivity, Btu/ft^2/hr/in./°F, 70°F: 336–360
Electrical resistivity, ohms/cmil/ft, 70°F: 300–1200

Short Time Tensile Properties

As Cast

Temperature, °F	T.S., psi	Y.S., psi, 0.2% offset	Elong., in 2 in., %	Hardness, Brinell
70	40,000–48,000	–	–	200–240

Table 1.1–6
GRAY IRONS – CAST

Class 60

Typical chemical composition, percent by weight: C, 2.7–3.0; Mn, 0.5–0.7; Fe, balance; S, 0.06–0.12; Si, 1.9–2.2; P, 0.1–0.2

Physical constants and thermal properties
Density, lb/in.^{3}: 0.27
Coefficient of thermal expansion, (70–200°F) in./in./°F × 10^{-6}: 6
Modulus of elasticity, psi: tension, 20 × 10^6
Melting point, °F: ~ 2150
Thermal conductivity, Btu/ft^2/hr/in./°F, 70°F: 336–360
Electrical resistivity, ohms/cmil/ft, 70°F: 300–1200

Short Time Tensile Properties

As Cast

Temperature, °F	T.S., psi	Y.S., psi, 0.2% offset	Elong., in 2 in., %	Hardness, Brinell
70	60,000–66,000	–	–	230–290

Table 1.1−7
MALLEABLE IRONS − CAST

Ferritic 32510

Typical chemical composition, percent by weight: C, 2.3−2.7; Mn, 0.55 max; Fe, balance; S, < 0.15; Si, 1.5−0.8; P, < 0.18

Physical constants and thermal properties
Density, lb/in.^{3}: 0.25−0.263
Coefficient of thermal expansion, (70−200°F) in./in./°F × 10^{-6}: 5.9
Modulus of elasticity, psi: tension, 25 × 10^{-6}
Poisson's ratio: 0.17
Melting point, °F: ∼ 2250
Thermal conductivity, Btu/ft^2/hr/in./°F, 70°F: 354
Electrical resistivity, ohms/cmil/ft, 70°F: 84.6
Curie temperature, °F: annealed, 2300

Short Time Tensile Properties

As Cast

Temperature, °F	T.S., psi	Y.S., psi, 0.2% offset	Elong., in 2 in., %	Hardness, Brinell
70	50,000	32,500	10	110−156

Fatigue Strength

As Cast

Test temperature, °F	Stress, psi
70	28,000 endurance limit

Impact Strength

As Cast

Test temperature, °F	Type test	Strength, ft-lb
−70	Charpy − V-notched[a]	16.5
+70	Charpy − unnotched	70−90

[a]0.394-in. square bar, 0.079-in. notch.

Table 1.1–8
MALLEABLE IRONS – CAST

Pearlitic 45010

Typical chemical composition, percent by weight: C,
2.35–2.50; Mn, 0.28–0.48; Fe, balance; S,
0.16–0.20; Si, 1.0–1.5; P, 0.06–0.10

Physical constants and thermal properties
Density, lb/in.^{3}: 0.265–0.268
Coefficient of thermal expansion, (70–200°F) in./in./
°F × 10^{-6}: 7.5
Modulus of elasticity, psi: tension, 26 × 10^{-6}
Poisson's ratio: 0.17
Thermal conductivity, Btu/ft^2/hr/in./°F, 70°F: 354
Electrical resistivity, ohms/cmil/ft,70°F: 95.4
Curie temperature, °F: annealed, 430

Short Time Tensile Properties

As Cast

Temperature, °F	T.S., psi	Y.S., psi, 0.2% offset	Elong., in 2 in., %	Hardness, Brinell
70	65,000	45,000	10	163–207

Fatigue Strength

As Cast

Test temperature, °F	Stress, psi
70	28,000 endurance limit

Impact Strength

As Cast

Test temperature, °F	Type test	Strength, ft-lb
–70	Charpy – V-notched[a]	14
+70	Charpy – unnotched	22–35

[a]0.394-in. square bar, 0.079-in notch.

Table 1.1–9
MALLEABLE IRONS – CAST

Pearlitic 48004

Typical chemical composition, percent by weight: C, 2.35–2.5; Mn, 0.28–0.48 Fe, balance; S, 0.16–0.20; Si, 1.0–1.5; P, 0.06–0.10

Physical constants and thermal properties
Density, lb/in.^{3}: 0.265–0.268
Coefficient of thermal expansion, (70–200°F) in./in./ °F × 10^{-6}: 7.5
Modulus of elasticity, psi: tension, 26.5 × 10^{-6}
Poisson's ratio: 0.17
Thermal conductivity, Btu/ft^2/hr/in./°F, 70°F: 354
Electrical resistivity, ohms/cmil/ft, 70°F: 95.4
Curie temperature, °F: annealed, 430

Short Time Tensile Properties

As Cast

Temperature, °F	T.S., psi	Y.S., psi, 0.2% offset	Elong., in 2 in., %	Hardness, Brinell
70	70,000	48,000	4	163–228

Fatigue Strength

As Cast

Test temperature, °F	Stress, psi
70	35,000 endurance limit

Impact Strength

As Cast

Test temperature, °F	Type test	Strength, ft-lb
–70	Charpy – V-notched[a]	14
+70	Charpy – unnotched	22–35

[a]0.394-in. square bar, 0.079-in. notch.

Table 1.1–10
MALLEABLE IRONS – CAST

Pearlitic 60003

Typical chemical composition, percent by weight: C, 2.35–2.5; Mn, 0.28–0.48; Fe, balance; S, 0.16–0.20; Si, 1.0–1.5; P, 0.06–0.10

Physical constants and thermal properties
Density, lb/in.^{3}: 0.265–0.268
Coefficient of thermal expansion, (70–200°F) in./in./ °F × 10^{-6}: 7.5
Modulus of elasticity, psi: tension, 26–28 × 10^{-6}
Poisson's ratio: 0.17
Thermal conductivity, Btu/ft^2/hr/in./°F, 70°F: 354
Electrical resistivity, ohms/cmil/ft, 70°F: 99
Curie temperature, °F: annealed, 170

Short Time Tensile Properties

As Cast

Temperature, °F	T.S., psi	Y.S., psi, 0.2% offset	Elong., in 2 in., %	Hardness, Brinell
70	80,000	60,000	3	197–255

Fatigue Strength

As Cast

Test temperature, °F	Stress, psi
70	39,000 endurance limit

Impact Strength

As Cast

Test temperature, °F	Type test	Strength, ft-lb
–70	Charpy – V-notched[a]	14
+70	Charpy – unnotched	22–35

[a]0.394-in. square bar, 0.079-in. notch.

Table 1.1—11
MALLEABLE IRONS — CAST

Pearlitic 80002

Typical chemical composition, percent by weight: C, 2.35—2.5; Mn, 0.28—0.48; Fe, balance; S, 0.16—0.20; Si, 1.0—1.5; P, 0.06—0.10

Physical constants and thermal properties
Density, lb/in.^{3}: 0.265—0.268
Coefficient of thermal expansion, (70—200° F) in./in./°F X 10^{-6}: 7.5
Modulus of elasticity, psi: tension, 26—28 X 10^{-6}
Poisson's ratio: 0.17
Thermal conductivity, Btu/ft^2/hr/in./°F, 70° F: 354
Electrical resistivity, ohms/cmil/ft, 70° F: 97.8
Curie temperature, °F: annealed, 170

Short Time Tensile Properties

As Cast

Temperature, °F	T.S., psi	Y.S., psi, 0.2% offset	Elong., in 2 in., %	Hardness, Brinell
70	100,000	80,000	2	241—269

Fatigue Strength

As Cast

Test temperature, °F	Stress, psi
70	35,000 endurance limit

Impact Strength

As Cast

Test temperature, °F	Type test	Strength, ft-lb
+70	Charpy — V-notched[a]	14
+70	Charpy — unnotched	22—35

[a]0.394-in. square bar, 0.079-in. notch.

Table 1.1–12
NODULAR OR DUCTILE IRONS

Cast 80-55-06

Typical chemical composition, percent by weight: Mn, 0.2–0.5; Fe, balance; Si, 2.0–3.0; Ni, 0–1.0; P, 0.06–0.08; Mg, 0.02–0.07

Physical constants and thermal properties
Density, lb/in.^{3} : 0.257
Coefficient of thermal expansion, (70–200°F) in./in./°F × 10^{-6} : 6.6
Modulus of elasticity, psi: tension, 22–25 × 10^6
Melting range, °F: 2050–2150
Thermal conductivity, Btu/ft^2/hr/in./°F, 70°F: 216

Electrical resistivity, ohms/cmil/ft, 70°F: 408

Heat treatments
Ferritic – heat in 1600–1650°F range, cool to 1300°F, hold 1–3 hr, furnace cool to 1100°F, air cool.
Pearlitic-ferritic – heat to 1600–1650°F, cool rapidly, reheat at 1100–1300°F.
Pearlitic – air cool from 1600–1650°F.
Tempered – quench from 1400–1650°F, temper at 800–1300°F.

Short Time Tensile Properties

As Cast

Temperature, °F	T.S., psi	Y.S., psi, 0.2% offset	Elong., in 2 in., %	Hardness, Brinell
70	90,000–110,000	60,000–75,000	3–10	179–255

Creep Strength
(Stress, psi, to Produce 0.0001%/hr Creep)

As Cast

Test temperature, °F	10,000 hr
1,000	1,750
1,200	480

Fatigue Strength

As Cast

Endurance ratio similar to other ferrous alloys. Notched endurance ratio from 30–40% for strong irons, up to 50–55% for soft irons.

Impact Strength

As Cast

Test temperature, °F	Type test	Strength, ft-lb
+70	Charpy – notched	2–5
+70	Charpy – unnotched	15–65

Table 1.1–13
NODULAR OR DUCTILE IRONS

Cast 100-70-03

Typical chemical composition, percent by weight: Mn, 0.3–0.6; Fe, balance; Si, 2.0–2.75; Ni, 0–2.5; Mo, 0–1.0; P, 0.08; Mg, 0.02–0.07

Physical constants and thermal properties
Density, lb/in.^{3} : 0.257
Coefficient of thermal expansion, (70–200°F) in./in./°F × 10^{-6} : 6.6
Modulus of elasticity, psi: tension, 22–25 × 10^6
Melting range, °F: 2050–2150

Heat treatments
Normalize, quench, and temper.
Ferritic – heat in 1600–1650°F range, cool to 1300°F, hold 1–3 hr, furnace cool to 1100°F, air cool.
Pearlitic-ferritic – heat to 1600–1650°F, cool rapidly, reheat at 1100–1300°F.
Pearlitic – air cool from 1600–1650°F.
Tempered – quench from 1400–1650°F; temper at 800–1300°F.

Short Time Tensile Properties

Temperature, °F	T.S., psi	Y.S., psi, 0.2% offset	Elong., in 2 in., %	Hardness, Brinell
70	100,000–120,000	75,000–90,000	6–10	229–285

Fatigue Strength

Endurance ratio similar to other ferrous alloys. Notched endurance ratio from 30–40% for strong irons, up to 50–55% for soft irons.

Impact Strength

Test temperature, °F	Type test	Strength, ft-lb
+70	Charpy – unnotched	35–50

Table 1.1−14
NODULAR OR DUCTILE IRONS − CAST

Heat Resistant

Typical chemical composition, percent by weight: Mn, 0.2−0.6; Fe, balance; Si, 2.5−6.0; Ni, 0−1.5; P, 0.08; Mg, 0.02−0.07

Physical constants and thermal properties
Density, lb/in.^{3}: 0.25
Modulus of elasticity, psi: tension, 22−25 × 10^6
Melting range, °F: 2050−2150

Heat treatments
Ferritic − heat in 1600−1650°F range, cool to 1300°F, hold 1−3 hr, furnace cool to 1100°F, air cool.
Pearlitic-ferritic − heat to 1600−1650°F, cool rapidly, reheat at 1100−1300°F.
Pearlitic − air cool from 1600−1650°F.
Tempered − quench from 1400−1650°F, temper at 800−1300°F.

Short Time Tensile Properties

As Cast

Temperature, °F	T.S., psi	Y.S., psi, 0.2% offset	Elong., in 2 in., %	Hardness, Brinell
70	60,000−100,000	45,000−75,000	0−20	140−300

Fatigue Strength

Endurance ratio similar to other ferrous alloys. Notched endurance ratio from 30−40% for strong irons, up to 50−55% for soft irons.

Impact Strength

Test temperature, °F	Type test	Strength, ft-lb
+70	Charpy − unnotched	5−115

Table 1.1–15
NODULAR OR DUCTILE IRONS – CAST

Austenitic

Typical chemical composition, percent by weight: Mn, 0.8–1.5; Fe, balance; Si, 2.0–3.2; Cr, 0–2.5; Ni, 18–22; P, 0.02

Physical constants and thermal properties
Density, lb/in.^{3}: 0.268
Coefficient of thermal expansion, (70–200°F) in./in./°F × 10^{-6}: 10.4
Modulus of elasticity, tension, 18.5 × 10^6
Melting range, °F: 2250
Electrical resistivity, ohms/cmil/ft, 70°F: 612

Short Time Tensile Properties

Temperature, °F	T.S., psi	Y.S., psi, 0.2% offset	Elong., in 2 in., %	Hardness, Brinell
70	58,000–68,000	32,000–38,000	7–40[a]	140–200

Creep Strength
(Stress, psi, to Produce 0.0001%/hr Creep)

Test temperature, °F	10,000 hr
1,000	13,000
1,200	5,700
1,400	2,000

Fatigue Strength

Endurance ratio similar to other ferrous alloys. Notched endurance ratio from 30–40% for strong irons, up to 50–55% for soft irons.

Impact Strength

Test temperature, °F	Type test	Strength, ft-lb
+70	Charpy – notched	10–28

[a]25–40% with 0% chromium.

Table 1.1–16
WHITE AND ALLOY IRONS – CAST;
WEAR AND ABRASION RESISTANT TYPES

White Iron

Typical chemical composition, percent by weight: C,
2.8–3.6; Mn, 0.4–0.9; Fe, balance; Si, 0.5–1.3

Physical constants and thermal properties
Density, lb/in.^{3}: 0.274–0.281
Coefficient of thermal expansion, (70–200°F) in./in./
°F × 10^{-6}: 5.0–5.3
Electrical resistivity, ohms/cmil/ft, 70°F: 30

Short Time Tensile Properties

Temperature, °F	T.S., psi	Y.S., psi, 0.2% offset	Elong., in 2 in., %	Hardness, Brinell
70	20,000–50,000	–	–	300–575

Impact Strength

Test temperature, °F	Type test	Strength, ft-lb
+70	Charpy	3.5–10.0

Table 1.1–17
WHITE AND ALLOY IRONS – CAST;
WEAR AND ABRASION RESISTANT TYPES

Ni-Hard

Typical chemical composition, percent by weight: C,
2.8–3.6; Mn, 0.4–0.9; Fe, balance; Si, 0.5–1.3

Physical constants and thermal properties
Density, lb/in.^{3}: 0.257–0.280
Coefficient of thermal expansion, (70–200°F) in./in./
°F × 10^{-6}: 4.5–5.0
Electrical resistivity, ohms/cmil/ft, 70°F: 480

Short Time Tensile Properties

Temperature, °F	T.S., psi	Y.S., psi, 0.2% offset	Elong., in 2 in., %	Hardness, Brinell
70	40,000–75,000	–	–	525–600

Impact Strength

Test temperature, °F	Type test	Strength, ft-lb
+70	Charpy	20–55

Table 1.1–18
WHITE AND ALLOY IRONS – CAST;
WEAR AND ABRASION RESISTANT TYPES

High Chromium

Typical chemical composition, percent by weight: C,
1.8–3.5; Mn, 0.3–1.0; Fe, balance; Si, 0.5–2.5; Cr,
10–35; Ni, 0–5.0; Mo, 0–3.0; Cu, 0–3.0

Physical constants and thermal properties
Density, lb/in.^{3}: 0.264–0.280
Coefficient of thermal expansion, (70–200°F) in./in./
°F × 10^{-6}: 5.2–5.5

Short Time Tensile Properties

Temperature, °F	T.S., psi	Y.S., psi, 0.2% offset	Elong., in 2 in., %	Hardness, Brinell
70	32,000–90,000	–	–	250–700

Impact Strength

Test temperature, °F	Type test	Strength, ft-lb
+70	Charpy	10–35

Table 1.1–19
WHITE AND ALLOY IRONS – CAST;
HEAT AND CORROSION RESISTANT TYPES

High Silicon (Duriron$^®$)

Typical chemical composition, percent by weight: C,
0.4–1.0; Mn, 0.4–1.0; Fe, balance; Si, 14–17; Mo,
0–3.5

Physical constants and thermal properties
Density, lb/in.^{3}: 0.252–0.254
Coefficient of thermal expansion, (70–200°F) in./in./
°F × 10^{-6}: 6.70

Short Time Tensile Properties

Temperature, °F	T.S., psi	Y.S., psi, 0.2% offset	Elong., in 2 in., %	Hardness, Brinell
70	13,000–18,000	–	–	450–500

Impact Strength

Test temperature, °F	Type test	Strength, ft-lb
+70	Charpy	2–4

Table 1.1–20
WHITE AND ALLOY IRONS – CAST;
HEAT AND CORROSION RESISTANT TYPES

High Silicon (Silal)

Typical chemical composition, percent by weight: C, 1.6–2.5; Mn, 0.4–0.8; Fe, balance; Si, 4.0–6.0

Physical constants and thermal properties
Density, lb/in.^{3}: 0.245–0.255
Coefficient of thermal expansion, (70–200°F) in./in./ °F × 10^{-6}: 6.0

Short Time Tensile Properties

Temperature, °F	T.S., psi	Y.S., psi, 0.2% offset	Elong., in 2 in., %	Hardness, Brinell
70	25,000–45,000	–	–	170–250

Impact Strength

Test temperature, °F	Type test	Strength, ft-lb
+70	Charpy	15–23

Table 1.1–21
WHITE AND ALLOY IRONS – CAST;
HEAT AND CORROSION RESISTANT TYPES

High Nickel (Ni-Resist)[a]

Typical chemical composition, percent by weight:
C, 1.8–3.0; Mn, 0.4–1.5; Fe, balance; Si, 1.0–2.75; Cr, 0.5–5.5; Ni, 14–30; Mo, 0–1.0: Cu, 0–7.0

Physical constants and thermal properties
Density, lb/in.^{3}: 0.264–0.270
Coefficient of thermal expansion, (70–200°F) in./in./ °F × 10^{-6}: 4.5–10.7

Short Time Tensile Properties

Temperature, °F	T.S., psi	Y.S., psi, 0.2% offset	Elong., in 2 in., %	Hardness, Brinell
70	25,000–45,000	–	–	130–250

Impact Strength

Test temperature, °F	Type test	Strength, ft-lb
+70	Charpy – unnotched	60–150

[a]High nickel alloy cast irons can be produced with either a low or high coefficient of thermal expansion. Special purpose grades are available which combine high electrical resistance with nonmagnetic properties.

Table 1.1–22
WHITE AND ALLOY IRONS – CAST;
HEAT AND CORROSION RESISTANT TYPES

High Aluminum

Typical chemical composition, percent by weight:
C, 1.3–1.7; Mn, 0.4–1.0; Fe, balance; Si, 1.3–6.0; Al, 18–25

Physical constants and thermal properties
Density, lb/in.^{3}: 0.200–0.232
Coefficient of thermal expansion, (70–200° F) in./in./ °F × 10^{-6}: 8.5

Short Time Tensile Properties

Temperature, °F	T.S., psi	Y.S., psi, 0.2% offset	Elong., in 2 in.,%	Hardness, Brinell
70	34,000–90,000	–	–	180–350

Table 1.1–23
IRON-BASE SUPERALLOYS – CAST, WROUGHT

Type A-286

Typical chemical composition, percent by weight:
C, 0.08; Mn, 1.35; Fe, balance; Si, 0.70; Cr, 15.00; Ni, 26.0; Mo, 1.25; Ti, 2.15; Al, 0.20; B, 0.003; V, 0.30

Physical constants and thermal properties
Density, lb/in.^{3}: 0.286
Coefficient of thermal expansion, (70–200° F) in./in./ °F × 10^{-6}: 10.3

Modulus of elasticity, psi: tension, 29.1 × 10^6
Melting range, °F: 2500–2600
Specific heat, Btu/lb/° F, 70° F: 0.10–0.11
Thermal conductivity, Btu/ft^2/hr/in./° F, 70° F: 164.4
Curie temperature, °F: age hardened, 1325 (18 hr)

Heat treatments
1 hr at 1800° F, oil quench, age 18 hr at 1325° F, air cool.

Short Time Tensile Properties

Temperature, °F	T.S., psi	Y.S., psi, 0.2% offset	Elong., in 2 in.,%	Hardness, Brinell
70	146,000	100,000	25	–
1200	104,000	88,000	13	–
1400	64,000	62,000	19	–
1500	37,000	–	69	–

Rupture Strength, 1000 hr

Test temperature, °F	Strength, psi	Elong., in 2 in.,%	Reduction of area,%
1350	21	–	–

Table 1.1–23 (continued)
IRON-BASE SUPERALLOYS – CAST, WROUGHT

Type A-286 (continued)

Creep Strength (Stress, psi, to Produce 0.0001%/hr Creep)

Test temperature, °F	10,000 hr
1200	30,000
1350	16,000

Fatigue Strength

Test temperature, °F	Stress, psi	Cycles to failure
1200	38,000	10^8

Impact Strength

Test Temperature, °F	Type test	Strength, ft-lb
+ 70	Charpy	64
1000	Charpy	46

Table 1.1–24
IRON-BASE SUPERALLOYS – CAST, WROUGHT

Type V-57

Typical chemical composition, percent by weight: C, 0.06; Mn 0.25; Fe, balance; Si, 0.55; Cr, 15.0; Ni, 25.5; Mo, 1.25; Ti, 3.0; Al, 0.25; B, 0.008; V, 0.02–5

Physical constants and thermal properties
Density, lb/in.³ : 0.286

Coefficient of thermal expansion, (70–200°F) in./in./°F × 10^{-6} : 10.5
Thermal conductivity, Btu/ft² /hr/in./°F, 70°F: 180

Heat treatments
2 hr at 1800°F, oil quench, age 16 hr at 1350°F, air cool.

Short Time Tensile Properties

Temperature, °F	T.S., psi	Y.S., psi, 0.2% offset	Elong., in 2 in.,%	Hardness, Brinell
70	172,000	119,000	24	–
1200	129,000	108,000	22	–
1400	95,000	89,000 (1350°F)	23 (1350°F)	–
1500	60,000	49,000	40	–

Rupture Strength, 1000 hr

Test temperature, °F	Strength, psi	Elong., in 2 in., %	Reduction of area,%
1350	25,000	–	–

Table 1.1—25
IRON-BASE SUPERALLOYS — CAST, WROUGHT

Type 16-25-6

Typical chemical composition, percent by weight: C, 0.10; Mn, 2.0; Fe, balance; Si, 1.0; Cr, 16.25; Ni, 25.5; Mo, 6.0

Modulus of elasticity, psi: tension, 28.2×10^6
Thermal conductivity, Btu/ft^2/hr/in./$^\circ$F, 70°F: 180

Physical constants and thermal properties
Density, lb/in.^{3}: 0.291
Coefficient of thermal expansion (70–200°F) in./in./$^\circ$F $\times 10^{-6}$: 9.4

Heat treatments
Bar solution treat 10 min at 2150°F, water quench, cold work 20%, stress relieve 4 hr at 1250°F.

Short Time Tensile Properties

Temperature, $^\circ$F	T.S., psi	Y.S., psi 0.2% offset	Elong., in 2 in., %	Hardness, Brinell
70	142,000	112,000	23	–
1200	90,000	75,000	12	–
1400	60,000	50,000	11	–
1500	47,000	37,000	9	–

Rupture Strength, 1000 hr

Test temperature, $^\circ$F	Strength, psi	Elong., in 2 in., %	Reduction of area, %
1350	21,000	–	–

Creep Strength (Stress, psi, to Produce 0.001%/hr Creep)

Test temperature, $^\circ$F	10,000 hr
1200	19,000
1350	13,000

Fatigue Strength

Test temperature, $^\circ$F	Stress, psi	Cycles to failure
1200	46,000	10^8

Impact Strength

Test temperature, $^\circ$F	Type test	Strength, ft-lb
+70	Charpy	15
1000	Charpy	50 (1500°F)

Table 1.1—26
IRON-BASE SUPERALLOYS — WROUGHT

Type 19-9DL

Typical chemical composition, percent by weight: C, 0.32; Mn, 1.15; Fe, balance; Si, 0.55; Cr, 18.5; Ni, 9.0; Mo, 1.40; W, 1.35; Cb + Ta, 0.40; Ti, 0.25; Cu, 0.15

Physical constants and thermal properties
Density, lb/in.^{3} : 0.287

Coefficient of thermal expansion, (70—200°F) in./in./°F $\times$ 10^{-6} : 10
Modulus of elasticity, psi: tension, 29.5 $\times$ 10^6
Melting range, °F: 2560—2615
Specific heat, Btu/lb/°F, 70°F: 0.10
Thermal conductivity, Btu/ft^2/hr/in./°F, 70°F: 146.4

Short Time Tensile Properties

Temperature, °F	T.S., psi	Y.S., psi, 0.2% offset	Hardness, Brinell
70	114,000	71,000	—
1000	79,000	55,000	—
1200	62,000	52,000	—
1400	50,000	40,000	—

Rupture Strength, 1000 hr

Test temperature, °F	Strength, psi	Elong., in 2 in., %	Reduction of area, %
1200	44,000	—	—

Fatigue Strength

Test temperature, °F	Stress, psi	Cycles to failure
70	81,000	10^8
1200	52,000	10^8

Impact Strength

Test temperature, °F	Type test	Strength, ft-lb
+70	Charpy	46

Specifications

SAE-AMS: 5526, 5527, 5720, 5722, 5721

Table 1.1–27
IRON-BASE SUPERALLOYS – WROUGHT

Type AMS 5700

Typical chemical composition, percent by weight: C, 0.45; Mn, 0.70; Fe, balance; S, 0.025; Si, 0.60; Cr, 14.00; Ni, 14.00; Mo, 0.50; W, 2.50; P, 0.030
Physical constants and thermal properties
  Density, $lb/in.^3$: 0.29

Coefficient of thermal expansion, $(70-200°F)$ $in./in./°F \times 10^{-6}$: 10
Modulus of elasticity, psi: tension, 30.0×10^6
Melting range, $°F$: 2550–2600
Thermal conductivity, $Btu/ft^2/hr/in./°F$, $70°F$: 126

Short Time Tensile Properties

Temperature, °F	T.S., psi	Y.S., psi, 0.2% offset	Elong., in 2 in., %	Hardness, Brinell
70	110,000	48,000	40	216
1000	–	–	–	–
1200	75,000	34,000	35	–
1400	46,000	31,000	47	–

Rupture Strength, 100 hr

Test temperature, °F	Strength, psi	Elong., in 2 in., %	Reduction of area, %
1350	19,000	–	–

Impact Strength

Test temperature, °F	Type test	Strength, ft-lb
+70	Charpy	38

Table 1.1–28
IRON-BASE SUPERALLOYS – CAST, WROUGHT

Multimet®, N-155

Typical chemical composition, percent by weight: C, 0.10; Mn, 1.50; Fe, balance; Si, 0.70; Cr, 20.75; Ni, 19.85; Co, 19.50; Mo, 2.95; W, 2.35; Cb + Ta, 1.15; Cu, 0.20

Coefficient of thermal expansion, (70–200°F) in./in./°F × 10^{-6}
Modulus of elasticity, psi: tension, 28.8 × 10^6
Melting range, °F: 2350–2470
Specific heat, Btu/lb/°F, 70°F: 0.104

Physical constants and thermal properties
Density, lb/in.³: 0.296

Heat treatments
Treat 1 hr at 2200 F°, rapid air cool.

Short Time Tensile Properties

Temperature, °F	T.S., psi	Y.S., psi, 0.2% offset	Elong., in 2 in., %	Hardness, Brinell
70	118,000	58,000	49	–
1000	94,000	40,000	54	–
1200	74,000	38,000	28	–
1400	59,000	36,000	12	–
1600	39,000	30,000	15	–

Rupture Strength, 1000 hr

Test temperature, °F	Strength, psi	Elong., in 2 in.,%	Reduction of area,%
1350	24,000	–	–

Creep Strength
(Stress, psi, to Produce 1% Creep)

Test temperature, °F	10,000 hr
1350	15,000

Fatigue Strength

Aged 50 hr at 1200°F

Test temperature, °F	Stress, psi	Cycles to failure
1200	66,000	10^8
1500	33,000	10^8

Impact Strength

Test temperature,°F	Type test	Strength, ft-lb
+ 70	Charpy	65

Table 1.1—29
IRON-BASE SUPERALLOYS — CAST, WROUGHT

Refractaloy® 26

Typical chemical composition, percent by weight: C, 0.03; Mn, 0.8; Fe, balance; S, 1.0; Cr, 18.0; Ni, 38.0; Co, 20.0; Mo, 3.2; Ti, 2.6; Al, 0.2

Physical constants and thermal properties
Density, lb/in.³: 0.296
Coefficient of thermal expansion, (70–200°F) in./in./°F × 10⁻⁶: 8.2
Modulus of elasticity, psi: tension, 30.6 × 10⁶

Melting range, °F: 2450
Specific heat, Btu/lb/°F, 70°F: 0.108
Curie temperature, °F: age hardened, 350–1500 (20–44 hr); 1500 (4–20 hr)

Heat treatments
Solution treat 1 hr at 1800–2100°F, oil quench, age 20–44 hr at 1350–1500°F and 4–20 hr at 1500°F, air cool.

Short Time Tensile Properties

Temperature, °F	T.S., psi	Y.S., psi, 0.2% offset	Elong., in 2 in., %	Hardness, Brinell
70	154,000	91,000	19	–
1000	143,000	85,000	18	–
1200	136,000	89,000	15	–
1460	71,000 (1500°F)	66,000 (1500°F)	29 (1500°F)	–
1600	48,000	47,000	49	–

Rupture Strength, 1000 hr

Test temperature, °F	Strength, psi	Elong., in 2 in., %	Reduction of area, %
1350	42,000	–	–

Creep Strength
(Stress, psi, to Produce 1% Creep)

Test temperature, °F	10,000 hr
1350	37,000

Fatigue Strength

Test temperature, °F	Stress, psi	Cycles to failure
1200	54,000	10⁸
1500	37,000	10⁸

Impact Strength

Test temperature, °F	Type test	Strength, ft-lb
+70	Charpy	18

Table 1.1–30
IRON-BASE SUPERALLOYS – CAST, WROUGHT

Discaloy®

Typical chemical composition, percent by weight: C, 0.08; Mn, 1.05; Fe, balance; Si, 0.7; Cr, 13.5; Ni, 25.5; Mo, 3; Ti, 1.7; Al, 0.35; B, 0.01; Cu, 0.5

Specific heat, Btu/lb/°F, 70°F: 0.113
Curie temperature, °F: age hardened, 1350°F (20 hr), then 1200°F (20 hr)

Physical constants and thermal properties
Density, lb/0.288
Coefficient of thermal expansion, (70–200°F) in./in./°F × 10⁻⁶ : 9.5
Modulus of elasticity, psi: tension, 28.4×10^6
Melting range, °F: 2516–2673

Heat treatments
Bar solution treat (1850°F, 2 hr, oil quench) then double age (1350°F, 20 hr air cool; 1200°F, 20 hr air cool). Disk forgings: heat treat (1875°F, 3 hr, oil quench and age).

Short Time Tensile Properties

Temperature, °F	T.S., psi	Y.S., psi, 0.2% offset	Elong., in 2 in., %	Hardness, Brinell
70	145,000	106,000	19	–
1200	104,000	91,000	19	–

Rupture Strength, 1000 hr

Test temperature, °F	Strength, psi	Elong., in 2 in., %	Reduction of area, %
1400	8,500	–	–

Fatigue Strength

Test temperature, °F	Stress, psi	Cycles to failure
1200	46,000–51,000	–

Impact Strength

Test temperature, °F	Type test	Strength, ft-lb
+70	Charpy	–

Table 1.1−31
ALLOY STEELS − CAST

Class 70,000

Chemical composition: total alloy content below 8%.

Physical constants and thermal properties
Density, lb/in.³ : 0.283
Coefficient of thermal expansion, (70−200°F) in./in./°F × 10^{-6}: 8.0−8.3
Modulus of elasticity, psi; tension, 29−30 × 10^6
Specific heat, Btu/lb/°F, 70°F: 0.10−0.11

Thermal conductivity, Btu/ft² /hr/in./°F, 70°F: 324
Electrical resistivity, ohms/cmil/ft, 70°F: 90−120
Curie temperature°F: annealed − about 200°F above critical range

Heat treatments
Normalized and tempered; quenching temperature about 100°F above critical range.

Short Time Tensile Properties

Temperature °F	T.S., psi	Y.S., psi, 0.2% offset	Elong., in 2 in., %	Hardness, Brinell
70	74,000	44,000	28	143

Fatigue Strength

Test temperature, °F	Stress, psi	Cycles to failure
70	35,000 endurance limit	−

Impact Strength

Test temperature, °F	Type test	Strength, ft-lb
− 40	Charpy − V-notched	22
+ 70	Charpy − V-notched	55

Table 1.1−32
ALLOY STEELS − CAST

Class 80,000

Chemical composition: total alloy content below 8%.

Physical constants and thermal properties
Density, lb/in.³ : 0.283
Coefficient of thermal expansion, (70−200°F) in./in./°F × 10^{-6}: 8.0−8.3
Modulus of elasticity, psi: tension 29−30 × 10^6
Specific heat, Btu/lb/°F, 70°F: 0.10−0.11

Thermal conductivity, Btu/ft² /hr/in./°F, 70°F: 324
Electrical resistivity, ohms/cmil/ft, 70°F 90−120
Curie temperature, °F: annealed − about 200°F above critical range

Heat treatments
Normalized and tempered; quenching temperature about 100°F above critical range.

Short Time Tensile Properties

Temperature, °F	T.S., psi	Y.S. psi, 0.2% offset	Elong., in 2 in.,%	Hardness, Brinell
70	86,000	54,000	24	170

Table 1.1–32 (continued)
ALLOY STEELS – CAST

Class 80,000 (continued)

Fatigue Strength

Test temperature, °F	Stress, psi	Cycles to failure
70	39,000 endurance limit	–

Impact Strength

Test temperature, °F	Type test	Strength, ft-lb
– 40	Charpy – V-notched	18
+ 70	Charpy – V-notched	48

Table 1.1–33
ALLOY STEELS–CAST

Class 105,000

Chemical composition: total alloy content below 8%.

Physical constants and thermal properties
Density, lb/in.^{3}: 0.283
Coefficient of thermal expansion, (70–200°F)
  in./in./°F × 10^{-6}: 8.0–8.3
Modulus of elasticity, psi: tension, 29–30 × 10^6
Specific heat, Btu/lb/°F, 70°F: 0.10–0.11

Thermal conductivity, Btu/ft^2/hr/in./°F, 70°F: 324
Electrical resistivity, ohms/cmil/ft, 70°F: 90–120
Curie temperature, °F: annealed – about 200°F above
  critical range

Heat treatments
  Normalized and tempered; quenching temperature
  about 100°F above critical range.

Short Time Tensile Properties

Temperature, °F	T.S., psi	Y.S., psi, 0.2% offset	Elong., in 2 in., %	Hardness, Brinell
70	110,000	91,000	21	217

Fatigue Strength

Test temperature, °F	Stress, psi	Cycles to failure
70	53,000 endurance limit	–

Impact Strength

Test temperature, °F	Type test	Strength, ft-lb
– 40	Charpy – V-notched	40
+ 70	Charpy – V-notched	58

Table 1.1–34
ALLOY STEELS – CAST

Class 120,000

Chemical composition: total alloy content below 8%.

Physical constants and thermal properties
Density, lb/in.³ : 0.283
Coefficient of thermal expansion, (70–200°F)
in./in./°F × 10⁻⁶ : 8.0–8.3
Modulus of elasticity, psi: tension, 29–30 × 10⁶
Specific heat, Btu/lb/°F, 70°F: 0.10–0.11

Thermal conductivity, Btu/ft²/hr/in./°F, 70°F: 324
Electrical resistivity, ohms/cmil/ft, 70°F: 90–120
Curie temperature,°F: annealed – About 200°F above
critical range

Heat treatments
Quenched and tempered about 100°F above critical
range.

Short Time Tensile Properties

Temperature, °F	T.S., psi	Y.S., psi, 0.2% offset	Elong., in 2 in., %	Hardness Brinell
70	128,000	112,000	16	262

Fatigue Strength

Test temperature, °F	Stress, psi	Cycles to failure
70	62,000 endurance limit	–

Impact Strength

Test temperature, °F	Type test	Strength, ft-lb
– 40	Charpy – V-notched	31
+ 70	Charpy – V-notched	45

Table 1.1–35
ALLOY STEELS–CAST

Class 150,000

Chemical composition: total alloy content below 8%.

Physical constants and thermal properties
Density, lb/in.³ : 0.283
Coefficient of thermal expansion, (70–200°F)
in./in./°F × 10⁻⁶ : 8.0–8.3
Modulus of elasticity, psi: tension, 29–30 × 10⁶
Specific heat, Btu/lb/°F, 70°F: 0.10–0.11

Thermal conductivity, Btu/ft²/hr/in./°F, 70°F: 324
Electrical resistivity, ohms/cmil/ft, 70°F: 90–120
Curie temperature, °F: annealed – about 200°F above
critical range

Heat treatments
Quenched and tempered; quenching temperature about
100°F above critical range.

Short Time Tensile Properties

Temperature, °F	T.S., psi	Y.S., psi, 0.2% offset	Elong., in 2 in., %	Hardness, Brinell
70	158,000	142,000	13	311

Table 1.1–35 (continued)
ALLOY STEELS – CAST

Class 150,000 (continued)

Fatigue Strength

Test temperature, °F	Stress, psi	Cycles to failure
70	74,000 endurance limit	–

Impact Strength

Test temperature, °F	Type test	Strength, ft-lb
– 40	Charpy – V-notched	17
+ 70	Charpy – V-notched	30

Table 1.1–36
ALLOY STEELS – CAST

Class 200,000

Chemical composition: total alloy content below 8%.

Physical constants and thermal properties
Density, lb/in.^{3}: 0.283
Coefficient of thermal expansion, (70–200°F) in./in./°F $\times$ 10^{-6}: 8.0–8.3
Modulus of elasticity, psi: tension, 29–30 $\times$ 10^6
Specific heat Btu/lb/°F, 70°F: 0.10–0.11

Thermal conductivity, Btu/ft^2/hr/in./°F, 70°F: 324
Electrical resistivity, ohms/cmil/ft, 70°F: 90–120
Curie temperature, °F: annealed – about 200°F above critical range

Heat treatments
Quenched and tempered; quenching temperature about 100°F above critical range.

Short Time Tensile Properties

Temperature, °F	T.S., psi	Y.S., psi, 0.2% offset	Elong., in 2 in., %	Hardness, Brinell
70	205,000	170,000	8	401

Fatigue Strength

Test temperature, °F	Stress, psi	Cycles to failure
70	88,000 endurance limit	–

Impact Strength

Test temperature, °F	Type test	Strength, ft-lb
–40	Charpy – V-notched	8
+70	Charpy – V-notched	14

Table 1.1–37
CARBON STEELS – CAST

Class 60,000

Physical constants and thermal properties
Density, lb/in.^{3}: 0.283
Coefficient of thermal expansion, (70–200°F) in./in./°F × 10^{-6}: 8.3
Modulus of elasticity, psi: tension, 30.1 × 10^6
Specific heat, Btu/lb/°F, 70°F: 0.10–0.11
Thermal conductivity, Btu/ft^2/hr/in./°F, 70°F: 324

Heat treatments
Annealed

Short Time Tensile Properties

Temperature, °F	T.S., psi	Y.S., psi, 0.2% offset	Elong., in 2 in., %	Hardness, Brinell
70	63,000	35,000	30	131

Fatigue Strength

Test temperature, °F	Stress, psi	Cycles to failure
70	30,000 endurance limit	–

Impact Strength

Test temperature, °F	Type test	Strength, ft-lb
–40	Charpy – V-notched	5
+70	Charpy – V-notched	12

Table 1.1—38
CARBON STEELS — CAST

Class 70,000

Physical constants and thermal properties
Density, lb/in.^{3} : 0.283
Coefficient of thermal expansion, (70–200°F)
  in./in./°F × 10^{-6} : 8.3
Modulus of elasticity, psi: tension, 30 × 10^6
Specific heat, Btu/lb/°F, 70°F: 0.10–0.11
Thermal conductivity, Btu/ft^2/hr/in./°F, 70°F: 324

Heat treatments
Normalized

Short Time Tensile Properties

Temperature, °F	T.S., psi	Y.S., psi, 0.2% offset	Elong., in 2 in., %	Hardness, Brinell
70	75,000	42,000	27	143

Fatigue Strength

Test temperature, °F	Stress, psi	Cycles to failure
70	35,000 endurance limit	—

Impact Strength

Test temperature, °F	Type test	Strength, ft-lb
−40	Charpy — V-notched	12
+70	Charpy — V-notched	30

Table 1.1–39
CARBON STEELS – CAST

Class 80,000

Physical constants and thermal properties
  Density, lb/in.³ : 0.283
  Coefficient of thermal expansion, (70–200°F)
    in./in./°F × 10⁻⁶ : 8.3
  Modulus of elasticity, psi: tension, 29.9 × 10⁶
  Specific heat, Btu/lb/°F, 70°F: 0.10–0.11
  Thermal conductivity, Btu/ft²/hr/in./°F, 70°F: 324

Heat treatments
  Normalized and tempered

Short Time Tensile Properties

Temperature, °F	T.S., psi	Y.S., psi, 0.2% offset	Elong., in 2 in., %	Hardness, Brinell
70	82,000	48,000	23	163

Fatigue Strength

Test temperature, °F	Stress, psi	Cycles to failure
70	37,000 endurance limit	–

Impact Strength

Test temperature, °F	Type test	Strength, ft-lb
–40	Charpy – V-notched	10
+70	Charpy – V-notched	35

Table 1.1–40
CARBON STEELS – CAST

Class 100,000

Physical constants and thermal properties
Density, lb/in.^{3}: 0.283
Coefficient of thermal expansion, (70–200°F) in./in./°F × 10^{-6} : 8.3
Modulus of elasticity, psi: tension, 29.7 × 10^6
Specific heat, Btu/lb/°F, 70°F: 0.10–0.11
Thermal conductivity, Btu/ft^2/hr/in./°F, 70°F: 324

Heat treatments
Quenched and tempered

Short Time Tensile Properties

Temperature, °F	T.S., psi	Y.S., psi, 0.2% offset	Elong., in 2 in., %	Hardness, Brinell
70	105,000	75,000	19	212

Fatigue Strength

Test temperature, °F	Stress, psi	Cycles to failure
70	45,000 endurance limit	–

Impact Strength

Test temperature, °F	Type test	Strength, ft-lb
–40	Charpy – V-notched	12
+70	Charpy – V-notched	40

Table 1.1–41
CARBON STEELS – CARBURIZING GRADES

AISI Type C1015

Physical constants and thermal properties
Density, lb/in.^{3} : 0.283
Coefficient of thermal expansion, (70–200°F)
  in./in./°F × 10^{-6} :8.4
Modulus of elasticity, psi: tension, 29-30 × 10^6
Melting range, °F: 2750–2775
Specific heat, Btu/lb/°F, 70°F: 0.10–0.11

Thermal conductivity, Btu/ft^2/hr/in./°F, 70°F: 324

Heat treatments
  1-in. rounds treated as follows: carburized at 1675°F
    for 8 hr, pot cooled, reheated to 1425°F, water
    quenched, tempered at 350°F.

Short Time Tensile Properties

Temperature, °F	T.S., psi	Y.S., psi, 0.2% offset	Elong., in 2 in., %	Hardness, Brinell
70	73,000	46,000	32	149 (core)

Fatigue Strength

Test temperature, °F	Stress, psi	Cycles to failure
70	30,000–35,000	–

Table 1.1–42
CARBON STEELS – CARBURIZING GRADES

AISI Type C1020

Physical Constants and Thermal Properties
Density, lb/in.^{3} : 0.283
Coefficient of thermal expansion, (70–200°F)
  in./in./°F × 10^{-6} : 8.4
Modulus of elasticity, psi: tension, 29–30 × 10^6
Melting range, °F: 2750–2775
Specific heat, Btu/lb/°F, 70°F: 0.10–0.11

Thermal conductivity, Btu/ ft^2/hr/in./°F, 70°F: 324

Heat Treatments
  1-in. rounds treated as follows: carburized at 1675°F
    for 8 hr, pot cooled, reheated to 1425°F, water
    quenched, tempered at 350°F.

Short Time Tensile Properties

Temperature, °F	T.S., psi	Y.S., psi, 0.2% offset	Elong., in 2 in., %	Hardness, Brinell
70	75,000	48,000	31	156 (core)

Fatigue Strength

Test temperature, °F	Stress, psi	Cycles to failure
70	30,000–37,000	–

Table 1.1–43
CARBON STEELS – CARBURIZING GRADES

AISI Type C1118

Physical constants and thermal properties
Density, lb/in.³: 0.283
Coefficient of thermal expansion, (70–200°F)
   in./in./°F × 10⁻⁶: 8.4
Modulus of elasticity, psi: tension, 29–30 × 10⁶
Melting range, °F: 2750–2775
Specific heat, Btu/lb°F, 70°F: 0.10–0.11

Thermal conductivity, Btu/ft²/hr/in./°F, 70°F: 324

Heat treatments
   1-in. rounds treated as follows: carburized at 1700°F
   for 8 hr, pot cooled, reheated to 1450°F, water
   quenched, tempered at 350°F.

Short Time Tensile Properties

Temperature, °F	T.S., psi	Y.S., psi, 0.2% offset	Elong., in 2 in., %	Hardness, Brinell
70	113,000	77,000	17	229 (core)

Fatigue Strength

Test temperature, °F	Stress, psi	Cycles to failure
70	45,000–56,000	–

Table 1.1–44
CARBON STEELS – HARDENING GRADES

AISI Type C1030

Physical constants and thermal properties
Density, lb/in.³: 0.283
Coefficient of thermal expansion, (70–200°F)
   in./in./°F × 10⁻⁶: 8.3
Modulus of elasticity, psi: tension, 29–30 × 10⁶
Melting range, °F: 2700–2750
Specific heat, Btu/lb/°F, 70°F: 0.10–0.11

Thermal conductivity, Btu/ft²/hr/in./°F, 70°F: 324
Electrical resistivity, ohms/cmil/ft, 70°F: 114

Heat treatments
   Normalized at 1700°F, reheated to 1600°F, quenched
   in water, tempered at 400–1300°F.

Short Time Tensile Properties

Temperature, °F	T.S., psi	Y.S., psi, 0.2% offset	Elong., in 2 in., %	Hardness, Brinell
70	122,000–75,000	93,000–58,000	18–33	495–179
400	123,000	94,000	17	495
600	116,000	90,000	19	401
800	106,000	84,000	23	302
1000	97,000	75,000	28	255
1200	85,000	64,000	32	207

Table 1.1–44 (continued)
CARBON STEELS – HARDENING GRADES

AISI Type C1030 (continued)

Rupture Strength, 1000 hr

Test temperature, °F	Strength, psi	Elong., in 2 in., %	Reduction of area, %
400	–	–	47
600	–	–	53
800	–	–	60
1000	–	–	65
1200	–	–	70

Table 1.1–45
CARBON STEELS – HARDENING GRADES

AISI Type C1050

Physical constants and thermal properties
Density, lb/in.^{3}: 0.283
Coefficient of thermal expansion, (70–200°F) in./in./°F $\times 10^{-6}$: 8.1
Modulus of elasticity, psi: tension, $29–30 \times 10^6$
Specific heat, Btu/lb/°F, 70°F: 0.10–0.11

Thermal conductivity, Btu/ft^2/hr/in./°F, 70°F: 324
Electrical resistivity, ohms/cmil/ft 70°F: 108

Heat treatments
Normalized at 1650°F, reheated to 1550°F, quenched in oil, tempered at 400–1300°F.

Short Time Tensile Properties

Temperature, °F	T.S., psi	Y.S., psi, 0.2% offset	Elong., in 2 in., %	Hardness, Brinell
70	143,000–96,000	108,000–61,000	10–30	321–192

Table 1.1–46
CARBON STEELS – HARDENING GRADES

AISI Type C1080

Physical constants and thermal properties
Density, lb/in.^{3}: 0.283
Coefficient of thermal expansion, (70–200°F) in./in./°F $\times 10^{-6}$: 8.1
Modulus of elasticity, psi: tension, $29–30 \times 10^6$
Specific heat, Btu/lb/°F, 70°F: 0.10–0.11

Thermal conductivity, Btu/ft^2/hr/in./°F, 70°F: 324
Electrical resistivity, ohms/cmil/ft, 70°F: 108

Heat treatments
Normalized at 1650°F, reheated to 1550°F, quenched in oil, tempered at 400–1300°F.

Table 1.1—46 (continued)
CARBON STEELS — HARDENING GRADES

AISI Type C1080 (continued)

Short Time Tensile Properties

Temperature, °F	T.S., psi	Y.S., psi, 0.2% offset	Elong., in 2 in., %	Hardness, Brinell
70	190,000–117,000	142,000–70,000	12–24	388–223
400	190,000	142,000	12	388
600	189,000	142,000	12	388
800	187,000	138,000	13	375
1000	164,000	117,000	16	321
1200	129,000	87,000	21	255

Rupture Strength, 1000 hr

Test temperature, °F	Strength, psi	Elong., in 2 in., %	Reduction of area, %
400	–	–	35
600	–	–	35
800	–	–	36
1000	–	–	40
1200	–	–	50

Table 1.1—47
CARBON STEELS — HARDENING GRADES

AISI Type C1095

Physical constants and thermal properties
Density, lb/in.^{3}: 0.283
Coefficient of thermal expansion, (70–200°F)
in./in./°F $\times 10^{-6}$: 8.1
Modulus of elasticity, psi: tension, 29–30 $\times 10^6$
Specific heat, Btu/lb/°F, 70°F: 0.10–0.11

Thermal conductivity, Btu/ft^2/hr/in./°F, 70°F: 324
Electrical resistivity, ohms/cmil/ft, 70°F: 108

Heat treatments
Normalized at 1650°F, reheated to 1475°F, quenched in oil, tempered at 400–1300°F.

Short Time Tensile Properties

Temperature, °F	T.S., psi	Y.S., psi, 0.2% offset	Elong., in 2 in., %	Hardness, Brinell
Oil Quenched				
70	188,000–190,000	120,000–74,000	10–26	401–229
400	187,000	120,000	10	401
600	183,000	118,000	10	375
800	176,000	112,000	12	363
1000	158,000	98,000	15	321
1200	130,000	80,000	21	269
Water Quenched				
400	216,000	152,000	10	601
600	212,000	150,000	11	534
800	199,000	139,000	13	388
1000	165,000	110,000	15	293
1200	122,000	85,000	20	235

Table 1.1–47 (continued)
CARBON STEELS – HARDENING GRADES

AISI Type C1095

Rupture Strength, 1000 hr

Test temperature, °F	Strength, psi	Elong., in 2 in., %	Reduction of area, %
		Oil Quenched	
400	–	–	30
600	–	–	30
800	–	–	32
1000	–	–	37
1200	–	–	47
		Water Quenched	
400	–	–	31
600	–	–	33
800	–	–	35
1000	–	–	40
1200	–	–	47

Table 1.1–48
CARBON STEELS – HARDENING GRADES

AISI Type C1141

Physical constants and thermal properties
Density, lb/in.^{3}: 0.282
Coefficient of thermal expansion, (70–200°F) in./in./°F × 10^{-6}: 7.5
Modulus of elasticity, psi: tension, 29–30 × 10^6
Specific heat, Btu/lb/°F, 70°F: 0.10–0.11

Thermal conductivity, Btu/ft^2/hr/in./°F, 70°F: 324
Electrical resistivity, ohms/cmil/ft, 70°F: 108

Heat treatments
½-in. round normalized at 1575°F, reheated to 1500°F, quenched in oil, tempered at 400–1300°F.

Short Time Tensile Properties

Temperature, °F	T.S., psi	Y.S., psi, 0.2% offset	Elong., in 2 in., %	Hardness, Brinell
70	237,000–94,000	188,000–68,000	7–28	461–192

Table 1.1–49
FREE-CUTTING CARBON STEELS – WROUGHT

AISI Type B1111

Typical chemical composition, percent by weight: C, 0.13; Mn, 0.60–0.90; Fe, balance; S, 0.08–0.15; P, 0.07–0.12

Physical constants and thermal properties
Density, lb/in.^{3}: 0.283
Coefficient of thermal expansion, (70–200°F) in./in./°F × 10^{-6}: 8.4

Modulus of elasticity, psi: tension, 29 × 10^6
Specific heat, Btu/lb/°F, 70°F: 0.10–0.11
Thermal conductivity, Btu/ft^2/hr/in./°F, 70°F: 324
Electrical resistivity, ohm./cmil/ft, 70°F: 85.8

Heat treatments
Tempering temperature, 300°F; case hardening temperature, 1450°F. Cold drawn.

Short Time Tensile Properties

1-in. Diameter

Temperature, °F	T.S., psi	Y.S., psi, 0.2% offset	Elong., in 2 in., %	Hardness, Brinell
70	80,000	70,000	12	163

Fatigue Strength

Notch sensitive as cold drawn.

Impact Strength

Relatively low impact strength at low temperatures; should not be used for shock loading applications at subzero temperatures.

Table 1.1–50
FREE-CUTTING CARBON STEELS – WROUGHT

AISI Type B1211

Typical chemical composition, percent by weight: C, 0.13; Mn, 0.60–0.90; Fe, balance; S, 0.08–0.15; P, 0.07–0.12

Physical constants and thermal properties
Density, lb/in.^{3}: 0.283
Coefficient of thermal expansion, (70–200°F) in./in./°F × 10^{-6}:8.4

Modulus of elasticity, psi: tension, 29 × 10^6
Specific heat, Btu/lb/°F, 70°F: 0.10–0.11
Thermal conductivity, Btu/ft^2/hr/in.°F, 70°F: 324
Electrical resistivity, ohms/cmil/ft, 70°F: 85.8

Heat treatments:
Tempering temperature, 300°F; case hardening temperature, 1700°F. Cold drawn.

Short Time Tensile Properties

1-in. Diameter

Temperature, °F	T.S., psi	Y.S., psi, 0.2% offset	Elong., in 2 in., %	Hardness, Brinell
70	105,000	90,000	22	229

Fatigue Strength

Notch sensitive as cold drawn.

Table 1.1–50 (continued)
FREE-CUTTING CARBON STEELS – WROUGHT

AISI Type B1211 (continued)

Impact Strength

Relatively low impact strength at low temperatures; should not be used for shock loading applications at subzero temperatures.

Table 1.1–51
FREE-CUTTING CARBON STEELS – WROUGHT

AISI Type B1112

Typical chemical composition, percent by weight: C, 0.13; Mn, 0.70–1.00; Fe, balance; S, 0.16–0.23; P, 0.07–0.12

Modulus of elasticity, psi: tension, 29×10^6
Specific heat, Btu/lb/$°$F 70$°$F: 0.10–0.11
Thermal conductivity, Btu/ft^2/hr/in./$°$F 70$°$F: 324
Electrical resistivity, ohms/cmil/ft, 70$°$F: 85.8

Physical constants and thermal properties
Density, lb/in.^{3}: 0.283
Coefficient of thermal expansion, (70–200$°$F) in./in./$°$F $\times$ 10^{-6}: 8.4

Heat treatments
Case hardening temperature 1450$°$F; tempering temperature 300$°$F.

Short Time Tensile Properties

1-in. Diameter

Temperature, $°$F	T.S., psi	Y.S., psi, 0.2% offset	Elong., in 2 in., %	Hardness, Brinell[a]
70	80,000	70,000	12	163

Fatigue Strength

Notch sensitive as cold drawn.

Impact Strength

Relatively low impact strength at low temperatures; should not be used for shock loading applications at subzero temperatures.

[a]Cold drawn.

Table 1.1–52
FREE-CUTTING CARBON STEELS – WROUGHT

AISI Type B1212

Typical chemical composition, percent by weight: C, 0.13; Mn, 0.70–1.00; Fe, balance; S, 0.16–0.23; P, 0.07–0.12

Physical constants and thermal properties
Density, lb/in.^{3}: 0.283
Coefficient of thermal expansion, (70–200°F) in./in./°F $\times 10^{-6}$:8.4

Modulus of elasticity, psi: tension, 29×10^6
Specific heat, Btu/lb/°F, 70°F: 0.10–0.11
Thermal conductivity, Btu/ft^2/hr/in./°F, 70°F: 324
Electrical resistivity, ohms/cmil/ft, 70°F: 85.8

Heat treatments
Case hardening temperature, 1700°F; tempering temperature, 300°F.

Short Time Tensile Properties

1-in. Diameter

Temperature, °F	T.S., psi	Y.S., psi, 0.2% offset	Elong., in 2 in., %	Hardness, Brinell[a]
70	105,000	90,000	22	229

Fatigue Strength

Notch sensitive as cold drawn.

Impact Strength

Relatively low impact strength at low temperatures; should not be used for shock loading applications at subzero temperatures.

[a]Cold drawn.

Table 1.1–53
HIGH STRENGTH STEELS – WROUGHT;
COLUMBIUM-BEARING CARBON STEELS

Type 50

Typical chemical composition, percent by weight:
C, 0.020; Mn, 0.50–1.00; Fe, balance; S, 0.05; Si, 0.10; Cb, 0.01; P, 0.04

Short Time Tensile Properties

Temperature, °F	T.S., psi	Y.S., psi, 0.2% offset	Elong., in 2 in., %	Hardness, Brinell
70	65,000	50,000	22	–

Specifications

ASTM: A-572

Table 1.1–54
HIGH STRENGTH STEELS – WROUGHT;
COLUMBIUM-BEARING CARBON STEELS

Type 60

Typical chemical composition, percent by weight:
C, 0.25; Mn, 1.50; Fe, balance; S, 0.05; Si, 0.10;
Cb, 0.01; P, 0.04

Short Time Tensile Properties

Temperature, °F	T.S., psi	Y.S., psi, 0.2% offset	Elong., in 2 in., %	Hardness, Brinell
70	75,000	60,000	18	–

Specifications

ASTM: A-572

Table 1.1–55
HIGH STRENGTH STEELS – WROUGHT;
VANADIUM-BEARING CARBON STEELS

Type 50

Typical chemical composition, percent by weight:
C, 0.22; Mn, 1.25; Fe, balance; S, 0.05; P, 0.04;
N, 0.015; V, 0.02

Short Time Tensile Properties

Temperature, °F	T.S., psi	Y.S., psi, 0.2% offset	Elong., in 2 in., %	Hardness, Brinell
70	70,000	50,000	18	156

Specifications

ASTM: A-572

Table 1.1–56
HIGH STRENGTH STEELS – WROUGHT;
VANADIUM-BEARING CARBON STEELS

Type 60

Typical chemical composition, percent by weight:
C, 0.22; Mn, 1.25; Fe, balance; S, 0.05; P, 0.04;
N, 0.015; V, 0.02

Short Time Tensile Properties

Temperature, °F	T.S., psi	Y.S., psi, 0.2% offset	Elong., in 2 in., %	Hardness, Brinell
70	75,000	60,000	16	183

Specifications

ASTM: A-572

Table 1.1–57
HIGH STRENGTH STEELS – WROUGHT;
LOW ALLOY, HIGH STRENGTH STEELS

ASTM Type A94

Typical chemical composition, percent by weight:
C, 0.33; Mn, 1.10–1.60; Fe, balance; S, 0.05;
Si, 0.15–0.30; P, 0.06; Cu, 0.20

Short Time Tensile Properties

Temperature, °F	T.S., psi	Y.S., psi, 0.2% offset	Elong., in 2 in., %	Hardness, Brinell
70	70,000–75,000	45,000–50,000	21–20	–

Table 1.1–58
HIGH STRENGTH STEELS – WROUGHT;
LOW ALLOY, HIGH STRENGTH STEELS

ASTM Type A242

Typical chemical composition, percent by weight:
C, 0.22; Mn, 1.25; Fe, balance; S, 0.05; Si,
0.20–0.90[a]; Cr, 0.30–1.25[a]; Zr, 0.10[a]; Cu,
0.25–0.55[a]

[a]Optional

Short Time Tensile Properties

Temperature, °F	T.S., psi	Y.S., psi, 0.2% offset	Elong., in 2 in., %	Hardness, Brinell
70	63,000–70,000	42,000–50,000	24–22	156

Table 1.1−59
HIGH STRENGTH STEELS − WROUGHT;
LOW ALLOY, HIGH STRENGTH STEELS

ASTM Type A440

Typical chemical composition, percent by weight:
C, 0.28; Mn, 1.10−1.60; Fe, balance; S, 0.05; Si, 0.30;
P, 0.06; Cu, 0.20

Short Time Tensile Properties

Temperature, °F	T.S., psi	Y.S., psi, 0.2% offset	Elong., in 2 in., %	Hardness, Brinell
70	63,000−70,000	42,000−50,000	24−22	−

Table 1.1−60
HIGH STRENGTH STEELS − WROUGHT;
LOW ALLOY, HIGH STRENGTH STEELS

ASTM Type A441

Typical chemical composition, percent by weight:
C, 0.22; Mn, 1.25; Fe, balance; S, 0.04; Si, 0.30;
P, 0.04; Cu, 0.20; V, 0.02

Short Time Tensile Properties

Temperature, °F	T.S., psi	Y.S., psi, 0.2% offset	Elong., in 2 in., %	Hardness, Brinell
70	63,000−70,000	42,000−50,000	24−22	−

Table 1.1−61
HIGH STRENGTH STEELS − WROUGHT;
LOW ALLOY, HIGH STRENGTH STEELS

ASTM Type A374

Typical chemical composition, percent by weight: C,
0.22; Mn, 1.25; Fe, balance; S, 0.05

Heat treatments
Cold rolled

Short Time Tensile Properties

Temperature, °F	T.S., psi	Y.S., psi, 0.2% offset	Elong., in 2 in., %	Hardness, Brinell
70	67,000−70,000	46,000−50,000	24−22	−

Table 1.1–62
HIGH STRENGTH STEELS – WROUGHT;
LOW ALLOY, HIGH STRENGTH STEELS

ASTM Type A375

Typical chemical composition, percent by weight: C, 0.22; Mn, 1.25; Fe, balance; S, 0.05

Heat treatments
Hot rolled

Short Time Tensile Properties

Temperature, °F	T.S., psi	Y.S., psi, 0.2% offset	Elong., in 2 in., %	Hardness, Brinell
70	67,000–70,000	46,000–50,000	24–22	–

Table 1.1–63
NITRIDING STEELS – WROUGHT

Type 135

Typical chemical composition, percent by weight: C, 0.30–0.40; Mn, 0.40–0.70; Fe, balance; Si, 0.20–0.40; Cr, 0.90–1.40; Mo, 0.15–0.25; Al, 0.85–1.20

Physical constants and thermal properties
Density, lb/in.^{3}: 0.283
Coefficient of thermal expansion, (70–200°F) in./in./ °F × 10^{-6}: 6.5
Modulus of elasticity, psi: tension, 29–30 × 10^6
Specific heat Btu/lb/°F, 70°F: 0.11–0.12

Thermal conductivity, Btu/ft^2/hr/in./°F, 70°F: 360
Electrical resistivity, ohms/cmil/ft, 70°F: 162–174
Curie temperature, °F: annealed, 1650–1700

Heat treatments
Quenching temperature 1700–1750°F. Tempering temperature 1100–1300°F. Nitriding temperature 930–1050°F for periods ranging to 100 hr; 24- to 48-hr treatments are most widely used. Core properties; oil quenched from 1700°F; tempered at 1200°F.

Short Time Tensile Properties

Temperature, °F	T.S., psi	Y.S., psi, 0.2% offset	Elong., in 2 in., %	Hardness, Brinell
70	138,000	120,000	20	280

Table 1.1–64
NITRIDING STEELS – WROUGHT

Type N

Typical chemical composition, percent by weight: C, 0.20–0.27; Mn, 0.40–0.70; Fe, balance; Si, 0.20–0.40; Cr, 1.00–1.50; Ni, 3.25–3.75; Mo, 0.20–0.30; Al, 0.85–1.20

Physical constants and thermal properties
Density, lb/in.^{3}: 0.283
Coefficient of thermal expansion, (70–200°F) in./in./°F × 10^{-6}: 6.5
Modulus of elasticity, psi: tension, 29–30 × 10^6
Specific heat, Btu/lb/°F, 70°F: 0.11–0.12
Thermal conductivity, Btu/ft^2/hr/in./°F, 70°F: 360

Electrical resistivity, ohms/cmil/ft, 70°F: 162–174
Curie temperature, °F: annealed, 1500–1550[a]

[a]Must be cooled rapidly below 1150°F to avoid precipitation hardening.

Heat treatments
Quenching temperature 1625–1675°F. Tempering temperature 1100–1300°F. Nitriding temperature 930–1050°F for periods ranging to 100 hr; 24- to 48-hr treatments are most widely used. Core properties; oil quenched from 1650°F, tempered at 1200°F before nitriding.

Short Time Tensile Properties

Temperature, °F	T.S., psi	Y.S., psi, 0.2% offset	Elong., in 2 in., %	Hardness, Brinell
70	132,000	114,000	22	277

Table 1.1–65
NITRIDING STEELS – WROUGHT

Type EZ

Typical chemical composition, percent by weight: C, 0.30–0.40; Mn, 0.50–1.10; Fe, balance; Si, 0.20–0.40; Cr, 1.00–1.50; Mo, 0.15–0.25; Al, 0.85–1.20; Se, 0.15–0.25

Physical constants and thermal properties
Density, lb/in.^{3}: 0.283
Coefficient of thermal expansion, (70–200°F) in./in./°F × 10^{-6}: 6.5
Modulus of elasticity, psi: tension, 29–30 × 10^6
Specific heat, Btu/lb/°F, 70°F: 0.11–0.12

Thermal conductivity, Btu/ft^2/hr/in./°F, 70°F: 360
Electrical resistivity, ohms/cmil/ft, 70°F: 162–174
Curie temperature, °F: annealed, 1650–1700

Heat treatments
Quenching temperature 1700–1750°F. Tempering temperature 1100–1300°F. Nitriding temperature 930–1050°F for periods ranging to 100 hr; 24- to 48-hr treatments are most widely used. Core properties; oil quenched from 1700°F, tempered at 1200°F.

Short Time Tensile Properties

Temperature, °F	T.S., psi	Y.S., psi, 0.2% offset	Elong., in 2 in., %	Hardness, Brinell
70	126,000	90,000	17	255

Table 1.1–66
AGE HARDENABLE STAINLESS STEELS – WROUGHT, CAST

Type Stainless W

Typical chemical composition, percent by weight: C, 0.08; Mn, 1.0; Fe, balance; Si, 1.0; Cr, 17.0; Ni, 7.0; Ti, 1.2; Al, 0.40

Physical constants and thermal properties
Density, lb/in.^{3} : 0.280
Coefficient of thermal expansion, (70–200°F) in./in./ °F × 10^{-6} : 5.5
Modulus of elasticity, psi: tension, 28 × 10^6

Thermal conductivity, Btu/ft^2/hr/in./°F, 70°F: 145.2
Electrical resistivity ohms/cmil/ft, 70°F: 510
Curie temperature, °F: annealed, 1850; age hardened, 950 (½ hr)

Heat treatments
Sheet solution annealed at 1850–1950°F, air cooled, aged ½ hr at 950°F, air cooled.

Short Time Tensile Properties

Temperature, °F	T.S., psi	Y.S., psi, 0.2% offset	Elong., in 2 in., %	Hardness, Brinell
70	195,000	180,000	3–15	–
800	146,000	135,000	13.5	–
1,000	94,000	54,000	22.5	–
1,200	41,000	24,000	40	–

Fatigue Strength

Test temperature, °F	Stress, psi	Cycles to failure
70	54,000–96,000	10–20 × 10^6

Table 1.1–67
AGE HARDENABLE STAINLESS STEELS – WROUGHT, CAST

Type Cast 17-4PH

Typical chemical composition, percent by weight: C, 0.07; Mn, 1.0; Fe, balance; Si, 1.0; Cr, 15–17; Ni, 3–5; Cu, 2.3–3.0

Physical constants and thermal properties
Density, lb/in.^{3} : 0.280
Coefficient of thermal expansion, (70–200°F) in./in./

°F × 10^{-6} : 6.0
Modulus of elasticity, psi: tension, 28.5 × 10^6
Thermal conductivity, Btu/ft^2/hr/in./°F, 70°F: 124.8
Electrical resistivity, ohms/cmil/ft 70°F: 588

Heat treatments
Solution treated and hardened.

Short Time Tensile Properties

Temperature, °F	T.S., psi	Y.S., psi, 0.2% offset	Elong., in 2 in., %	Hardness, Brinell
70	170,000	140,000	6	269–420
800	158,000	138,000	–	–

Rupture Strength, 1,000 hr

Test temperature, °F	Strength, psi	Elong., in 2 in., %	Reduction of area, %
800	89,000	–	–

Table 1.1–67 (continued)
AGE HARDENABLE STAINLESS STEELS – WROUGHT, CAST

Type Cast 17-4PH (continued)

Creep Strength
(Stress, psi, to Produce 1% Creep)

Test temperature, °F	10,000 hr	100,000 hr
700	120,000	–

Impact Strength

Test temperature, °F	Type test	Strength, ft-lb
+70	Charpy – unnotched	17

Table 1.1–68
AGE HARDENABLE STAINLESS STEELS – WROUGHT, CAST

Type PH 15-7 Mo

Typical chemical composition, percent by weight: C, 0.09; Mn, 1.0; Fe, balance; Si, 1.0; Cr, 15.0; Ni, 7.0; Mo, 2.5; Al, 1.0

Physical constants and thermal properties
Density, lb/in.^{3}: 0.277
Coefficient of thermal expansion, (70–200°F) in./in./°F × 10^{-6} : 5–6
Modulus of elasticity, psi: tension, 29 × 10^6

Thermal conductivity, Btu/ft^2/hr/in./°F, 70°F: 111.6
Electrical resistivity, ohms/cmil/ft, 70°F: 492
Curie temperature, °F: annealed, 1950; age hardened, 1050 (90 min)

Heat treatments
Sheet solution annealed at 1950°F, austenitized 90 min at 1400°F, air cooled to 50–60°F, held 30 min, aged 90 min at 1050°F.

Short Time Tensile Properties

Temperature, °F	T.S., psi	Y.S., psi, 0.2% offset	Elong., in 2 in., %	Hardness, Rockwell
70	210,000	200,000	7	B90–C48
800	160,000	150,000	9.5	–
1,000	110,000	105,000	21	–

Rupture Strength, 1,000 hr

Test temperature, °F	Strength, psi	Elong., in 2 in., %	Reduction of area, %
800	137,000	–	–

Impact Strength

Test temperature, °F	Type test	Strength, ft-lb
+70	Charpy – unnotched	4

Specifications

ASTM: A-461-564; SAE-AMS: 5520, 5657, 5812, 5813

Table 1.1–69
AGE HARDENABLE STAINLESS STEELS – WROUGHT, CAST

Type 17-4 PH

Typical chemical composition, percent by weight: C, 0.07; Mn, 1.0; Fe, balance; Si, 1.0; Cr, 16.5; Ni, 4.0; Cb + Ta, 0.30; Cu, 4.0

Physical constants and thermal properties
Density, $lb/in.^3$: 0.281
Coefficient of thermal expansion, (70–200°F) in./in./°F × 10^{-6} : 6.0
Modulus of elasticity, psi: tension, 28.5 × 10^6

Thermal conductivity, $Btu/ft^2/hr/in./°F$, 70°F: 124.8
Electrical resistivity, ohms/cmil/ft, 70°F: 462
Curie temperature, °F: annealed, 1900; age hardened, 900 (1 hr)

Heat treatments
1. H 900 – bar solution annealed at 1900°F, oil quench or air cool, aged 1 hr at 900°F, air cooled;
2. H 1150 – 4 hr at 1150°F, air cooled.

Short Time Tensile Properties

Temperature, °F	T.S., psi	Y.S., psi, 0.2% offset	Elong., in 2 in., %	Hardness, Rockwell
70	195,000	180,000	13	C30–45
800	157,000	138,000	10	–
1,000	100,000	77,000	15	–
1,200	59,000	42,000	15	–

Rupture Strength, 1,000 hr

Test temperature, °F	Strength, psi	Elong., in 2 in., %	Reduction of area, %
800	128,000	–	–

Creep Strength
(Stress, psi, to Produce 1% Creep)

Test temperature, °F	10,000 hr	100,000 hr
800	50,000	–

Fatigue Strength

Test temperature, °F	Stress, psi	Cycles to failure
70	90,000	10–20 × 10^6

Impact Strength

Test temperature, °F	Type test	Strength, ft-lb
1,000	Charpy – unnotched	19

Specifications

ASTM: A-461, 564

Table 1.1–70
AGE HARDENABLE STAINLESS STEELS – WROUGHT, CAST

Type 17-7 PH

Typical chemical composition, percent by weight: C, 0.09; Mn, 1.0; Fe, balance; Si, 1.0; Cr, 17.0; Ni, 7.1; Al, 1.0

Electrical resistivity, ohms/cmil/ft, 70°F: 492
Curie temperature, °F: annealed, 1950; age hardened, 1050 (90 min)

Physical constants and thermal properties
Density, lb/in.^{3}: 0.276
Coefficient of thermal expansion, (70–200°F) in./in./°F × 10^{-6}: 5.6
Modulus of elasticity, psi: tension, 29 × 10^6
Thermal conductivity, Btu/ft^2/hr/in./°F, 70°F: 116.4

Heat treatments
1. TH 1050 – sheet solution annealed at 1950°F, austenitized 90 min at 1400°F, air cooled to 50–60°F, held 30 min, aged 90 min at 1050°F.
2. RH 950 – solution annealed at 1950°F, austenitized 10 min at 1750°F, air cooled, treated 8 hr at –100°F, aged 1 hr at 950°F.

Short Time Tensile Properties

Temperature, °F	T.S., psi	Y.S., psi, 0.2% offset	Elong., in 2 in., %	Hardness, Brinell
70	200,000	185,000	9	290
800	143,000	129,000	6.5	–

Rupture Strength, 1,000 hr

Test temperature, °F	Strength, psi	Elong., in 2 in., %	Reduction of area, %
800	90,000	–	–

Creep Strength
(Stress, psi, to Produce 1% Creep)

Test temperature, °F	10,000 hr	100,000 hr
800	40,000	–

Fatigue Strength

Test temperature, °F	Stress, psi	Cycles to failure
70	75,000	10–20 × 10^6

Impact Strength

Test temperature, °F	Type test	Strength, ft-lb
+70	Charpy – unnotched	6

Specifications

ASTM: A-461, 564

Table 1.1–71
AGE HARDENABLE STAINLESS STEELS – WROUGHT, CAST

Type AM-350

Typical chemical composition, percent by weight: C, 0.10; Mn, 0.80; Fe, balance; Si, 0.25; Cr, 16.5; Ni, 4.3; Mo, 2.75; N, 0.10

Physical constants and thermal properties
Density, lb/in.^{3} : 0.282
Coefficient of thermal expansion, (70–200°F)

in./in./°F × 10^{-6} : 6.3
Modulus of elasticity, psi: tension, 29.4 × 10^6
Thermal conductivity, Btu/ft^2/hr/in./°F, 70°F: 106.44
Electrical resistivity, ohms/cmil/ft, 70°F: 472.8

Heat treatments
Solution treated and hardened

Short Time Tensile Properties

Temperature, °F	T.S., psi	Y.S., psi, 0.2% offset	Elong., in 2 in., %	Hardness, Rockwell
70	206,000	173,000	13.5	B-93,C-45
600	189,000	136,000	7.0	–
800	186,000	125,000	9.5	–
1000	106,000	85,000	16.0	–

Rupture Strength, 1000 hr

Test temperature, °F	Strength, psi	Elong., in 2 in., %	Reduction of area, %
800	182,000	–	–

Creep Strength,
(Stress, psi, to Produce 1% Creep)

Test temperature, °F	10,000 hr	100,000 hr
800	91,000	–

Impact Strength

Test temperature, °F	Type test	Strength, ft-lb
70	Charpy – unnotched	14
212	Charpy – unnotched	24

Specifications

SAE-AMS 5546, 5548

Table 1.1–72
AGE HARDENABLE STAINLESS STEELS – WROUGHT, CAST

Type AM-355

Typical chemical composition, percent by weight: C, 0.13; Mn, 0.95; Fe, balance; Si, 0.25; Cr, 15.5; Ni, 4.3; Mo, 2.75; N, 0.10

Physical constants and thermal properties
Density, lb/in.^{3}: 0.282
Coefficient of thermal expansion, (70–200°F) in./in./°F × 10^{-6}: 6.4
Modulus of elasticity, psi: tension, 29.3 × 10^6
Thermal conductivity, Btu/ft^2/hr/in./°F, 70°F: 110.16
Electrical resistivity, ohms/cmil/ft, 70°F: 454.2

Short Time Tensile Properties

Temperature, °F	T.S., psi	Y.S., psi, 0.2% offset	Elong., in 2 in., %	Hardness, Brinell
70	186,000	171,000	19	402–477
600	159,000	143,000	14	–
800	140,000	128,000	15	–
1000	115,000	96,000	19	–

Rupture Strength, 1000 hr

Test temperature, °F	Strength, psi	Elong., in 2 in., %	Reduction of area, %
800	132,000	–	38.5

Impact Strength

Test temperature, °F	Type test	Strength, ft-lb
+70	Charpy – unnotched	33
212	Charpy – unnotched	46

Specifications

ASTM: A-461, 564

Table 1.1–73
AGE HARDENABLE STAINLESS STEELS – WROUGHT, CAST

Type AM-362

Typical chemical composition, percent by weight: C,
0.03; Mn, 0.30; Fe, balance; Si, 0.20; Cr, 14.5; Ni, 6.5;
Ti, 0.80

Physical constants and thermal properties
Density, lb/in.^{3} : 0.281
Coefficient of thermal expansion, (70–200°F)
    in./in./°F × 10^{-6} :5.7
Modulus of elasticity, psi: tension, 28.9 × 10^6
Electrical resistivity, ohms/cmil/ft, 70°F: 540

Short Time Tensile Properties

Temperature, °F	T.S., psi	Y.S., psi, 0.2% offset	Elong., in 2 in., %	Hardness, Brinell
70	165,000	160,000	16	–

Impact Strength

Test temperature, °F	Type test	Strength, ft-lb
+70	Charpy – unnotched	28–32
212	Charpy – unnotched	42

Table 1.1–74
AGE HARDENABLE STAINLESS STEELS – WROUGHT, CAST

Type PH14-8 Mo

Typical chemical composition, percent by weight: C,
0.05; Mn, 1.00; Fe, balance; Si, 1.00; Cr, 13.75–15.00;
Ni, 7.50–8.75; Mo, 2.00–3.00; Al, 0.75–1.50

Physical constants and thermal properties
Density, lb/in.^{3} : 0.278
Coefficient of thermal expansion, (70–200°F)
    in./in./°F × 10^{-6} : 5.3
Curie temperature, °F: annealed, 1825; age hardened,
    950 (1 hr)

Heat treatments
Annealed (1825°F, austenitized (1700°F, 1 hr), air
    cooled, refrigerated within 1 hr (–100°F, 8 hr), aged
    (950°F, 1 hr).

Table 1.1−74 (continued)
AGE HARDENABLE STAINLESS STEELS − WROUGHT, CAST

Type PH14-8 Mo (continued)

Short Time Tensile Properties

Temperature, °F	T.S., psi	Y.S., psi, 0.2% offset	Elong., in 2 in., %	Hardness, Rockwell
70	235,000	220,000	5	B-90,C-48
800	205,000	185,000	5	−
1000	185,000	160,000	7	−
1200	132,000	110,000	16	−

Specifications

SAE-AMS: 5601, 5603

Table 1.1−75
AGE HARDENABLE STAINLESS STEELS − WROUGHT, CAST

Type 15-5PH

Typical chemical composition, percent by weight: C, 0.07; Mn, 1.00; Fe, balance; Si, 1.00; Cr, 14.00−15.00; Ni, 3.50−5.50; Cb + Ta, 0.15−0.45; Cu, 2.50−4.50

Modulus of elasticiity, psi: tension, 28.5×10^6
Electrical resistivity, ohms/cmil/ft, 70°F: 462
Curie temperature, °F: annealed, 1900; age hardened, 900 (1 hr)

Physical constants and thermal properties
Density, lb/in.^{3} : 0.282
Coefficient of thermal expansion, (70−200°F) in./in./°F $\times$ 10^{-6}: 6.2

Heat treatments
Annealed (1900°F, air or oil cooled), aged (900°F 1 hr) air cooled.

Short Time Tensile Properties

Temperature, °F	T.S., psi	Y.S., psi, 0.2% offset	Elong., in 2 in., %	Hardness, Brinell
70	190,000	170,000	10	−

Impact Strength

Test temperature, °F	Type test	Strength, ft-lb
+ 70	Charpy − unnotched	15

Table 1.1–76
AGE HARDENABLE STAINLESS STEELS – WROUGHT, CAST

Type Custom 450

Typical chemical composition, percent by weight: C, 0.05; Fe, balance; Cr, 15.5; Ni, 6.75; Mo, 0.75; Cb, 8 × C; Cu, 1.5

Physical constants and thermal properties
Density, lb/in.^{3} : 0.28

Coefficient of thermal expansion, (70–200°F) in./in./°F × 10^{-6} : 5.6
Curie temperature, °F: annealed, 1900

Heat treatments
Hardened 1900°F (1/2 hr), water quenched, 900°F (4 hr) air cooled.

Short Time Tensile Properties

Temperature, °F	T.S., psi	Y.S., psi, 0.2% offset	Elong., in 2 in., %	Hardness, Rockwell
70	144,000	117,000	14	C 26–44

Impact Strength

Test temperature, °F	Type test	Strength, ft-lb
+ 70	Charpy – unnotched	60

Table 1.1–77
AGE HARDENABLE STAINLESS STEELS – WROUGHT, CAST

Type Custom 455

Typical chemical composition, percent by weight: C, 0.03; Mn, 0.50; Fe, balance; Si, 0.50; Cr, 11.75 Ni, 9.0; Cb + Ta, 0.30; Ti, 1.20; Cu, 2.25

Physical constants and thermal properties
Density, lb/in.^{3} : 0.28
Coefficient of thermal expansion, (70–200°F) in./in./°F × 10^{-6} : 5.9

Electrical resistivity, ohms/cmil/ft, 70°F: 450
Curie temperature, °F: annealed, 1500°F; Age hardened, 900 (4 hr)

Heat treatments
Annealed (1500°F, ½ hr, water quenched) aged (900°F, 4 hr) air cooled.

Short Time Tensile Properties

Temperature, °F	T.S., psi	Y.S., psi, 0.2% offset	Elong., in 2 in., %	Hardness, Rockwell
70	140,000–260,000	115,000–250,000	3–18	C25–54

Specifications

SAE-AMS 5578, 5617,5672, 5860

Impact Strength

Test temperature, °F	Type test	Strength, ft-lb
70	Charpy – unnotched	10

Table 1.1–78
AGE HARDENABLE STAINLESS STEELS – WROUGHT, CAST

Type AFC-77

Typical chemical composition, percent by weight: C, 0.12–0.17; Fe, balance; Cr, 13.5–14.5; Co, 13.0–14.0; Mo, 4.5–5.5; V, 0.10–0.30

Physical constants and thermal properties
Density, lb/in.^{3} : 0.284
Coefficient of thermal expansion, (70–200°F) in./in./°F × 10^{-6} : 5.87

Electrical resistivity, ohms/cmil/ft, 70°F: 441

Heat treatments
1. 700°F temper: austenitized (1900°F, 1 hr, oil quenched), refrigerated (–100°F, 1 hr minimum), double temper (700°F, 2 + 2 hr).
2. 1100°F temper: austenitize (1900°F, 1 hr, oil quenched), double temper (1100°F, 2 + 2 hr).

Short Time Tensile Properties

Temperature, °F	T.S., psi	Y.S., psi, 0.2% offset	Elong., in 2 in., %	Hardness, Rockwell
70	252,000	200,000	17	C 39–53
800	242,000	151,000	17	–

Impact Strength

Test temperature, °F	Type test	Strength, ft-lb
+70	Charpy – unnotched	10

Specifications

SAE-AMS: 5748

Table 1.1–79
AUSTENITIC STAINLESS STEELS–WROUGHT

AISI Type 201

Typical chemical composition, percent by weight:
C, 0.15; Mn, 5.5–10.0; Fe, balance; Cr, 16.0–19.0; Ni, 3.5–6.0; N, 0.25

Physical constants and thermal properties
Density, lb/in.³: 0.28
Coefficient of thermal expansion, (70–200°F) in./in./°F × 10^{-6}: 8.7

Modulus of elasticity, psi: tension, 28.0 × 10^6
Specific heat, Btu/lb/°F, 70°F: 0.12
Electrical resistivity, ohms/cmil/ft, 70°F: 414
Curie temperature, °F: annealed, 1850–2050
Thermal conductivity, Btu/ft²/hr/in./°F, 70°F: 112.8
Heat treatments
Forging temperature (start) 2100–2250°F; annealing temperature 1850–2050°F.

Short Time Tensile Properties

Temperature, °F	T.S., psi	Y.S., psi, 0.2% offset	Elong., in 2 in., %	Hardness, Rockwell
70	1,157,000	55,000	55	B 90

Impact Strength

Test temperature, °F	Type test	Strength, ft-lb
+70	Izod	110–120
–300		38–70

Table 1.1–80
AUSTENITIC STAINLESS STEELS – WROUGHT

AISI Type 301

Typical chemical composition, percent by weight:
C, 0.15; Mn, 2; Fe, balance; S, 0.03; Si, 1; Cr, 16–18; Ni, 6–8; P, 0.045

Physical constants and thermal properties
Density, lb/in.³: 0.29
Coefficient of thermal expansion, (70–200°F) in./in./°F × 10^{-6}: 9.4
Modulus of elasticity, psi: tension: 28.0 × 10^6

Melting range, °F: 2550–2590
Specific heat, Btu/lb/°F, 70°F: 0.12
Thermal conductivity, Btu/ft²/hr/in./°F, 70°F: 112.8
Electrical resistivity, ohms/cmil/ft, 70°F: 432
Curie temperature, °F: annealed, 1850–2050

Heat treatments
Forging temperature (start) 2100–2300°F; annealing temperature 1850–2050°F.

Short Time Tensile Properties

Temperature, °F	T.S., psi	Y.S., psi, 0.2% offset	Elong., in 2 in., %	Hardness, Rockwell
70	110,000	40,000	60	B 85
+32	155,000	43,000	53	–
–40	180,000	48,000	42	–
–320	275,000	75,000	30	–

Table 1.1–80 (continued)
AUSTENITIC STAINLESS STEELS – WROUGHT

AISI Type 301 (continued)

Creep Strength (Stress, psi, to Produce 1% Creep)

Test temperature, °F	10,000 hr	100,000 hr
1,000	19,000	–
1,100	12,500	–
1,200	8,000	–
1,300	9,500	–
1,500	1,800	–

Fatigue Strength

Test temperature, °F	Stress, psi	Cycles to failure
70	35,000 endurance limit	–

Impact Strength

Test temperature, °F	Type test	Strength, ft-lb
+ 32	Izod	110
+ 70	Izod	110
– 40	Izod	110
– 320	Izod	110

Table 1.1–81
AUSTENITIC STAINLESS STEELS – WROUGHT

AISI Type 302

Typical chemical composition, percent by weight:
C, 0.15; Mn, 2; Fe, balance; S, 0.03; Si, 1; Cr, 17–19; Ni, 8–10; P, 0.045

Physical constants and thermal properties
Density, lb/in.^{3}: 0.29
Coefficient of thermal expansion, (70–200°F) in./in./°F $\times 10^{-6}$: 9.6
Modulus of elasticity, psi: tension, 28.0×10^6

Melting range, °F: 2550–2590
Specific heat, Btu/lb/°F, 70°F: 0.12
Thermal conductivity, Btu/ft^2/hr/in./°F, 70°F: 112.8
Electrical resistivity, ohms/cmil/ft., 70°F: 432
Curie temperature, °F: annealed, 1850–2050

Heat treatments
Forging temperature (start) 2100–2300°F; annealing temperature 1850–2050°F.

Short Time Tensile Properties

Annealed

Temperature, °F	T.S., psi	Y.S., psi, 0.2% offset	Elong., in 2 in., %	Hardness, Rockwell
70	90,000	37,000	68	B 85
+32	122,000	40,000	65	–
–40	145,000	48,000	60	–
–320	219,000	68,000	46	–
–423	250,000	125,000	41	–

Table 1.1–81 (continued)
AUSTENITIC STAINLESS STEELS – WROUGHT

AISI Type 302 (continued)

Creep Strength (Stress, psi, to Produce 1% Creep)

Test temperature, °F	10,000 hr	100,000 hr
1,000	17,000	–
1,300	4,000	–
1,500	1,200	–

Impact Strength

Test temperature, °F	Type test	Strength, ft-lb
+32	Izod	110
+70	Izod	110
–40	Izod	110
–320	Izod	110

Table 1.1–82
AUSTENITIC STAINLESS STEELS – WROUGHT

AISI Type 304

Typical chemical composition, percent by weight:
C, 0.08; Mn, 2; Fe, balance; S, 0.030; Si, 1; Cr, 18–20; Ni, 8–12; P, 0.045

Physical constants and thermal properties
Density, lb/in.^{3}: 0.29
Coefficient of thermal expansion, (70–200°F) in/in./°F $\times 10^{-6}$: 9.6
Modulus of elasticity, psi: tension, 28.0 $\times 10^6$

Melting range, °F: 2550–2650
Specific heat, Btu/lb/°F, 70°F: 0.12
Thermal conductivity, Btu/ft^2/hr/in./°F, 70°F: 112.8
Electrical resistivity, ohms/cmil/ft, 70°F: 432
Curie temperature, °F: annealed, 1850–2050

Heat treatments
Annealing temperature 1850–2050°F; forging temperature (start) 2100–2300°F.

Short Time Tensile Properties

Temperature, °F	T.S., psi	Y.S., psi, 0.2% offset	Elong., in 2 in., %	Hardness, Brinell
70	84,000	42,000	65	149
+32	130,000	34,000	55	–
–40	155,000	34,000	47	–
–320	221,000	39,000	40	–
–423	243,000	50,000	40	–

Creep Strength (Stress, psi, to Produce 1% Creep)

Test temperature, °F	1,000 hr	10,000 hr	100,000 hr
1,000	17,000	20,000	–
1,100	–	12,000	–
1,200	–	7,000	–
1,300	4,000	4,500	–
1,500	1,200	2,000	–

Table 1.1—82 (continued)
AUSTENITIC STAINLESS STEELS — WROUGHT

AISI Type 304

Impact Strength

Annealed

Test temperature, °F	Type test	Strength, ft-lb
+ 32	Izod	110
+ 70	Izod	110
− 40	Izod	110
− 32	Izod	110

Table 1.1—83
AUSTENITIC STAINLESS STEELS — WROUGHT

AISI Type 304 L

Typical chemical composition, percent by weight: C, 0.030; Mn, 2; Fe, balance; S, 0.030; Si, 1; Cr, 18–20; Ni, 8–12; P, 0.045

Physical constants and thermal properties
Density, lb/in.³: 0.29
Coefficient of thermal expansion, (70–200°F) in./in./°F × 10⁻⁶: 9.6
Modulus of elasticity, psi: tension, 28.0 × 10⁶

Melting range, °F: 2550–2650
Specific heat, Btu/lb/°F, 70°F: 0.12
Thermal conductivity, Btu/ft²/hr/in./°F, 70°F: 112.8
Electrical resistivity, ohms/cmil/ft, 70°F: 432
Curie temperature, °F: annealed, 1850–2050

Heat treatments
Annealing temperature 1850–2050°F; forging temperature 2100–2300°F.

Short Time Tensile Properties

Temperature, °F	T.S., psi	Y.S., psi, 0.2% offset	Elong., in 2 in., %	Hardness, Brinell
70	81,000	39,000	65	143
+32	130,000	34,000	55	–
−40	155,000	34,000	47	–
−320	221,000	39,000	40	–
−423	243,000	50,000	40	–

Creep Strength
(Stress, psi, to Produce 1% Creep)

Test temperature, °F	10,000 hr	100,000 hr
1000	20,000	–
1100	12,000	–
1200	7,500	–
1300	4,500	–
1500	2,000	–

Impact Strength

Annealed

Test temperature, °F	Type test	Strength, ft-lb
+32	Izod	110
+70	Izod	110
−40	Izod	110
−32	Izod	110

Table 1.1—84

AUSTENITIC STAINLESS STEELS — WROUGHT

AISI Type 308

Typical chemical composition, percent by weight: C, 0.08; Mn, 2; Fe, balance; S, 0.030; Si, 1; Cr, 19—21; Ni, 10—12; P, 0.45

Physical constants and thermal properties
Density, lb/in.^{3} : 0.29
Coefficient of thermal expansion, (70—200°F) in./in./°F × 10^{-6}: 9.6
Modulus of elasticity, psi: tension, 28.0 × 10^6

Melting range, °F: 2550—2590
Specific heat, Btu/lb/°F, 70°F: 0.12
Thermal conductivity, Btu/ft^2/hr/in./°F, 70°F: 105.6
Electrical resistivity, ohms/cmil/ft, 70°F: 432
Curie temperature, °F: annealed, 1850—2050

Heat treatments
Annealing temperature 1850—2050°F; forging temperature (start) 2100—2300°F.

Short Time Tensile Properties

Temperature, °F	T.S. psi	Y.S., psi, 0.2% offset	Elong., in 2 in., %	Hardness, Brinell
70	85,000	35,000	50	150

Impact Strength

Annealed

Test temperature, °F	Type test	Strength, ft-lb
+70	Izod	110

Table 1.1—85

AUSTENITIC STAINLESS STEELS — WROUGHT

AISI Type 316

Typical chemical composition, percent by weight: C, 0.08; Mn, 2; Fe, balance; S, 0.03; Si, 1; Cr, 16—18; Ni, 10—14; Mo, 2—3; P, 0.045

Physical constants and thermal properties
Density, lb/in.^{3} : 0.29
Coefficient of thermal expansion, (70—200°F) in./in./°F × 10^{-6}: 8.9
Modulus of elasticity, psi: tension, 28.0 × 10^6

Melting range, °F: 2500—2550
Specific heat, Btu/lb/°F, 70°F: 0.12
Thermal conductivity, Btu/ft^2/hr/in./°F, 70°F: 112.8
Electrical resistivity, ohms/cmil/ft, 70°F: 444
Curie temperature, °F: annealed, 1850—2050

Heat treatments
Annealing temperature 1850—2050°F; forging temperature (start) 2100—2300°F.

Short Time Tensile Properties

Temperature, °F	T.S., psi	Y.S., psi, 0.2% offset	Elong., in 2 in., %	Hardness, Brinell
70	84,000	42,000	65	149
+32	90,000	39,000	60	—
—40	104,000	41,000	59	—
—320	185,000	75,000	59	—
—423	210,000	84,000	52	—

Table 1.1–85 (continued)
AUSTENITIC STAINLESS STEELS – WROUGHT

AISI Type 316 (continued)

Creep Strength
(Stress, psi, to Produce 1% Creep)

Test temperature, °F	10,000 hr	100,000 hr
1000	25,000	–
1100	17,400	–
1200	11,600	–
1300	7,500	–
1500	2,400	–

Fatigue Strength

Test temperature, °F	Stress, psi	Cycles to failure
70	39,000 endurance limit	–

Impact Strength

Annealed

Test temperature, °F	Type test	Strength, ft-lb
+32	Izod	110
+70	Izod	110
–40	Izod	110
–80	Izod	110

Table 1.1–86
AUSTENITIC STAINLESS STEELS – WROUGHT

AISI Type 316L

Typical chemical composition percent by weight:
C, 0.030; Mn, 2; Fe, balance; S, 0.030; Si, 1.00; Cr, 16–18; Ni, 10–14; Mo, 2–3; P, 0.045

Physical constants and thermal properties
Density, lb/in.^{3}: 0.29
Coefficient of thermal expansion, (70–200° F) in./in./° F × 10^{-6}: 8.9
Melting range, °F: 2500–2550

Specific heat, Btu/lb/° F, 70° F: 0.12
Thermal conductivity, Btu/ft^2/hr/in./° F, 70° F: 112.8
Electrical resistivity, ohms/cmil/ft, 70° F: 444
Curie temperature, °F: annealed, 1850–2050

Heat treatments
Forging temperature (start) 2100–2300° F; annealing temperature 1850–2050° F.

Short Time Tensile Properties

Temperature, °F	T.S., psi	Y.S., psi, 0.2% offset	Elong., in 2 in.,%	Hardness, Brinell
70	84,000	42,000	50	146
+32	90,000	39,000	60	–
–40	104,000	41,000	59	–
–320	185,000	75,000	59	–
–423	210,000	84,000	52	–

Creep Strength (Stress, psi, to Produce 1% Creep)

Test temperature, °F	10,000 hr	100,000 hr
1000	25,000	–
1100	17,400	–
1200	11,600	–
1300	7,500	–
1500	2,400	–

Impact Strength

Annealed

Test temperature, °F	Type test	Strength, ft-lb
+32	Izod	110
+70	Izod	110
–40	Izod	110
–80	Izod	110

Table 1.1–87
AUSTENITIC STAINLESS STEELS – WROUGHT

AISI Type 321

Typical chemical composition, percent by weight: C, 0.08; Mn, 2; Fe, balance; S, 0.03; Si, 1; Cr, 17–19; Ni, 9–12; Ti, 5 × C; P, 0.04

Physical constants and thermal properties
Density, lb/in.^{3} : 0.29
Coefficient of thermal expansion, (70–200°F) in./in./
  °F × 10^{-6} : 9.3
Modulus of elasticity, psi: tension, 28.0 × 10^6

Melting range, °F: 2550–2600
Specific heat, Btu/lb/°F, 70°F: 0.12
Thermal conductivity, Btu/ft^2 /hr/in./°F, 70°F: 111.6
Electrical resistivity, ohms/cmil/ft, 70°F: 432
Curie temperature, °F: annealed, 1750–2050

Heat treatments
Forging temperature (start) 2100–2300°F.

Short Time Tensile Properties

Temperature, °F	T.S., psi	Y.S., psi, 0.2% offset	Elong., in 2 in., %	Hardness, Brinell
70	90,000	35,000	62	160
+32	99,000	38,000	58	
–40	117,000	44,000	58	
–320	208,000	64,000	44	
–423	238,000	92,000	35	

Creep Strength
(Stress, psi, to Produce 1% Creep)

Test temperature, °F	10,000 hr	100,000 hr
1000	18,000	–
1300	4,500	–
1500	850	–

Impact Strength

Annealed

Test temperature, °F	Type test	Strength, ft-lb
+32	Izod	110
+70	Izod	110
–40	Izod	115
–80	Izod	117

Table 1.1–88
AUSTENITIC STAINLESS STEELS – WROUGHT

AISI Type 384

Typical chemical composition, percent by weight: C, 0.08; Mn, 2; Fe, balance; S, 0.030; Si, 1.00; Cr, 15–17; Ni, 17–19; P, 0.045

Physical constants and thermal properties
Density, lb/in.^{3} : 0.29
Coefficient of thermal expansion, (70–200°F) in./in./
  °F × 10^{-6} : 9.6
Modulus of elasticity, psi: tension, 28 × 10^6

Melting range, °F: 2550–2650
Specific heat, Btu/lb/°F, 70°F: 0.12
Thermal conductivity, Btu/ft^2 /hr/in./°F, 70°F: 112.8
Curie temperature, °F: annealed 1900–2100

Heat treatments
Forging temperature (start) 2100–2250°F; annealing temperature 1900–2100°F.

Table 1.1−88 (continued)
AUSTENITIC STAINLESS STEELS − WROUGHT

AISI Type 384

Short Time Tensile Properties

Temperature, °F	T.S., psi	Y.S., psi, 0.2% offset	Elong., in 2 in., %	Hardness, Rockwell
70	75,000	35,000	55	B 70

Table 1.1−89
FERRITIC STAINLESS STEELS − WROUGHT

AISI Type 405

Typical chemical composition, percent by weight: C, 0.08; Mn, 1.00; Fe, balance; S, 0.030; Si, 1.00; Cr, 11.5−14.5; Al, 0.10−0.30; P, 0.040

Physical constants and thermal properties
Density, lb/in.^{3} : 0.28
Coefficient of thermal expansion, (70−200°F) in./in./ °F × 10^{-6} : 6.0
Modulus of elasticity, psi: tension, 29 × 10^6

Melting range, °F: 2700−2790
Specific heat, Btu/lb/°F, 70°F: 0.11
Thermal conductivity, Btu/ft^2 /hr/in./°F, 70°F: 180
Electrical resistivity, ohms/cmil/ft, 70°F: 360
Curie temperature, °F: annealed, 1350−1500

Heat treatments
Forging temperature (start) 1950−2050°F; annealing temperature 1350−1500°F.

Short Time Tensile Properties

Temperature, °F	T.S., psi	Y.S., psi, 0.2% offset	Elong., in 2 in., %	Hardness, Rockwell
70	65,000	35,000	25	B75-90

Creep Strength
(Stress, psi, to Produce 1% Creep)

Test temperature, °F	10,000 hr	100,000 hr
1000	8,400	−

Impact Strength

Annealed

Test temperature, °F	Type test	Strength, ft-lb
+70	Izod	20−35

Table 1.1–90
FERRITIC STAINLESS STEELS – WROUGHT

AISI Type 430

Typical chemical composition, percent by weight: C, 0.12; Mn, 1.00; Fe, balance; S, 0.030; Si, 1.00; Cr, 14.0–18.0; P, 0.040

Physical constants and thermal properties
Density, lb/in.^{3} : 0.28
Coefficient of thermal expansion, (70–200°F) in./in./°F × 10^{-6} : 5.8
Modulus of elasticity, psi: tension, 29 × 10^6

Melting range, °F: 2600–2750
Specific heat, Btu/lb/°F, 70°F: 0.11
Thermal conductivity, Btu/ft^2/hr/in./°F, 70°F: 181.2
Electrical resistivity, ohms/cmil/ft, 70°F: 360
Curie temperature, °F: annealed 1400–1500

Heat treatments
Forging temperature (start) 1900–2050°F; annealing temperature 1400–1500°F.

Short Time Tensile Properties

Temperature, °F	T.S., psi	Y.S., psi, 0.2% offset	Elong., in 2 in., %	Hardness, Rockwell
70	65,000–75,000	40,000	25–30	B80
+32	69,000	40,000	37	–
–40	76,000	41,000	36	–
–320	90,000	87,000	2	–

Creep Strength
(Stress, psi, to Produce 1% Creep)

Test temperature, °F	10,000 hr	100,000 hr
1000	8,500	–
1100	4,700	–
1200	2,600	–
1300	1,400	–

Impact Strength

Annealed

Test temperature, °F	Type test	Strength, ft-lb
+32	Izod	20
+70	Izod	35
–40	Izod	10
–320	Izod	2

Table 1.1—91
FERRITIC STAINLESS STEELS — WROUGHT

AISI Type 434

Typical chemical composition, percent by weight: C, 0.12; Mn, 1.00; Fe, balance; S, 0.030; Si, 1.00; Cr, 16–18; Mo, 0.75–1.25; P, 0.040

Physical constants and thermal properties
Density, lb/in.^{3}: 0.28
Coefficient of thermal expansion, (70–200° F) in./in./ °F × 10^{-6} : 6.6
Modulus of elasticity, psi: tension, 29 × 10^6

Melting range, °F: 2600–2750
Specific heat, Btu/lb/°F, 70°F: 0.11
Thermal conductivity, Btu/ft^2/hr/in./°F, 70°F: 182.4
Electrical resistivity, ohms/cmil/ft, 70°F: 360
Curie temperature, °F: annealed, 1450

Heat treatments
Forging temperature (start) 1900–2050°F; annealing temperature 1450–1550°F.

Short Time Tensile Properties

Temperature, °F	T.S., psi	Y.S., psi, 0.2% offset	Elong., in 2 in., %	Hardness, Rockwell
70	77,000	53,000	23	B 83

Table 1.1—92
FERRITIC STAINLESS STEELS — WROUGHT

AISI Type 446

Typical chemical composition, percent by weight: C, 0.20; Mn, 1.50; Fe, balance; S, 0.030; Si, 1.00; Cr, 23.0–27.0; P, 0.040; N, 0.25

Physical constants and thermal properties
Density, lb/in.^{3}: 0.27
Coefficient of thermal expansion, (70–200° F) in./in./ °F × 10^{-6}: 5.8
Modulus of elasiticty, psi: tension, 29 × 10^6

Melting range, °F: 2600–2750
Specific heat, Btu/lb/°F, 70°F: 0.12
Thermal conductivity, Btu/ft^2/hr/in./°F, 70°F: 145.2
Electrical resistivity, ohms/cmil/ft, 70°F: 402
Curie temperature, °F: annealed, 1450–1600

Heat treatments
Forging temperature (start) 1950–2050°F; annealing temperature 1450–1600°F.

Short Time Tensile Properties

Temperature, °F	T.S., psi	Y.S., psi, 0.2% offset	Elong., in 2 in., %	Hardness, Brinell
70	80,000	50,000	20	83

Creep Strength
(Stress, psi, to Produce 1% Creep)

Test temperature, °F	10,000 hr	100,000 hr
1,000	6,000	4,200
1,100	2,900	–
1,200	1,400	–
1,300	600	–
1,500	400	–

Table 1.1–92 (continued)
FERRITIC STAINLESS STEELS – WROUGHT

AISI Type 446 (continued)

Impact Strength

Annealed

Test temperature, °F	Type test	Strength, ft-lb
+70	Izod	2–10

Table 1.1–93
MARTENSITIC STAINLESS STEELS – WROUGHT

AISI Type 403

Typical chemical composition, percent by weight: C, 0.15; Mn, 1.00; Fe, balance; S, 0.030; Si, 0.50; Cr, 11.5–13; P, 0.040

Specific heat Btu/lb/°F, 70°F: 0.11
Thermal conductivity, Btu/ft²/hr/in./°F, 70°F: 172.8
Electrical resistivity, ohms/cmil/ft, 70°F: 342
Curie temperature, °F: annealed, 1500–1650

Physical constants and thermal properties
Density, lb/in.³: 0.28
Coefficient of thermal expansion, (70–200°F) in./in./
°F × 10⁻⁶: 5.5
Modulus of elasticity, psi: tension, 29 × 10⁶
Melting range, °F: 2700–2790°F

Heat treatments
Annealing temperature 1500–1650°F; hardening temperature 1700–1850°F; tempering temperature 400–1400°F; forging temperature (start) 2000–2200°F.

Short Time Tensile Properties

Temperature, °F	T.S., psi	Y.S., psi, 0.2% offset	Elong., in 2 in., %	Hardness, Brinell
70	70,000–110,000	40,000–85,000	25–35	155
+32	115,000	89,000	24	–
–40	122,000	90,000	23	–
–320	158,000	148,000	10	–

Creep Strength
(Stress, psi, to Produce 1% Creep)

Test temperature, °F	10,000 hr	100,000 hr
1,000	11,000	–
1,100	4,500	–
1,200	2,000	–
1,300	1,400	–

Fatigue Strength

Test temperature, °F	Stress, psi	Cycles to failure
70	40,000 endurance limit	–

Table 1.1–93 (continued)
MARTENSITIC STAINLESS STEELS – WROUGHT

AISI Type 403 (continued)

Impact Strength

Annealed

Test temperature, °F	Type test	Strength, ft-lb
+32	Izod	40
+70	Izod	85
–80	Izod	25
–320	Izod	5

Table 1.1–94
MARTENSITIC STAINLESS STEELS – WROUGHT

AISI Type 410

Typical chemical composition, percent by weight: C, 0.15; Mn, 1.00; Fe, balance; S, 0.030; Si, 1; Cr, 11.5–13.5; P, 0.040

Specific heat, Btu/lb/°F, 70°F: 0.11
Thermal conductivity, Btu/ft²/hr/in./°F, 70°F: 172.8
Electrical resistivity, ohms/cmil/ft, 70°F: 420
Curie temperature, °F: annealed, 1500–1650

Physical constants and thermal properties
Density, lb/in.³: 0.28
Coefficient of thermal expansion, (70–200°F) in./in./°F ×10⁻⁶: 5.5
Modulus of elasticity, psi: tension, 29 × 10⁶
Melting range, °F: 2700–2790

Heat treatments
Annealing temperature 1500–1650°F; hardening temperature 1700–1850°F; tempering temperature 400–1400°F; forging temperature (start) 2000–2200°F.

Short Time Tensile Properties

Annealed

Temperature, °F	T.S., psi	Y.S., psi, 0.2% offset	Elong., in 2 in., %	Hardness, Brinell
70	65,000–110,000	35,000–85,000	25–35	155
+32	115,000	89,000	24	–
–40	122,000	90,000	23	–
–320	158,000	148,000	10	–

Creep Strength
(Stress, psi, to Produce 1% Creep)

Test temperature, °F	10,000 hr	100,000 hr
1,000	9,200	–
1,100	4,300	–
1,200	2,000	–
1,300	1,500	–

Table 1.1–94 (continued)
MARTENSITIC STAINLESS STEELS – WROUGHT

AISI Type 410 (continued)

Fatigue Strength

Test temperature, °F	Stress, psi	Cycles to failure
70	40,000 endurance limit	–

Impact Strength

Annealed

Test temperature, °F	Type test	Strength, ft-lb
+32	Izod	80
+70	Izod	85
−80	Izod	25
−320	Izod	5

Table 1.1–95
MARTENSITIC STAINLESS STEELS – WROUGHT

AISI Type 420

Typical chemical composition, percent by weight: C, 0.15; Mn, 1.00; Fe, balance; S, 0.030; Si, 1.00; Cr, 12–14; P, 0.040

Physical constants and thermal properties
Density, lb/in.³: 0.28
Coefficient of thermal expansion, (70–200°F) in./in./°F × 10⁻⁶: 5.7
Modulus of elasticity, psi: tension, 29.0 × 10⁶
Melting range, °F: 2650–2750

Specific heat, Btu/lb/°F, 70°F: 0.11
Thermal conductivity, Btu/ft²/hr/in./°F, 70°F: 172.8
Electrical resistivity, ohms/cmil/ft, 70°F: 330
Curie temperature, °F: annealed, 1550–1650

Heat treatments
Annealing temperature 1550–1650°F; hardening temperature 1800–1900°F; tempering temperature 300–700°F; forging temperature 2000–2200°F.

Short Time Tensile Properties

Temperature, °F	T.S., psi	Y.S., psi, 0.2% offset	Elong., in 2 in., %	Hardness, Brinell
70	95,000–230,000	50,000–195,000	25–28	195–500

Creep Strength
(Stress, psi, to Produce 1% Creep)

Test temperature, °F	10,000 hr	100,000 hr
1,000	9,200	–
1,100	4,200	–
1,200	2,000	–
1,300	1,000	–

Table 1.1–95 (continued)
MARTENSITIC STAINLESS STEELS – WROUGHT

AISI Type 420 (continued)

Fatigue Strength

Test temperature, °F	Stress, psi	Cycles to failure
70	40,000 endurance limit	–

Impact Strength (AISI Type 431)

Annealed

Test temperature, °F	Type test	Strength, ft-lb
+32	Izod	10
+70	Izod	10
–40	Izod	8
–80	Izod	7

Table 1.1–96
MARTENSITIC STAINLESS STEELS – WROUGHT

AISI Type 440A

Typical chemical composition, percent by weight: C, 0.60–0.75; Mn, 1.00; Fe, balance; S, 0.030; Si, 1.00; , Cr, 16–18; Mo, 0.75; P, 0.040

Physical constants and thermal properties
Density, lb/in.³: 0.28
Coefficient of thermal expansion, (70–200°F) in./in./ °F × 10⁻⁶: 5.6
Modulus of elasticity, psi: tension 29.0 × 10⁶
Melting range, °F: 2500–2750

Specific heat, Btu/lb/°F, 70°F: 0.11
Thermal conductivity, Btu/ft²/hr/in./°F, 70°F: 168
Electrical resistivity, ohms/cmil/ft, 70°F: 360
Curie temperature, °F: annealed, 1550–1650

Heat treatments
Annealing temperature 1550–1650°F; hardening temperature 1850–1950°F; tempering temperature 300–800°F; forging temperature (start) 1900–2200°F.

Short Time Tensile Properties

Temperature, °F	T.S., psi	Y.S., psi, 0.2% offset	Elong., in 2 in., %	Hardness, Brinell
70	125,000	60,000	20	215

Fatigue Strength

Test temperature, °F	Stress, psi	Cycles to failure
70	40,000 endurance limit	–

Impact Strength

Test temperature °F	Type test	Strength, ft-lb
70	Izod	2

Table 1.1–97
MARTENSITIC STAINLESS STEELS – WROUGHT

AISI Type 501

Typical chemical composition, percent by weight: C, 0.10; Mn, 1.00; Fe, balance; S, 0.030; Si, 1.00; Cr, 4–6; Mo, 0.40–0.65; P, 0.40

Specific heat, Btu/lb/°F, 70°F: 0.11
Thermal conductivity, Btu/ft²/hr/in./°F, 70°F: 254.4
Electrical resistivity, ohms/cmil/ft, 70°F: 240
Curie temperature, °F: annealed, 1525–1600

Physical constants and thermal properties
Density, lb/in.³: 0.28
Coefficient of thermal expansion, (70–200°F) in./in./°F × 10^{-6}:6.2
Modulus of elasticity, psi: tension, 29 × 10^6
Melting range, °F: 2700–2800

Heat treatments
Annealing temperature 1525–1600°F; hardening temperature 1600–1700°F; tempering temperature 400–1400°F; forging temperature (start) 2100–2200°F.

Short Time Tensile Properties

Temperature, °F	T.S., psi	Y.S., psi, 0.2% offset	Elong., in 2 in., %	Hardness, Brinell
70	70,000	30,000	28	160

Table 1.1–98
SPECIALTY STAINLESS STEELS – WROUGHT

Type Flo 302 HQ

Typical chemical composition, percent by weight: C, 0.08; Fe, balance; Cr, 18; Ni, 9; Cu, 3.5

Physical constants and thermal properties
Density, lb/in.³: 0.29
Coefficient of thermal expansion, (70–200°F) in./in./°F × 10^{-6}: 9.6

Modulus of elasticity, psi: tension, 28 × 10^6
Specific heat, Btu/lb/°F, 70°F: 0.12
Thermal conductivity, Btu/ft²/hr/in./°F, 70°F: 78
Electrical resistivity, ohms/cmil/ft, 70°F: 432

Heat treatments
Rod, cold worked 50%.

Short Time Tensile Properties

Temperature, °F	T.S., psi	Y.S., psi, 0.2% offset	Elong., in 2 in., %	Hardness, Rockwell
70	73,000	27,000	65	B 80

Table 1.1–99
SPECIALTY STAINLESS STEELS – WROUGHT

Type JS 700

Typical chemical composition, percent by weight: C, 0.03; Fe, balance; Cr, 21; Ni, 25; Mo, 4.5; Cb, 0.30

Physical constants and thermal properties
Density, lb/in.^{3}: 0.29
Coefficient of thermal expansion, (70–200°F) in./in./
°F × 10^{-6}: 9.15
Modulus of elasticity, psi: tension, 29 × 10^6
Specific heat, Btu/lb/°F, 70°F: 0.12
Thermal conductivity, Btu/ft^2/hr/in./°F, 70°F: 102

Short Time Tensile Properties

Temperature, °F	T.S., psi	Y.S., psi, 0.2% offset	Elong., in 2 in., %	Hardness, Brinell
70	85,000	39,000	45	170

Table 1.1–100
SPECIALTY STAINLESS STEELS – WROUGHT

Type MF-1

Typical chemical composition, percent by weight: C, 0.045; Fe, balance; Cr, 11; Ti, 0.5

Physical constants and thermal properties
Density, lb/in.^{3}: 0.276
Coefficient of thermal expansion, (70–200°F) in./in./
°F × 10^{-6}: 6.5
Specific heat, Btu/lb/°F, 70°F: 0.11

Short Time Tensile Properties

Temperature, °F	T.S., psi	Y.S., psi, 0.2% offset	Elong., in 2 in., %	Hardness, Rockwell
70	65,000	34,000	32	B 72

Table 1.1–101
SPECIALTY STAINLESS STEELS – WROUGHT

Pyromet® 355

Typical chemical composition, percent by weight: C, 0.12; Fe, balance; Cr, 15.5; Ni, 4.5; Mo, 3.0; N, 0.1

Physical constants and thermal properties
Density, lb/in.³ : 0.286
Coefficient of thermal expansion, (70–200°F) in./in./°F × 10⁻⁶ : 8.3

Modulus of elasticity, psi: tension, 29.3×10^6
Specific heat, Btu/lb/°F, 70°F: 0.12
Electrical resistivity, ohms/cmil/ft, 70°F: 450

Heat treatments
Strip 1750°F, rapid cool, -100°F 3 hr, tempered 850°F.

Short Time Tensile Properties

Temperature, °F	T.S., psi	Y.S., psi, 0.2% offset	Elong., in 2 in., %	Hardness, Rockwell
70	186,000	55,000	29	B 100

Table 1.1–102
SPECIALTY STAINLESS STEELS – WROUGHT

Uniloy® 326

Typical chemical composition, percent by weight: C, 0.05; Fe, balance; Cr, 26; Ni, 6.5; Ti, 0.2

Physical constants and thermal properties
Density, lb/in.³ : 0.279
Coefficient of thermal expansion, (70–200°F) in./in./°F × 10⁻⁶ : 5.9
Modulus of elasticity, psi: tension, 27×10^6
Specific heat, Btu/lb/°F, 70°F: 0.102
Thermal conductivity, Btu/ft² /hr/in./°F, 70°F: 135.6
Electrical resistivity, ohms/cmil/ft, 70°F: 312

Short Time Tensile Properties

Temperature, °F	T.S., psi	Y.S., psi, 0.2% offset	Elong., in 2 in., %	Hardness, Brinell
70	100,000	75,000	32	95

Table 1.1−103
SPECIALTY STAINLESS STEELS − WROUGHT

Type 18 SR

Typical chemical composition, percent by weight: C, 0.05; Fe, balance; Cr, 18; Ti, 0.4; Al, 2.0

Physical constants and thermal properties
Density, lb/in.^{3} : 0.27
Coefficient of thermal expansion, (70−200°F) in./in./°F × 10^{-6} : 5.9
Electrical resistivity ohms/cmil/ft, 70°F: 660

Heat treatments
Annealed

Short Time Tensile Properties

Temperature, °F	T.S., psi	Y.S., psi, 0.2% offset	Elong., in 2 in., %	Hardness, Brinell
70	85,000	65,000	27	90

Table 1.1−104
SPECIALTY STAINLESS STEELS − WROUGHT

Type 18-2 Mn

Typical chemical composition, percent by weight: C, 0.10; Mn, 12; Fe, balance; Cr, 18; Ni, 1.6; N, 0.35

Physical constants and thermal properties
Density, lb/in.^{3} : 0.281
Coefficient of thermal expansion, (70−200°F) in./in./°F × 10^{-6} : 10.3
Modulus of elasticity, psi: tension, 29 × 10^6

Heat treatments
Annealed; forging temperature 2100−2200°F.

Short Time Tensile Properties

Temperature, °F	T.S., psi	Y.S., psi, 0.2% offset	Elong., in 2 in., %	Hardness, Brinell
70	120,000	65,000	60	98

Table 1.1–105
SPECIALTY STAINLESS STEELS – WROUGHT

Type 18-18-2

Typical chemical composition, percent by weight: C, 0.06; Fe, balance; Si, 1.9; Cr, 18; Ni, 18

Physical constants and thermal properties
Density, lb/in.^{3} : 0.284
Coefficient of thermal expansion, (70–200°F) in./in./°F $\times$ 10^{-6} : 7.6
Electrical resistivity, ohms/cmil/ft, 70°F: 516

Heat treatments
Annealed

Short Time Tensile Properties

Temperature, °F	T.S., psi	Y.S., psi, 0.2% offset	Elong., in 2 in., %	Hardness, Brinell
70	81,000	36,000	54	–

Table 1.1–106
SPECIALTY STAINLESS STEELS – WROUGHT

Type 211

Typical chemical composition, percent by weight: C, 0.05; Mn, 6; Fe, balance; Cr, 17; Ni, 5.5; Cu, 1.5

Physical constants and thermal properties
Density, lb/in.^{3} : 0.284
Coefficient of thermal expansion, (70–200°F) in./in./°F $\times$ 10^{-6} : 9.4
Modulus of elasticity, psi: tension, 28.6 $\times$ 10^6
Electrical resistivity, ohms/cmil/ft, 70°F: 438

Heat treatments
Annealed; forging temperature, 2200–2300°F.

Short Time Tensile Properties

Temperature, °F	T.S., psi	Y.S., psi, 0.2% offset	Elong., in 2 in., %	Hardness, Brinell
70	87,000	31,000	60	74

Table 1.1–107
SPECIALTY STAINLESS STEELS – WROUGHT

Type 21-6-9

Typical chemical composition, percent by weight: C, 0.08; Mn, 9; Fe, balance; Cr, 20; Ni, 6.5; N, 0.30

Physical constants and thermal properties
Density, lb/in.^{3} : 0.283
Coefficient of thermal expansion, (70–200°F) in./in./°F × 10^{-6} : 8.5
Modulus of elasticity, psi: tension, 28.5 × 10^6
Thermal conductivity, Btu/ft^2/hr/in./°F, 70°F: 96
Electrical resistivity, ohms/cmil/ft, 70°F: 420

Short Time Tensile Properties

Temperature, °F	T.S., psi	Y.S., psi, 0.2% offset	Elong., in 2 in., %	Hardness, Rockwell
70	111,000	64,000	42	B 93

Table 1.1–108
SPECIALTY STAINLESS STEELS – WROUGHT

Type 22-13-5

Typical chemical composition, percent by weight: C, 0.06; Mn, 5; Fe, balance; Cr, 22; Ni, 12.5; Mo, 2.25; Cb, 0.20; N, 0.30; V, 0.20

Physical constants and thermal properties
Density, lb/in.^{3} : 0.285
Coefficient of thermal expansion, (70–200°F) in./in./°F × 10^{-6} : 9.0
Thermal conductivity, Btu/ft^2/hr/in./°F, 70°F: 108
Electrical resistivity, ohms/cmil/ft, 70°F: 480

Heat treatments
Annealed; forging temperature 2150–2250°F.

Short Time Tensile Properties

Temperature, °F	T.S., psi	Y.S., psi, 0.2% offset	Elong., in 2 in. %	Hardness, Rockwell
70	121,000	65,000	46	B 98

Impact Strength

Test temperature, °F	Type test	Strength, ft-lb
+70	Charpy – V-notched	170

Table 1.1–109
SPECIALTY STAINLESS STEELS – WROUGHT

Type 303 Plus-X$^{®}$

Typical chemical composition, percent by weight: C, 0.15; Mn, 3.5; Fe, balance; S, 0.15; Cr, 18; Ni, 8.5

Physical constants and thermal properties
Density, lb/in.^{3} : 0.286
Coefficient of thermal expansion, (70–200°F) in./in./°F × 10^{-6} : 9.6

Modulus of elasticity, psi: tension, 28 × 10^6
Specific heat, Btu/lb/°F, 70°F: 0.12
Thermal conductivity, Btu/ft^2/hr/in./°F, 70°F: 112.8
Electrical resistivity, ohms/cmil/ft, 70°F: 432

Heat treatments
Annealed; forging temperature 2100–2350°F.

Short Time Tensile Properties

Temperature, °F	T.S., psi	Y.S., psi, 0.2% offset	Elong., in 2 in., %	Hardness, Rockwell
70	90,000	35,000	3	B 88

Impact Strength

Test temperature, °F	Type test	Strength, ft-lb
+70	Izod	80

Table 1.1–110
SPECIALTY STAINLESS STEELS – WROUGHT

Type 304 + B

Typical chemical composition, percent by weight: C, 0.08; Fe, balance; Cr, 19; Ni, 13.5; B, 2

Physical constants and thermal properties
Density, lb/in.^{3} : 0.29
Coefficient of thermal expansion, (70–200°F) in./in./°F × 10^{-6} : 10.5

Modulus of elasticity, psi: tension, 30 × 10^6
Specific heat, Btu/lb/°F, 70°F: 0.12
Electrical resistivity, ohms/cmil/ft, 70°F: 444

Heat treatments
Annealed; forging temperature 2000–2200°F.

Short Time Tensile Properties

Temperature, °F	T.S., psi	Y.S., psi, 0.2% offset	Elong., in 2 in., %	Hardness, Rockwell
70	100,000	50,000	10	B 95

Impact Strength

Test temperature, °F	Type test	Strength, ft-lb
+70	Charpy – V-notched	5

Table 1.1–111
SPECIALTY STAINLESS STEELS – WROUGHT

Type 305 H

Typical chemical composition, percent by weight: C, 0.08; Fe, balance; Cr, 16; Ni, 18

Physical constants and thermal properties
Density, lb/in.^{3} : 0.29
Coefficient of thermal expansion, (70–200°F) in./in./°F $\times$ 10^{-6} : 9.6
Specific heat, Btu/lb/°F, 70°F: 0.12
Thermal conductivity, Btu/ft^2/hr/in./°F, 70°F: 112.8
Electrical resistivity, ohms/cmil/ft, 70°F: 474

Heat treatments
Annealed; forging temperature 2100–2300°F.

Short Time Tensile Properties

Temperature, °F	T.S., psi	Y.S., psi, 0.2% offset	Elong., in 2 in., %	Hardness, Brinell
70	80,000	40,000	55	156

Impact Strength

Test temperature, °F	Type test	Strength, ft-lb
+70	Izod	92

Table 1.1–112
SPECIALTY STAINLESS STEELS – WROUGHT

Type 305 MH

Typical chemical composition, percent by weight: C, 0.08; Fe, balance; Cr, 12.5; Ni, 15

Physical constants and thermal properties
Density, lb/in.^{3} : 0.29
Coefficient of thermal expansion, (70–200°F) in./in./°F $\times$ 10^{-6} : 10.4
Modulus of elasticity, psi: tension, 29 $\times$ 10^6
Specific heat, Btu/lb/°F, 70°F: 0.12
Electrical resistivity, ohms/cmil/ft, 70°F: 444

Heat treatments
Annealed; forging temperature 2000–2200°F.

Short Time Tensile Properties

Temperature, °F	T.S., psi	Y.S., psi, 0.2% offset	Elong., in 2 in., %	Hardness, Brinell
70	78,000	46,000	35	163

Table 1.1–113
SPECIALTY STAINLESS STEELS – WROUGHT

Type 329

Typical chemical composition, percent by weight: C, 0.15; Fe, balance; Cr, 27.5; Ni, 4.5; Mo, 1.5

Physical constants and thermal properties
Density, lb/in.^{3}: 0.28
Coefficient of thermal expansion, (70–200°F) in./in./°F × 10^{-6}: 8.0
Specific heat, Btu/lb/°F, 70°F: 0.11
Electrical resistivity, ohms/cmil/ft, 70°F: 450

Heat treatments
Forging temperature 1900–2100°F.

Short Time Tensile Properties

Temperature, °F	T.S., psi	Y.S., psi, 0.2% offset	Elong., in 2 in., %	Hardness, Brinell
70	105,000	80,000	25	–

Table 1.1–114
SPECIALTY STAINLESS STEELS – WROUGHT

Type 404

Typical chemical composition, percent by weight: C, 0.05; Fe, balance; Cr, 11.75; Ni, 1.6; N, 0.03

Physical constants and thermal properties
Density, lb/in.^{3}: 0.28
Coefficient of thermal expansion, (70–200°F) in./in./°F × 10^{-6}: 4.8
Modulus of elasticity, psi: tension, 29 × 10^6
Specific heat, Btu/lb/°F, 70°F: 0.11
Electrical resistivity, ohms/cmil/ft, 70°F: 420

Heat treatments
Forging temperature 2000–2100°F; worked 1750°F, oil quenched, tempered 700°F; cold worked.

Short Time Tensile Properties

Temperature, °F	T.S., psi	Y.S., psi, 0.2% offset	Elong., in 2 in., %	Hardness, Rockwell
70	165,000	135,000	18	C 32

Table 1.1–115
SPECIALTY STAINLESS STEELS – WROUGHT

Type 410 Cb

Typical chemical composition, percent by weight: C, 0.15; Fe, balance; Cr, 12.5; Cb, 0.25

Physical constants and thermal properties
Density, lb/in.^{3} : 0.28
Coefficient of thermal expansion, (70–200°F) in./in./°F × 10^{-6} : 6.5
Modulus of elasticity, psi: tension, 29 × 10^6

Specific heat, Btu/lb/°F, 70°F: 0.11
Thermal conductivity, Btu/ft^2 /hr/in./°F, 70°F: 172.8
Electrical resistivity, ohms/cmil/ft, 70°F: 342

Heat treatments
Forging temperature 1600–2100°F; worked 1850°F, oil quenched, tempered 700°F, 4 hr; cold worked.

Short Time Tensile Properties

Temperature, °F	T.S., psi	Y.S., psi, 0.2% offset	Elong., in 2 in., %	Hardness, Rockwell
70	194,000	162,000	16	C 43

Table 1.1–116
STAINLESS STEELS – CAST

ACI Type CA-15

Typical chemical composition, percent by weight: C, 0.15; Mn, 1.0; Fe, balance; S, 0.04; Si, 1.5; Cr, 11.5–14.0; Ni, 1.0; Mo, 0.5; P, 0.04

Physical constants and thermal properties
Density, lb/in.^{3} : 0.275
Coefficient of thermal expansion, (70–200°F) in./in./°F × 10^{-6} : 6.4
Modulus of elasticity, psi: tension, 29 × 10^6
Melting range, °F: 2750
Specific heat, Btu/lb/°F, 70°F: 0.11
Thermal conductivity, Btu/ft^2 /hr/in./°F, 70°F: 174

Electrical resistivity, ohms/cmil/ft, 70°F: 468
Curie temperature, °F: annealed, 1450–1650° (furnace cooled); age hardened, 1800–1850° (air cooled or oil quenched)
Permeability, 70°F, 200 Oe: annealed – ferromagnetic

Heat treatments
Air cooled from 1800°F, tempered at 1450°F; temperature less than 600 or 1100–1500°F. Highest strength and corrosion resistance by tempering below 600°F.

Short Time Tensile Properties

Hardened and Tempered

Temperature, °F	T.S., psi	Y.S., psi, 0.2% offset	Elong., in 2 in., %	Hardness, Brinell
70	200,000, 100,000	150,000, 75,000	7, 30	390, 185

Impact Strength

Hardened and Tempered

Test temperature, °F	Type test	Strength, ft-lb
+70	Charpy, keyhole notched	15, 35

Table 1.1–117
STAINLESS STEELS – CAST

ACI Type CA-40

Typical chemical composition, percent by weight: C, 0.20–0.40; Mn, 1.0; Fe, balance; S, 0.04; Si, 1.50; Cr, 11.5–14.0; Ni, 1.0; Mo, 0.5 (not intentionally added); P, 0.04

Physical constants and thermal properties
Density, lb/in.³: 0.275
Coefficient of thermal expansion, (70–200°F) in./in./°F × 10⁻⁶: 6.4
Modulus of elasticity, psi: tension, 29 × 10⁶
Melting range; °F: 2725
Specific heat, Btu/lb/°F, 70°F: 0.11

Thermal conductivity, Btu/ft²/hr/in./°F, 70°F: 174
Electrical resistivity, ohms/cmil/ft, 70°F: 456
Curie temperature, °F: annealed, 1450–1650 (furnace cooled); age hardened, 1800–1850 (air cooled or oil quenched)
Permeability, 70°F, 200 Oe: annealed – ferromagnetic

Heat treatments
Air cooled from 1800°F, tempered at 1400°F; tempering temperature less than 600 or 1100–1500°F. Highest strength and corrosion resistance by tempering below 600°F.

Short Time Tensile Properties

Hardened and Tempered

Temperature, °F	T.S., psi	Y.S., psi, 0.2% offset	Elong., in 2 in., %	Hardness, Brinell
70	220,000, 110,000	165,000, 67,000	1, 18	470, 212

Impact Strength

Hardened and Tempered

Test temperature, °F	Type test	Strength, ft-lb
+70	Charpy, keyhole notched	1, 3

Table 1.1–118
STAINLESS STEELS – CAST

ACI Type CE-30

Typical chemical composition, percent by weight: C, 0.30; Mn, 1.50; Fe, balance; S, 0.04; Si, 2.0; Cr, 26–30; Ni, 8–11; P, 0.04

Physical constants and thermal properties
Density, lb/in.³: 0.277

Coefficient of thermal expansion, (70–200°F) in./in./°F × 10⁻⁶: 9.6
Modulus of elasticity, psi: tension, 25 × 10⁶
Melting range, °F: 2650
Specific heat, Btu/lb/°F, 70°F: 0.14
Electrical resistivity, ohms/cmil/ft, 70°F: 510
Permeability, 70°F, 200 Oe: annealed – less than 1.5

Short Time Tensile Properties

Temperature, °F	T.S., psi	Y.S., psi, 0.2% offset	Elong., in 2 in., %	Hardness, Brinell
70	97,000	63,000	18	170

Table 1.1–119
STAINLESS STEELS – CAST

ACI Type CF-8

Typical chemical composition, percent by weight: C, 0.08; Mn, 1.50; Fe, balance; S, 0.04; Si, 2.0; Cr, 18–21; Ni, 8–11; P, 0.04

Physical constants and thermal properties
Density, lb/in.^{3} : 0.280
Coefficient of thermal expansion, (70–200°F) in./in./°F × 10^{-6} : 10.0
Modulus of elasticity, psi: tension, 28 × 10^6
Melting range, °F: 2600
Specific heat, Btu/lb/°F, 70°F: 0.12
Thermal conductivity, Btu/ft^2/hr/in./°F, 70°F: 110.4
Electrical resistivity, ohms/cmil/ft, 70°F: 456
Permeability, 70°F, 200 Oe: annealed – 1.0–1.3 after heat treatment

Short Time Tensile Properties

Temperature, °F	T.S., psi	Y.S., psi, 0.2% offset	Elong., in 2 in., %	Hardness, Brinell
70	77,000	37,000	55	140

Impact Strength

Test temperature, °F	Type test	Strength, ft-lb
+70	Charpy, keyhole notched	74

Table 1.1–120
STAINLESS STEELS – CAST

ACI Type CF-3M

Typical chemical composition, percent by weight: C, 0.03; Mn, 1.5; Fe, balance; Si, 1.5; Cr, 17–21; Ni, 9–13; Mo, 2–3

Physical constants and thermal properties
Density, lb/in.^{3} : 0.280

Coefficient of thermal expansion, (70–200°F) in./in./°F × 10^{-6} : 9.7
Modulus of elasticity, psi: tension, 27 × 10^6
Melting range, °F: 2550
Specific heat, Btu/lb/°F, 70°F: 0.12
Thermal conductivity, Btu/ft^2/hr/in./°F, 70°F: 112.8
Electrical resistivity, ohms/cmil/ft, 70°F: 492
Permeability, 70°F, 200 Oe: annealed 1.5–2.5

Short Time Tensile Properties

Temperature, °F	T.S., psi	Y.S., psi, 0.2% offset	Elong., in 2 in., %	Hardness, Brinell
70	80,000	42,000	50	156–170

Impact Strength

Test temperature, °F	Type test	Strength, ft-lb
+70	Charpy, keyhole notched	70

Table 1.1–121
STAINLESS STEELS – CAST

ACI Type CF-8C

Typical chemical composition, percent by weight: C, 0.08; Mn, 1.5; Fe, balance; S, 0.04; Si, 2.0; Cr, 18–21, Ni, 9–12; Cb, 1.0; P, 0.04

Physical constants and thermal properties
Density, 1b/in.^{3}: 0.280
Coefficient of thermal expansion, (70°–200°F)
  in./in./°F × 10^{-6}: 10.3
Modulus of elasticity, psi: tension, 28 × 10^6
Melting range, °F: 2600
Specific heat, Btu/lb/°F, 70°F: 0.12
Thermal conductivity, Btu/ft^2/hr/in./°F, 70°F: 111.6
Electrical resistivity, ohms/cmil/ft, 70°F: 426
Permeability, 70°F, 200 Oe: annealed, 1.20–1.80

Short Time Tensile Properties

Temperature, °F	T.S., psi	Y.S., psi, 0.2% offset	Elong., in 2 in., %	Hardness, Brinell
70	77,000	38,000	39	149

Impact Strength

Test temperature, °F	Type test	Strength, ft-lb
+70	Charpy, keyhole notched	30

Table 1.1–122
STAINLESS STEELS – CAST

ACI Type CG-8M

Typical chemical composition, percent by weight: C, 0.08; Mn, 1.5; Fe, balance; Si, 1.5; Cr, 18–21; Ni, 9–13; Mo, 3–4

Physical constants and thermal properties
Density, lb/in.^{3}: 0.281
Coefficient of thermal expansion, (70–200°F) in./in./°F × 10^{-6}: 9.7

Modulus of elasticity, psi: tension, 28 × 10^6
Melting range, °F: 2550
Specific heat, Btu/lb/°F, 70°F: 0.12
Thermal conductivity, Btu/ft^2/hr/in./°F, 70°F: 112.8
Electrical resistivity, ohms/cmil/ft, 70°F: 492
Permeability, 70°F, 200 Oe: annealed, 1.5–2.5

Heat treatments
Water quenched from 2000°F.

Short Time Tensile Properties

Temperature, °F	T.S., psi	Y.S., psi, 0.2% offset	Elong., in 2 in., %	Hardness, Brinell
70	82,000	43,000	50	170

Table 1.1–122 (continued)
STAINLESS STEELS – CAST

ACI Type CG-8M (continued)

Impact Strength

Test temperature, °F	Type test	Strength, ft-lb
+70	Charpy, keyhole notched	70

Table 1.1–123
STAINLESS STEELS – CAST

ACI Type CH-20

Typical chemical composition, percent by weight: C, 0.20; Mn, 1.5; Fe, balance; S, 0.04; Si, 2.0; Cr, 22–26; Ni, 12–15; P, 0.04

Physical constants and thermal properties
Density, lb/in.^{3}: 0.279
Coefficient of thermal expansion, (70–200°F) in./in./°F $\times$ 10^{-6}: 9.6

Modulus of elasticity, psi: tension, 28 $\times$ 10^6
Melting range, °F: 2600
Specific heat, Btu/lb/°F, 70°F: 0.12
Thermal conductivity, Btu/ft^2/hr/in./°F, 70°F: 98.4
Electrical resistivity, ohms/cmil/ft, 70°F: 504
Permeability, 70°F, 200 Oe: annealed, 1.71

Heat treatments
Water quenched from 2000°F.

Short Time Tensile Properties

Temperature, °F	T.S., psi	Y.S., psi, 0.2% offset	Elong., in 2 in., %	Hardness, Brinell
70	88,000	50,000	38	190

Impact Strength

Test temperature, °F	Type test	Strength, ft-lb
+70	Charpy, keyhole notched	30

Table 1.1–124
LOW TEMPERATURE STEELS – WROUGHT;
FINE GRAIN C-Mn-Si CARBON STEELS

Type 40-50

Typical chemical composition, percent by weight: C, 0.14 or 0.2; Mn, 0.7–1.35; Fe, balance; S, 0.04 or 0.05[a]; Si, 0.15–0.5; P, 0.035 or 0.4

[a]Higher value is flange quality; lower value is firebox quality.

Heat treatments
Normalized

Short Time Tensile Properties

Temperature, °F	T.S., psi	Y.S., psi, 0.2% offset	Elong., in 2 in., %	Hardness, Brinell
70	58,000–90,000	40,000–50,000	19–30	120–170

Impact Strength

Test temperature, °F	Type test	Strength, ft-lb
–100	Charpy – V-notched	10–16
+70	Charpy – V-notched	40–75

Table 1.1–125
LOW TEMPERATURE STEELS – WROUGHT;
FINE GRAIN C-Mn-Si CARBON STEELS

Type 55-60

Typical chemical composition, percent by weight: C, 0.2; Mn, 0.7–1.35; Fe, balance; S, 0.04 or 0.05; Si, 0.15–0.3; P, 0.35 or 0.04[a]

[a]Higher value is flange quality; lower value is firebox quality.

Heat treatments
Quenched and tempered.

Short Time Tensile Properties

Temperature, °F	T.S., psi	Y.S., psi, 0.2% offset	Elong., in 2 in., %	Hardness, Brinell
70	75,000–100,000	55,000–65,000	23–30	–

Table 1.1−125 (continued)
LOW TEMPERATURE STEELS − WROUGHT;
FINE GRAIN C-Mn-Si CARBON STEELS

Type 55-60 (continued)

Impact Strength

Test temperature, °F	Type test	Strength, ft-lb
−100	Charpy − V-notched	5−16
+70	Charpy − V-notched	40−75

Table 1.1−126
LOW TEMPERATURE STEELS − WROUGHT;
Cr-Ni-Mo LOW ALLOY STEELS

Type 80-100

Typical chemical composition, percent by weight: C, 0.18−0.2; Mn, 0.1−0.4; Fe, balance; S, 0.025; Si, 0.15−0.35; Cr, 1−1.8; Ni, 2−3.25; Mo, 0.2−0.6; Ti, 0.02; P, 0.025; V, 0.3; Cu, 0.25

Heat treatments
    Quenched and tempered.

Short Time Tensile Properties

Temperature, °F	T.S., psi	Y.S., psi, 0.2% offset	Elong., in 2 in., %	Hardness, Brinell
70	>100,000	80,000−100,000	19−20	205−240

Impact Strength

Test temperature, °F	Type test	Strength, ft-lb
−120	Charpy − V-notched	30−50

TABLE 1.1–127
LOW TEMPERATURE STEELS – WROUGHT;
Cr-Ni-Mo LOW ALLOY STEELS

Type 100-120

Typical chemical composition, percent by weight: C, 0.2;
Mn, 0.25; Fe, balance; S, 0.02; Si, 0.25; Cr, 1–1.8; Ni,
2.25–3.5; Mo, 0.2–0.6; P, 0.02

Heat treatments
Quenched and tempered.

Short Time Tensile Properties

Temperature, °F	T.S. psi	Y.S., psi, 0.2% offset	Elong., in 2 in., %	Hardness, Brinell
70	>120,000	100,000–120,000	18	–

Impact Strength

Test temperature, °F	Type test	Strength, ft-lb
–120	Charpy – V-notched	30–50

TABLE 1.1–128
LOW TEMPERATURE STEELS – WROUGHT;
NICKEL STEELS

Type 5 Ni

Typical chemical composition, percent by weight: C,
0.14; Mn, 0.3–0.6; Fe, balance; S, 0.04; Si, 0.35; Ni,
4.75–5.25; P, 0.04

Heat treatments:
Normalized tubing.

Short Time Tensile Properties

Temperature, °F	T.S., psi	Y.S., psi, 0.2% offset	Elong., in 2 in., %	Hardness, Brinell
70	80,000–100,000	60,000–75,000	30–50	145–210

Impact Strength

Test temperature, °F	Type test	Strength, ft-lb
–100	Charpy – keyhole notched; transition fracture	49
+70	Charpy – keyhole notched	55

Table 1.1–129
LOW TEMPERATURE STEELS – WROUGHT; NICKEL STEELS

Type 9 Ni

Typical chemical composition, percent by weight: C,
0.12–0.13; Mn, 0.3–0.9; Fe, balance; S, 0.03 or 0.04;
Si, 0.15–0.32; Ni, 8–10; P, 0.03–0.035

Short Time Tensile Properties

Temperature, °F	T.S., psi	Y.S., psi, 0.2% offset	Elong., in 2 in., %	Hardness, Brinell
70	90,000–120,000	60,000–75,000	20–22	–

Impact Strength

Test temperature, °F	Type test	Strength, ft-lb
–320	Charpy – V-notched	15–25

Table 1.1–130
HEAT RESISTANT ALLOYS – CAST

ACI Type HH (Ferritic)

Typical chemical composition, percent by weight: C,
0.20–0.50; Mn, 2.00; Fe, balance; S, 0.04; Si, 2.00;
Cr, 24–28; Ni, 11–14; Mo, 0.5; P, 0.04; N, 0.2

Specific heat, Btu/lb/°F, 70°F: 0.12
Thermal conductivity, Btu/ft²/hr/in./°F, 70°F: 98.4
Electrical resistivity, ohms/cmil/ft, 70°F: 450–510
Curie temperature, °F: annealed, 1900; age hardened, 1400
Permeability, 70°F, 200 Oe: annealed, 1.0–1.9

Physical constants and thermal properties
Density, lb/in.³ : 0.279
Coefficient of thermal expansion, (70–200°F) in./
in./°F × 10⁻⁶ : 10.5
Modulus of elasticity, psi: tension, 27 × 10⁶
Melting range, °F: 2500

Heat treatments
As cast and heat treated (aged 24 hr at 1400°F,
furnace cooled); 12 hr at 1900°F may improve
life.

Short Time Tensile Properties

Temperature, °F	T.S., psi	Y.S., psi, 0.2% offset	Elong., in 2 in., %	Hardness, Brinell
70	80,000–86,000	50,000–55,000	25, 11	185, 200
1400	33,000	17,000	18	–
1600	18,500	13,500	30	–
1800	9,000	6,300	45	–

Rupture Strength, 1000 hr

Test temperature, °F	Strength, psi	Elong., in 2 in., %	Reduction of area, %
1400	6500	–	–
1600	3800	–	–
2000	–	–	–

Table 1.1–130 (continued)
HEAT RESISTANT ALLOYS – CAST

ACI Type HH (Ferritic) (continued)

Creep Strength (Stress, psi, to Produce 0.0001%/hr Creep)

Test temperature, °F	1000 hr	100,000 hr
1400	3000	–
1600	1700	–
2000	300	–

Table 1.1–131
HEAT RESISTANT ALLOYS – CAST

ACI Type HH (Austenitic)

Typical chemical composition, percent of weight: C, 0.20–0.50; Mn, 2.00; Fe, balance; S, 0.04; Si, 2.00; Cr, 24–28; Ni, 11–14; Mo, 0.5[a], P, 0.04; N, 0.2

[a]Molybdenum not intentionally added.

Physical constants and thermal properties
Density, lb/in.^{3}: 0.279
Coefficient of thermal expansion, (70–200°F) in./in./°F $\times$ 10^{-6}: 10.5
Modulus of elasticity, psi: tension, 27 $\times$ 10^6
Melting range, °F: 2500

Specific heat, Btu/lb/°F, 70°F: 0.12
Thermal conductivity, Btu/ft^2/hr/in./°F, 70°F: 98.4
Electrical resistivity, ohms/cmil/ft, 70°F: 450–510
Curie temperature, °F: annealed, 1900; age hardened, 1400
Permeability, 70°F, 200 Oe: annealed, 1.0–1.9

Heat treatments
As cast and heat treated (aged 24 hr at 1400°F, furnace cooled); 12 hr at 1900°F may improve life.

Short Time Tensile Properties

Temperature, °F	T.S., psi	Y.S., psi, 0.2% offset	Elong., in 2 in., %	Hardness, Brinell
70	85,000–92,000	40,000–45,000	15–8	180, 200
1400	35,000	18,000	12	–
1600	22,000	14,000	16	–
1800	11,000	7,000	30	–

Rupture Strength, 1000 hr

Test temperature, °F	Strength, psi	Elong., in 2 in., %	Reduction of area, %
1400	10,000	–	–
1600	4,700	–	–
2000	1,200	–	–

Creep Strength (Stress, psi, to Produce 0.0001%/hr Creep)

Test temperature, °F	1000 hr	100,000 hr
1400	7000	–
1600	4000	–
2000	800	–

Table 1.1–132
HEAT RESISTANT ALLOYS – CAST

ACI Type HK

Typical chemical composition, percent by weight: C, 0.20–0.60; Mn, 2.00; Fe, balance; S, 0.04; Si, 2.00; Cr, 24–28; Ni, 18–22; Mo, 0.5[a]; P, 0.04

[a]Molybdenum not intentionally added.

Physical constants and thermal properties
  Density, lb/in.^{3}: 0.280
  Coefficient of thermal expansion, (70–200°F) in./in./°F × 10^{-6}: 10.0
  Modulus of elasticity, psi: tension, 29 × 10^6

Melting range, °F: 2550
Specific heat, Btu/lb/°F, 70°F: 0.12
Thermal conductivity, Btu/ft^2 /hr.,/in./°F, 70°F: 98.4
Electrical resistivity, ohms/cmil/ft, 70°F: 540
Curie temperature, °F: age hardened, 1400
Permeability, 70°F, 200 Oe: annealed, 1.02; age hardened

Heat treatments
  As cast and heat treated (aged 24 hr at 1400°F, furnace cooled).

Short Time Tensile Properties

Temperature, °F	T.S., psi	Y.S., psi, 0.2% offset	Elong., in 2 in., %	Hardness, Brinell
70	75,000–85,000	50,000	17–10	170, 190
1600	23,000	–	21	–

Rupture Strength, 1000 hr

Test temperature, °F	Strength, psi	Elong., in 2 in., %	Reduction of area, %
1400	9000	–	–
1600	5000	–	–

Creep Strength (Stress, psi, to Produce 0.0001%/hr Creep)

Test temperature, °F	1000 hr	100,000 hr
1400	6800	–
1600	4200	–
2000	1000	–
2150	200	–

Table 1.1–133
HEAT RESISTANT ALLOYS – CAST

ACI Type HA

Typical chemical composition, percent by weight: C, 0.20; Mn, 0.35–0.65; Fe, balance; S, 0.04; Si, 1.00; Cr 8–10; Mo, 0.90–1.20; P, 0.04

Physical constants and thermal properties
Density, lb/in.^{3}: 0.279
Coefficient of thermal expansion, (70–200°F) in./in./°F × 10^{-6}: 7.5
Modulus of elasticity, psi: tension, 29 × 10^6
Melting range, °F: 2750

Specific heat, Btu/lb/°F, 70°F: 0.11
Thermal conductivity, Btu/ft^2/hr/in./°F: 182.4
Electrical resistivity, ohms/cmil/ft, 70°F: 420
Curie temperature, °F: annealed 1625
Permeability, 70°F, 200 Oe: age hardened, ferromagnetic

Heat treatments
Normalized at 1825°F, tempered at 1250°F.

Short Time Tensile Properties

Temperature, °F	T.S., psi	Y.S., psi, 0.2% offset	Elong., in 2 in., %	Hardness, Brinell
70	95,000	65,000	23	180
1000	67,000	42,000	–	–
1100	44,000	32,000	36	–

Rupture Strength, 1000 hr

Test temperature, °F	Strength, psi	Elong., in 2 in., %	Reduction of area, %
1000	27,000	–	–

Creep Strength (Stress, psi, to Produce 0.0001%/hr Creep)

Test temperature, °F	1000 hr	100,000 hr
1000	16,000	–
1100	7,200	–
1200	3,100	–

Table 1.1–134
HEAT RESISTANT ALLOYS – CAST

ACI Type HF

Typical chemical composition, percent by weight: C, 0.20–0.40; Mn, 2.00; Fe, balance; S, 0.04; Si, 2.00; Cr, 19–23; Ni, 9–12; Mo, 0.5[a]; P, 0.04

[a]Molybdenum not intentionally added.

Physical constants and thermal properties
Density, lb/in.³: 0.280
Coefficient of thermal expansion, (70–200°F) in./in./°F × 10⁻⁶: 10.1
Modulus of elasticity, psi: tension, 28 × 10⁶
Melting range, °F: 2550

Specific heat, Btu/lb/°F, 70°F: 0.12
Thermal conductivity, Btu/ft²/hr/in./°F, 70°F: 108
Electrical resistivity, ohms/cmil/ft, 70°F: 480
Curie temperature, °F: annealed, 1900; age hardened, 1400 (24 hr)
Permeability, 70°F, 200 Oe: annealed, 1.0

Heat Treatments
Annealing temperature (1900°F) before cyclic temperature service; 6 hr at 1900°F may improve life. Aged 24 hr at 1400°F, furnace cooled.

Short Time Tensile Properties

Temperature, °F	T.S., psi	Y.S., psi, 0.2% offset	Elong., in 2 in., %	Hardness, Brinell
70	85,000	45,000	35	165
1200	57,000	–	16	–
1400	35,000	21,000	20	–
1600	22,000	–	22	–

Rupture Strength, 1000 hr

Test temperature, °F	Strength, psi	Elong., in 2 in., %	Reduction of area, %
1200	17,000	–	–
1400	8,000	–	–
1600	3,800	–	–

Creep Strength (Stress, psi, to Produce 0.0001%/hr Creep)

Test temperature, °F	1000 hr	100,000 hr
1200	13,000	–
1400	6,000	–
1600	3,200	–

Table 1.1–135
HEAT RESISTANT ALLOYS – CAST

ACI Type HT

Typical chemical composition, percent by weight: C, 0.35-0.75; Mn, 2.00, Fe,balance; S, 0.04; Si, 2.50; Cr, 13–17; Ni, 33–37: Mo, 0.5[a]; P, 0.04

[a]Molybdenum not intentionally added.

Physical constants and thermal properties
Density, lb/in.^{3}: 0.286
Coefficient of thermal expansion, (70–200°F) in./in./°F $\times$ 10^{-6}: 9.8
Modulus of elasticity, psi: tension, 27 $\times$ 10^6
Melting range, °F: 2450

Specific heat, Btu/lb/°F, 70°F: 0.11
Thermal conductivity, Btu/ft^2/hr/in./°F, 70°F: 92.4
Electrical resistivity, ohms/cmil/ft, 70°F: 600
Curie temperature, °F: annealed, 1900; age hardened, 1400 (24 hr)
Permeability, 70°F, 200 Oe: annealed, 1.10–2.00

Heat treatments
Annealing temperature (1900°F) before cyclic temperature service; 6 hr at 1900°F may improve life. Aged 24 hr at 1800°F, air cooled.

Short Time Tensile Properties

Temperature, °F	T.S., psi	Y.S., psi, 0.2% offset	Elong., in 2 in., %	Hardness, Brinell
70	70,000	40,000	10	180
1400	35,000	26,000	10	–
1600	18,800	15,000	26	–
1800	11,000	8,000	28	–

Rupture Strength, 1000 hr

Test temperature, °F	Strength, psi	Elong., in 2 in., %	Reduction of area, %
1400	12,500	–	–
1600	7,000	–	–
2000	1,800	–	–

Creep Strength (Stress, psi, to Produce 0.0001%/hr Creep)

Test temperature, °F	1000 hr	100,000 hr
1400	8000	–
1600	4500	–
2000	500	–
2150	150	–

Table 1.1–136
HEAT RESISTANT ALLOYS – CAST

ACI Type HX

Typical chemical composition, percent by weight: C, 0.35–0.75; Mn, 2.00; Fe, balance; S, 0.04; Si, 2.50; Cr, 15–19; Ni, 64–68; Mo, 0.5[a]; P, 0.04

[a]Molybdenum not intentionally added.

Physical constants and thermal properties
    Density, lb/in.^{3}: 0.294
    Coefficient of thermal expansion, (70–200°F) in/in/°F × 10^{-6}: 9.2

Modulus of elasticity, psi: tension, 25 × 10^6
Melting range, °F: 2350
Specific heat, Btu/lb/°F, 70°F: 0.11
Curie temperature, °F: age hardened, 1400 (24 hr)
Permeability, 70°F, 200 Oe: annealed, 2.0

Heat treatments
    Aged 24 hr at 1400°F, air cooled.

Short Time Tensile Properties

Temperature, °F	T.S., psi	Y.S., psi, 0.2% offset	Elong., in 2 in., %	Hardness, Brinell
70	65,000	36,000	9	176
1600	20,500	17,500	48	–
1800	10,700	6,900	40	–

Rupture Strength, 1000 hr

Test temperature, °F	Strength, psi	Elong., in 2 in., %	Reduction of area, %
1600	4000	–	–
2000	900	–	–

Creep Strength (Stress, psi, to Produce 0.0001%/hr Creep)

Test temperature, °F	1000 hr	100,000 hr
1400	6400	–
1600	3200	–
2000	600	–

Table 1.1–137
HIGH TEMPERATURE STEELS – WROUGHT

Type 1415 NW (Greek Ascoloy)

Typical chemical composition, percent by weight: C, 0.17; Mn, 0.40; Fe, balance; Si, 0.30; Cr, 12.75; Ni, 1.95; Mo, 0.15; W, 3.0; Cu, 0.13

Physical constants and thermal properties
Density, lb/in.^{3}: 0.284
Coefficient of thermal expansion, (70–200°F) in./in./°F × 10^{-6}: 6.3
Modulus of elasticity, psi: tension – 29 × 10^6 (70°F);

21.5 × 10^6 (1000°F)
Melting range, °F: 2660–2670

Heat treatments
Mechanical properties: austenitized ½ hr at 1800°F, oil quenched, tempered 2 hr at 1050°F, air cooled; rupture strength applies to material tempered 1½ hr at 1260°F. Hot working temperature 1700–2200°F.

Short Time Tensile Properties

Temperature, °F	T.S., psi	Y.S., psi, 0.2% offset	Elong., in 2 in., %	Hardness, Brinell
70	170,000	150,000	13.3	–
800	135,000	122,000	13.3	–
1000	103,000	98,000	17.1	–

Rupture Strength, 1000 hr

Test temperature, °F	Strength, psi	Elong., in 2 in., %	Reduction of area, %
1000	36,000	–	–

Fatigue Strength

Test temperature, °F	Stress, psi	Cycles to failure
1000	53,000	10^7

Impact Strength

Test temperature, °F	Type test	Strength, ft-lb
+70	Charpy – V-notched	19.5

Table 1.1–138
HIGH TEMPERATURE STEELS – WROUGHT

Type 1430 MV (Lapelloy)

Typical chemical composition, percent by weight: C, 0.30; Mn, 1.05; Fe, balance; Si, 0.30; Cr, 11.80; Ni, 0.25; Mo, 2.80; V, 0.25

22×10^6 (1000°F)
Melting range, °F: 2700–2750
Thermal conductivity, Btu/ft^2/hr/in./°F, 70°F: 189.6

Physical constants and thermal properties
Density, lb/in.^{3}: 0.281
Coefficient of thermal expansion, (70–200°F) in./in./°F $\times 10^{-6}$: 6.5
Modulus of elasticity, psi: tension – 30×10^6 (70°F);

Heat treatments
Mechanical properties: austenitized at 2000°F, oil quenched, tempered 2 hr at 1200°F, air cooled; rupture strength applies to material tempered at 1300°F. Hot working temperature 1700–2200°F.

Short Time Tensile Properties

Temperature, °F	T.S., psi	Y.S., psi 0.2% offset	Elong., in 2 in., %	Hardness, Brinell
70	157,000	125,000	12	–
900	123,000	94,000	10	–
1100	89,000	81,000	28	–

Rupture Strength, 1000 hr

Test temperature, °F	Strength, psi	Elong., in 2 in., %	Reduction of area, %
1000	58,000	–	–

Fatigue Strength

Test temperature, °F	Stress, psi	Cycles to failure
900	53,000	10^7
1100	36,000	10^7

Impact Strength

Test temperature, °F	Type test	Strength, ft-lb
+70	Charpy – unnotched	10.0
1000	Charpy – unnotched	15.0

Table 1.1–139
HIGH TEMPERATURE STEELS – WROUGHT

Type 14 CVM (Chromoloy)

Typical chemical composition, percent by weight: C, 0.20; Mn, 0.50; Fe, balance; Si, 0.75; Cr, 1.0; Mo, 1.0; V, 0.10

Physical constants and thermal properties
Density, lb/in.^{3}: 0.285
Coefficient of thermal expansion, (70–200°F) in./in./°F × 10^{-6}: 7.9

Modulus of elasticity, psi: tension – 31.6 × 10^6 (70°F); 25.4 × 10^6 (1000°F)

Heat treatments
Mechanical properties: normalized 1 hr at 1650–1750°F, air cooled, tempered 2 hr at 1200°F, air cooled. Hot working temperature 1500–2100°F.

Short Time Tensile Properties

Temperature, °F	T.S., psi	Y.S., psi, 0.2% offset	Elong., in 2 in., %	Hardness, Brinell
70	139,000	117,000	8	–
800	119,000	96,000	–	–
1000	103,000	85,000	–	–

Rupture Strength, 1000 hr

Test temperature, °F	Strength, psi	Elong., in 2 in., %	Reduction of area, %
1000	52,000	–	–

Table 1.1–140
HIGH TEMPERATURE STEELS – WROUGHT

Type 17-22AS (14 MV)

Typical chemical composition, percent by weight: C, 0.30; Mn, 0.55; Fe, balance; Si, 0.70; Cr, 1.30, Mo, 0.50; V, 0.25

Physical constants and thermal properties
Density, lb/in.^{3}: 0.283

Coefficient of thermal expansion, (70–200°F) in./in./°F × 10^{-6}: 7.8
Modulus of elasticity, psi: tension – 29.5 × 10^6 (70°F); 20 × 10^6 (1000°F)
Melting range, °F: 2700–2750
Thermal conductivity, Btu/ft^2/hr/in./°F, 70°F: 207.6

Short Time Tensile Properties

Temperature, °F	T.S., psi	Y.S., psi, 0.2% offset	Elong., in 2 in., %	Hardness, Brinell
70	150,000	127,000	16.5	–
800	121,000	101,000	18	–
1000	91,000	78,000	18	–

Table 1.1–140 (continued)
HIGH TEMPERATURE STEELS — WROUGHT

Type 17-22AS (14 MV) (continued)

Rupture Strength, 1000 hr

Test temperature, °F	Strength, psi	Elong., in 2 in., %	Reduction of area, %
1000	55,000	—	—

Impact Strength

Test temperature, °F	Type test	Strength, ft-lb
+70	Charpy – unnotched	31
1000	Charpy – unnotched	37

Table 1.1–141
ULTRA HIGH STRENGTH STEELS — WROUGHT

Type Modified H-11

Typical chemical composition, percent by weight: C, 0.40; Mn, 0.35; Fe, balance; Si, 1.0; Cr, 5.0; Mo, 1.4; V, 0.45

Physical constants and thermal properties
Density, lb/in.^{3}: 0.281
Coefficient of thermal expansion, (70–200°F)

in./in./°F $\times 10^{-6}$
Modulus of elasticity, psi: tension, 30×10^6 (70°F); $27.6–27.8 \times 10^6$ (400°F); $21.9–26.6 \times 10^6$ (800°F)
Thermal conductivity, Btu/ft^2/hr/in./°F, 70°F: 204
Heat treatments
Hot working temperature 1700–2100°F

Short Time Tensile Properties

Temperature, °F	T.S., psi	Y.S., psi, 0.2% offset	Elong., in 2 in., %	Hardness, Brinell
70	295,000–311,000	241,000–247,000	6.6–12.0	—
500	270,000–281,000	220,000–221,000	9.8–9.9	—
800	252,000–259,000	199,000–207,000	10.8–12.0	—
1000	216,000–220,000	172,000–173,000	11.8–15.0	—

Rupture Strength, 1000 hr

Test temperature, °F	Strength, psi	Reduction of area, %
900	140,000–150,000	70°F: 27.0–39.9
		500°F: 33.0–42.1
		800°F: 35.2–42.2
		1000°F: 42.5–43.0

Fatigue Strength

Test temperature, °F	Stress, psi	Cycles to failure
70	130,000–135,000	10^6

Table 1.1–141 (continued)
ULTRA HIGH STRENGTH STEELS – WROUGHT

Type Modified H-11 (continued)

Impact Strength

Test temperature, °F	Type test	Strength, ft-lb
–200	Charpy – unnotched	10
+70	Charpy – unnotched	15–22
500	Charpy – unnotched	31

Table 1.1–142
ULTRA HIGH STRENGTH STEELS – WROUGHT

Type 300-M

Typical chemical composition, percent by weight: C, 0.40; Mn, 0.75; Fe, balance; Si, 1.60; Cr, 0.85; Mo, 0.40; V, 0.08

Thermal conductivity, Btu/ft² /hr/in./°F, 70°F: 260.4

Physical constants and thermal properties
Coefficient of thermal expansion, (70–200°F) in./in./°F × 10⁻⁶ : 7.61 (0–600°F)

Heat treatments
Normalized at 1700°F, austenitized at 1600°F, oil quenched, tempered at 600°F. Hot working temperature 1700–2250°F.

Short Time Tensile Properties

Temperature, °F	T.S., psi	Y.S., psi, 0.2% offset	Elong., in 2 in., %	Hardness, Brinell
70	289,000	242,000	10.0	–
500	270,000	200,000	13.3	–
700	232,000	178,000	–	–
800	–	–	15.0	–

Rupture Strength, 1000 hr

Test temperature, °F	Strength, psi	Reduction of area, %
–	–	70°F: 38
		500°F: 35
		700°F: 52

Fatigue Strength

Test temperature, °F	Stress, psi	Cycles to failure
70	116,000	10⁶

Impact Strength

Test temperature, °F	Type test	Strength, ft-lb
–200	Charpy – unnotched	11
+70	Charpy – unnotched	22
500	Charpy – unnotched	23

Table 1.1–143
ULTRA HIGH STRENGTH STEELS – WROUGHT

Type D-6A

Typical chemical composition, percent by weight: C, 0.46; Mn, 0.75; Fe, balance; Si, 0.22; Cr, 1.0; Ni, 0.55; Mo, 1.0

Modulus of elasticity, psi: tension – 30×10^6 ($70°F$); 24.4×10^6 ($400°F$); 23.7×10^6 ($800°F$)

Physical constants and thermal properties
Density, lb/in.^{3}: 0.283
Coefficient of thermal expansion, ($70–200°F$) in./in./$°F \times 10^{-6}$: 7.3 ($80–400°F$)

Heat treatments
Normalized at $1650°F$, oil quenched, tempered at $500–700°F$. Hot working temperature range $1800–2250°F$.

Short Time Tensile Properties

Temperature, $°F$	T.S., psi	Y.S., psi, 0.2% offset	Elong., in 2 in., %	Reduction of area, %
70	284,000	250,000	7.5	26.8
500	267,000	188,000	15.2	55.0
800	185,000[a]	159,000[a]	15.2[a]	55.0[a]
1000	139,000[b]	121,000[b]	19.8[b]	64.5[b]

[a]Tempered at $850°F$.
[b]Tempered at $1050°F$.

Rupture Strength, 100 hr

Test temperature, $°F$	Strength, psi
900	144,000

Fatigue Strength

Test temperature, $°F$	Stress, psi	Cycles to failure
70	110,000	10^6

Impact Strength

Test temperature, $°F$	Type test	Strength, ft-lb
−200	Charpy – unnotched	10
+70	Charpy – unnotched	14
500	Charpy – unnotched	26

Table 1.1—144
ULTRA HIGH STRENGTH STEELS — WROUGHT

Type 4130

Typical chemical composition, percent by weight: C, 0.3; Mn, 0.5; Fe, balance; Si, 0.3; Cr, 0.95; Mo, 0.2

Modulus of elasticity, psi: tension — 30×10^6 ($70°$F); 27.5×10^6 ($400°$F); 24×10^6 ($800°$F)
Thermal conductivity, Btu/ft^2/hr/in./$°$F, $70°$F: 300

Physical constants and thermal properties
Density, lb/in.^{3}: 0.283
Coefficient of thermal expansion, ($70-200°$F) in./in./$°$F $\times 10^{-6}$: 7.5 ($80-600°$F)

Heat treatments
Bar austenitized at $1575°$F, water quenched and tempered (1 hr) at $800°$F. Hot working temperature range $1850-2250°$F.

Short Time Tensile Properties

Temperature, $°$F	T.S., psi	Y.S., psi, 0.2% offset	Elong., in 2 in., %	Reduction of area, %
70	200,000	170,000	20	50

Rupture Strength, 100 hr

Test temperature, $°$F	Strength, psi
900	65,000

Impact Strength

Test temperatture, $°$F	Type test	Strength, ft-lb
+70	Charpy — unnotched	25

Table 1.1—145
ULTRA HIGH STRENGTH STEELS — WROUGHT

Type 4330 V-modified

Typical chemical composition, percent by weight: C, 0.3; Mn, 0.9; Fe, balance; Si, 0.3; Cr, 0.8; Ni, 1.8; Mo, 0.4; V, 0.07

Modulus of elasticity, psi: tension, 30×10^6
Thermal conductivity, Btu/ft^2/hr/in./$°$F, $70°$F: 348

Physical constants and thermal properties
Density, lb/in.^{3}: 0.283
Coefficient of thermal expansion, ($70-200°$F) in./in./$°$F $\times 10^{-6}$: 7.5 ($80-600°$F)

Heat treatments
1-in. vacuum melted plate austenitized $1600°$F (1 hr), oil quenched, double tempered (2 hr each) at $600°$F, air cooled. Hot working temperature range $1850-2200°$F.

Short Time Tensile Properties

Temperature, $°$F	T.S., psi	Y.S., psi, 0.2% offset	Elong., in 2 in., %	Reduction of area, %
70	235,000	205,000	12	50

Impact Strength

Test temperature, $°$F	Type test	Strength, ft-lb
+70	Charpy — unnotched	20

Table 1.1–146
ULTRA HIGH STRENGTH STEELS – WROUGHT

Type 4340

Typical chemical composition, percent by weight: C, 0.40; Mn, 0.85; Fe, balance; Si, 0.20; Cr, 0.75; Ni, 1.80; Mo, 0.25

Physical constants and thermal properties
Density, lb/in.^{3} : 0.283

Coefficient of thermal expansion, (70–200°F) in./in./°F × 10^{-6} : 6.3
Modulus of elasticity, psi: tension, 30 × 10^6

Heat treatments
Austenitized at 1550°F, oil quenched, tempered at 400°F.

Short Time Tensile Properties

Temperature, °F	T.S., psi	Y.S., psi, 0.2% offset	Elong., in 2 in., %	Reduction of area, %
70	287,000	270,00	11	39

Fatigue Strength

Test temperature, °F	Stress, psi	Cycles to failure
70	107,000	10^6

Table 1.1–147
ULTRA HIGH STRENGTH STEELS – WROUGHT

Type 18Ni

Typical chemical composition, percent by weight: C, 0.026; Mn, 0.1; Fe, balance; Si, 0.11; Ni, 18.5; Co, 7.0; Mo, 4.5; Ti, 0.22; B, 0.003

Physical constants and thermal properties
Density, lb/in.^{3} : 0.290
Coefficient of thermal expansion, (70–200°F) in./in./°F × 10^{-6} : 5.6

Modulus of elasticity, psi: tension, 26.5 × 10^6
Curie temperature, °F: annealed, 1500°F (1 hr); age hardened, 900°F (3 hr)

Heat treatments
Annealed 1 hr at 1500°F, air cooled, aged 3 hr at 900°F. Hot working temperature range 1700–2300°F.

Short Time Tensile Properties

Temperature, °F	T.S., psi	Y.S., psi, 0.2% offset	Elong., in 2 in., %	Reduction of area, %
70	275,000	268,000	11	48
800	221,000	209,000	12	56
1000	154,000	138,000	24	74

Impact Strength

Test temperature, °F	Type test	Strength, ft-lb
–200	Charpy–unnotched	20
+70	Charpy–unnotched	23

Table 1.1–148
ULTRA HIGH STRENGTH STEELS – WROUGHT

Type 9Ni-4Co-0.20C

Typical chemical composition, percent by weight: C, 0.20; Mn, 0.25; Fe, balance; Si, 0.1; Cr, 0.75; Ni, 9.0; Co, 4.5; Mo, 1.0; V, 0.08; P, 0.01

Physical constants and thermal properties
Density, lb/in.^{3}: 0.283
Coefficient of thermal expansion, (70–200°F)

in./in./°F $\times$ 10^{-6}: 6.4
Modulus of elasticity, psi: tension, 28 $\times$ 10^6

Heat treatments
Austenitized at 1500–1550°F, water or oil quenched, double tempered at 1000°F. Hot working temperature range 1700–2050°F.

Short Time Tensile Properties

Temperature, °F	T.S., psi	Y.S., psi 0.2% offset	Elong., in 2 in., %	Reduction of area, %
70	195,000–220,000	173,000–194,000	12–19	45–65
800	159,000	139,000	16	72
1000	139,000	107,000	18	80

Fatigue Strength

Test temperature, °F	Stress, psi	Cycles to failure
70	105,000–110,000	10^6

Impact Strength

Test temperature, °F	Type test	Strength, ft-lb
–200	Charpy – unnotched	40
+70	Charpy – unnotched	40–60

Table 1.1–149
ULTRA HIGH STRENGTH STEELS – WROUGHT

Type 9Ni-4Co-0.25C

Typical chemical composition, percent by weight: C, 0.27; Mn, 0.25; Fe, balance; Si, 0.1; Ni, 8.25; Co, 4.0; Mo, 0.45; V, 0.09; P, 0.01

Physical constants and thermal properties
Density, lb/in.^{3} : 0.283
Coefficient of thermal expansion, (70–200°F)

in./in./°F × 10^{-6} : 6.4
Modulus of elasticity, psi: tension, 28.4 × 10^6

Heat treatments
Austenitized at 1500–1550°F, water or oil quenched, double tempered at 1000°F. Hot working temperature range 1700–2050°F.

Short Time Tensile Properties

Temperature, °F	T.S., psi	Y.S., psi, 0.2% offset	Elong., in 2 in.,%	Reduction of area, %
70	195,000–240,000	178,000–198,000	10–18	49–65
800	155,000	140,000	18	80
1000	120,000	90,000	19	87

Fatigue Strength

Test temperature, °F	Stress, psi	Cycles to failure
70	105,000–110,000	10^6

Impact Strength

Test temperature, °F	Type test	Strength, ft-lb
–200	Charpy–unnotched	30
+70	Charpy–unnotched	17–40

Table 1.1–150
ULTRA HIGH STRENGTH STEELS – WROUGHT

Type 9Ni-4Co-0.30C

Typical chemical composition, percent by weight: C, 0.31; Mn, 0.25; Fe, balance; S, 0.1; Si, 0.1; Ni, 7.5; Co, 4.5; Mo, 1.0; V, 0.11

Physical constants and thermal properties
Density, lb/in.³ : 0.283
Coefficient of thermal expansion, (70–200°F)

in./in./°F × 10⁻⁶ : 6.2
Modulus of elasticity, psi: tension, 28.6 × 10⁶

Heat treatments
Austenitized at 1500–1550°F, water or oil quenched, double tempered at 1000°F. Hot working temperature range 1700–2050°F.

Short Time Tensile Properties

Temperature, °F	T.S., psi	Y.S., psi, 0.2% offset	Elong., in 2 in.,%	Reduction of area, %
70	220,000–260,000	190,000–210,000	10–16	45–60
800	185,000	178,000	16	70
1000	155,000	140,000	–	–

Fatigue Strength

Test temperature, °F	Stress, psi	Cycles to failure
70	110,000–120,000	10⁶

Impact Strength

Test temperature, °F	Type test	Strength, ft-lb
+70	Charpy–unnotched	15–30

Table 1.1–151

APPROXIMATE EQUIVALENT HARDNESS NUMBERS[a] FOR STEEL

| Brinell Indentation Dia, mm | Brinell Hardness No.[b] 10-mm Ball, 3000-kg Load | | | Diamond Pyramid Hardness No. | Rockwell Hardness No.[c] | | | | Rockwell Superficial Hardness No. Superficial Brale Penetrator | | | Shore Scleroscope Hardness No. | Tensile Strength (Approximate) in 1000 psi |
| | Standard Ball | Hultgren Ball | Tungsten-Carbide Ball | | A-Scale, 60-kg Load, Brale Penetrator | B-Scale, 100kg Load, 1/16-in. Dia Ball | C-Scale, 150kg Load, Brale Penetrator | D-Scale, 100kg Load, Brale Penetrator | 15-N Scale, 15-kg Load | 30-N Scale, 30-kg Load | 45-N Scale, 45-kg Load | | |
Col. 1	Col. 2	Col. 3	Col. 4	Col. 5	Col. 6	Col. 7	Col. 8	Col. 9	Col. 10	Col. 11	Col. 12	Col. 13	Col. 14
2.25	—	—	—	940	85.6	—	68.0	76.9	93.2	84.4	75.4	97	—
	—	—	—	920	85.5	—	67.5	76.5	93.0	84.0	74.8	96	—
	—	—	767	900	85.0	—	67.0	76.1	92.9	83.6	74.2	95	—
	—	—	757	880	84.7	—	66.4	75.7	92.7	83.1	73.6	93	—
	—	—	745	860	84.4	—	65.9	75.3	92.5	82.7	73.1	92	—
2.30	—	—	733	840	84.1	—	65.3	74.8	92.3	82.2	72.2	91	—
	—	—	722	820	83.8	—	64.7	74.3	92.1	81.7	71.8	90	—
	—	—	712	800	83.4	—	64.0	73.8	91.8	81.1	71.0	88	—
2.35	—	—	710	780	83.0	—	63.3	73.3	91.5	80.4	70.2	87	—
	—	—	698	760	82.6	—	62.5	72.6	91.2	79.7	69.4	86	—
2.40	—	—	684	740	82.2	—	61.8	72.1	91.0	79.1	68.6	84	—
	—	—	682	737	82.2	—	61.7	72.0	91.0	79.0	68.5	83	—
	—	—	670	720	81.8	—	61.0	71.5	90.7	78.4	67.7	—	—
	—	—	656	700	81.3	—	60.1	70.8	90.3	77.6	66.7	—	—
	—	—	653	697	81.2	—	60.0	70.7	90.2	77.5	66.5	81	—
2.45	—	—	647	690	81.1	—	59.7	70.5	90.1	77.2	66.2	—	—
	—	—	638	680	80.8	—	59.2	70.1	89.8	76.8	65.7	80	—
	—	—	630	670	80.6	—	58.8	69.8	89.7	76.4	65.3	79	—
	—	—	627	667	80.5	—	58.7	69.7	89.6	76.3	65.1	—	—
2.50	—	601	601	677	80.7	—	59.1	70.0	89.8	76.8	65.7	77	—
	—	—	—	640	79.8	—	57.3	68.7	89.0	75.1	63.5	—	—
2.55	—	578	578	640	79.8	—	57.3	68.7	89.0	75.1	63.5	75	—
	—	—	—	615	79.1	—	56.0	67.7	88.4	73.9	62.1	—	—
2.60	—	555	555	607	78.8	—	55.6	67.4	88.1	73.5	61.6	73	—
	—	—	—	591	78.4	—	54.7	66.7	87.8	72.7	60.6	—	298
2.65	—	534	534	579	78.0	—	54.0	66.1	87.5	72.0	59.8	71	292
	—	—	—	569	77.8	—	53.5	65.8	87.2	71.6	59.2	—	288
2.70	—	514	514	553	77.1	—	52.5	65.0	86.7	70.7	58.0	70	278
	—	—	—	547	76.9	—	52.1	64.7	86.5	70.3	57.6	—	274
2.75	495	495	495	539	76.7	—	51.6	64.3	86.3	69.9	56.9	—	269
	—	—	—	530	76.4	—	51.1	63.9	86.0	69.5	56.2	—	265
	—	—	—	528	76.3	—	51.0	63.8	85.9	69.4	56.1	68	264
2.80	477	477	477	516	75.9	—	50.3	63.2	85.6	68.7	55.2	—	258
	—	—	—	508	75.6	—	49.6	62.7	85.3	68.2	54.5	66	252
	—	—	—	508	75.6	—	49.6	62.7	85.3	68.2	54.5	—	252
2.85	461	461	461	495	75.1	—	48.8	61.9	84.9	67.4	53.5	—	244
	—	—	—	491	74.9	—	48.5	61.7	84.7	67.2	53.2	65	242
	—	—	—	491	74.9	—	48.5	61.7	84.7	67.2	53.2	—	242
2.90	444	444	444	474	74.3	—	47.2	61.0	84.1	66.0	51.7	—	231
	—	—	—	472	74.2	—	47.1	60.8	84.0	65.8	51.5	63	230
	—	—	—	472	74.2	—	47.1	60.8	84.0	65.8	51.5	—	230

Table 1.1–151 (continued)

APPROXIMATE EQUIVALENT HARDNESS NUMBERS[a] FOR STEEL

Brinell Indentation Dia, mm	Brinell HN[b] Standard Ball	Brinell HN[b] Hultgren Ball	Brinell HN[b] Tungsten-Carbide Ball	Diamond Pyramid Hardness No.	Rockwell[c] A-Scale, 60-kg Load, Brale Penetrator	Rockwell[c] B-Scale, 100-kg Load, 1/16-in. Dia Ball	Rockwell[c] C-Scale, 150-kg Load, Brale Penetrator	Rockwell[c] D-Scale, 100-kg Load, Brale Penetrator	Superficial 15-N Scale, 15-kg Load	Superficial 30-N Scale, 30-kg Load	Superficial 45-N Scale, 45-kg Load	Shore Scleroscope Hardness No.	Tensile Strength (Approximate) in 1000 psi
Col. 1	Col. 2	Col. 3	Col. 4	Col. 5	Col. 6	Col. 7	Col. 8	Col. 9	Col. 10	Col. 11	Col. 12	Col. 13	Col. 14
2.95	429	429	429	455	73.4	—	45.7	59.7	83.4	64.6	49.9	61	219
3.00	415	415	415	440	72.8	—	44.5	58.8	82.8	63.5	48.4	59	212
3.05	401	401	401	425	72.0	—	43.1	57.8	82.0	62.3	46.9	58	202
3.10	388	388	388	410	71.4	—	41.8	56.8	81.4	61.1	45.3	56	193
3.15	375	375	375	396	70.6	—	40.4	55.7	80.6	59.9	43.6	54	184
3.20	363	363	363	383	70.0	—	39.1	54.6	80.0	58.7	42.0	52	177
3.25	352	352	352	372	69.3	(110.0)	37.9	53.8	79.3	57.6	40.5	51	171
3.30	341	341	341	360	68.7	(109.0)	36.6	52.8	78.6	56.4	39.1	50	164
3.35	331	331	331	350	68.1	(108.5)	35.5	51.9	78.0	55.4	37.8	48	159
3.40	321	321	321	339	67.5	(108.0)	34.3	51.0	77.3	54.3	36.4	47	154
3.45	311	311	311	328	66.9	(107.5)	33.1	50.0	76.7	53.3	34.4	46	149
3.50	302	302	302	319	66.3	(107.0)	32.1	49.3	76.1	52.2	33.8	45	146
3.55	293	293	293	309	65.7	(106.0)	30.9	48.3	75.5	51.2	32.4	43	141
3.60	285	285	285	301	65.3	(105.5)	29.9	47.6	75.0	50.3	31.2	—	138
3.65	277	277	277	292	64.6	(104.5)	28.8	46.7	74.4	49.3	29.9	41	134
3.70	269	269	269	284	64.1	(104.0)	27.6	45.9	73.7	48.3	28.5	40	130
3.75	262	262	262	276	63.6	(103.0)	26.6	45.0	73.1	47.3	27.2	39	127
3.80	255	255	255	269	63.0	(102.0)	25.4	44.2	72.5	46.2	26.0	38	123
3.85	248	248	248	261	62.5	(101.0)	24.2	43.2	71.7	45.1	24.5	37	120
3.90	241	241	241	253	61.8	100.0	22.8	42.0	70.9	43.9	22.8	36	116
3.95	235	235	235	247	61.4	99.0	21.7	41.4	70.3	42.9	21.5	35	114
4.00	229	229	229	241	60.8	98.2	20.5	40.5	69.7	41.9	20.1	34	111
4.05	223	223	223	234	—	97.3	(18.8)	—	—	—	—	—	—
4.10	217	217	217	228	—	96.4	(17.5)	—	—	—	—	33	105
4.15	212	212	212	222	—	95.5	(16.0)	—	—	—	—	—	102
4.20	207	207	207	218	—	94.6	(15.2)	—	—	—	—	32	100
4.25	201	201	201	212	—	93.8	(13.8)	—	—	—	—	31	98
4.30	197	197	197	207	—	92.8	(12.7)	—	—	—	—	30	95
4.35	192	192	192	202	—	91.9	(11.5)	—	—	—	—	29	93
4.40	187	187	187	196	—	90.7	(10.0)	—	—	—	—	—	90
4.45	183	183	183	192	—	90.0	(9.0)	—	—	—	—	28	89
4.50	179	179	179	188	—	89.0	(8.0)	—	—	—	—	27	87
4.55	174	174	174	182	—	87.8	(6.4)	—	—	—	—	—	85
4.60	170	170	170	178	—	86.8	(5.4)	—	—	—	—	26	83
4.65	163	163	167	175	—	86.0	(4.4)	—	—	—	—	—	81
4.70	163	163	163	171	—	85.0	(3.3)	—	—	—	—	25	79
4.80	156	156	156	163	—	82.9	—	—	—	—	—	—	76
4.90	149	149	149	156	—	80.8	(0.9)	—	—	—	—	23	73

Table 1.1–151 (continued)
APPROXIMATE EQUIVALENT HARDNESS NUMBERS[a] FOR STEEL

Brinell Indentation Dia, mm	Brinell Hardness No.,[b] 10-mm Ball, 3000-kg Load			Diamond Pyramid Hardness No.	Rockwell Hardness No.[c]				Rockwell Superficial Hardness No. Superficial Brale Penetrator			Shore Sclero-scope Hardness No.	Tensile Strength (Approximate) in 1000 psi
	Standard Ball	Hultgren Ball	Tungsten-Carbide Ball		A-Scale, 60-kg Load, Brale Penetrator	B-Scale, 100-kg Load, 1/16-in. Dia Ball	C-Scale, 150-kg Load, Brale Penetrator	D-Scale, 100-kg Load, Brale Penetrator	15-N Scale, 15-kg Load	30-N Scale, 30-kg Load	45-N Scale, 45-kg Load		
Col. 1	Col. 2	Col. 3	Col. 4	Col. 5	Col. 6	Col. 7	Col. 8	Col. 9	Col. 10	Col. 11	Col. 12	Col. 13	Col. 14
5.00	143	143	143	150	—	78.7	—	—	—	—	—	22	71
5.10	137	137	137	143	—	76.4	—	—	—	—	—	21	67
5.20	131	131	131	137	—	74.0	—	—	—	—	—	—	65
5.30	126	126	126	132	—	72.0	—	—	—	—	—	20	63
5.40	121	121	121	127	—	69.8	—	—	—	—	—	19	60
5.50	116	116	116	122	—	67.6	—	—	—	—	—	18	58
5.60	111	111	111	117	—	65.7	—	—	—	—	—	15	56

[a]The values in this table shown in **boldface type** correspond to the values shown in the corresponding joint SAE-ASM-ASTM Committee on Hardness Conversions as printed in ASTM E 140, Table 3.

[b]Brinell numbers are based on the diameter of impressed indentation. If the ball distorts (flattens) during test, Brinell numbers will vary in accordance with the degree of such distortion when related to hardnesses determined with a Vickers Diamond Pyramid, Rockwell Brale, or other penetrator which does not sensibly distort. At high hardnesses, therefore, the relationship between Brinell and Vickers or Rockwell scales is affected by the type of ball used. Steel balls (Standard or Hultgren) tend to flatten slightly more than carbide balls, resulting in larger indentation and lower Brinell number than shown by a carbide ball. Thus, on a specimen of 640 Vickers, a Hultgren ball will leave a 2.55-mm impression (578 Bhn), and the carbide ball a 2.50-mm impression (601 Bhn). Conversely, identical impression diameters for both types of ball will correspond to different Vickers or Rockwell values. Thus, if both impressions are 2.55 mm (578 Bhn), material tested with a Hultgren ball has a Vickers Hardness 640, while material tested with a carbide ball has a Vickers Hardness 615.

[c]Values in parentheses are beyond normal range and are given for information only.

From Hardness Tests and Hardness Number Conversions, SAE Information Report J417b. Copyright © Society of Automotive Engineers, Inc., 1974. All rights reserved. Reprinted with permission.

1.2 LIGHT METALS

W. F. Simmons
Battelle/Columbus Laboratories

MAGNESIUM BASE ALLOYS

Nomenclature

Magnesium alloy designations are based on chemical composition. They consist of two letters representing the two alloying elements specified in the greatest amount, arranged in decreasing percentages or alphabetically if of equal percentage. The letters are followed by the respective percentages rounded off to whole numbers with a serial letter at the end. The serial letter indicates some variation in composition.

The letters used to designate various alloying elements include:

A—Aluminum	M—Manganese
E—Rare earths	Q—Silver
H—Thorium	S—Silicon
K—Zirconium	T—Tin
L—Lithium	Z—Zinc

Temper designations for magnesium alloys are separated from the alloy designations by a dash. The following designations are used to denote tempers of magnesium mill products:

F	— as fabricated
T4	— solution heat treated
T5,T51	— artificially aged
T6,T61	— solution heat treated and artificially aged
T7	— solution heat treated and stabilized
T8	— solution heat treated and artificially aged
O	— annealed
H24,H26	— strain hardened, then partially annealed

Key to mechanical property tests:

a — Permanent mold or sand castings, ½-in. diameter section
b — 0.2% offset
c — 500-kg, 10-mm ball
d — 3/16-in. diameter pin
e — Rotating beam
f — Axial load
g — Flexure

Table 1.2–1
AM100A

AM100A is used for pressure-tight sand and permanent mold castings. It contains aluminum and a small amount of manganese. It has a good combination of room temperature properties, similar to AZ92A, but has less tendency to crack when used for permanent mold castings.

Typical chemical composition, percent by weight: Al, 10; Mn, 0.10 minimum; Mg, balance

Physical constants and thermal properties
Density at $68°F$, lb/cu in.$^{-1}$: 0.0651
Coefficient of thermal expansion, $°F^{-1} \times 10^{-6}$, $65-212°F$: 14.5
Modulus of elasticity, psi: 6.5×10^6
Poisson's ratio: 0.35
Melting range, $°F$: 867–1,101
Specific heat, Btu/lb^{-1}/$°F^{-1}$, $70°F$: 0.31
Electrical resistivity, microhm-cm, $70°F$: 15.0

Thermal conductivity, Btu/hr^{-1}/ft^{-2}/$°F^{-1}$/ft, $70°F$: 32.5; $300°F$: 42.0; $500°F$: 47.0

Table 1.2–1 (continued)
AM100A

Heat Treatments

Treatment	Temperature, °F	Time, hours	Cooling
Solution[a]	790	20	Strong air blast
Artificial aging (partial)	325	12	Still air
Artificial aging (complete)	450	5	Still air or oven
Stabilizing[b]	500	4	Still air

[a]An atmosphere with at least 0.5% SO_2 is required.
[b]Used to minimize growth in castings to be used at elevated temperatures.

Specifications

	Ingot	Sand castings	Permeable mold castings
ASM			4483
ASTM	B93	B80	B199
SAE			502
Federal			QQ-M-55

Typical Room Temperature Mechanical Properties

Temper	TS, ksi	YS, ksi	CYS, ksi	Elong., %	Hardness Bhn	Hardness RE	Shear strength, ksi	Charpy impact strength, ft-lb
F	20/22	10/12	12	2	53	64	18	0.6
T4	34/40	10/13	13	10	52	62	20	2.0
T61	34/40	17/22	19	1	69	80	21	0.7
T5	22	16	16	2	58	70		
T7	38	18	18	1	67	78		
T6	40	16		4				

Temper	Bearing, ksi Ultimate	Bearing, ksi Yield	Fatigue strength, ksi 10^5 cycles	10^6 cycles	10^7 cycles	10^8 cycles	5×10^8 cycles
F							10
T4	69.0	45.0					11
T61	81.0	68.0					10

Table 1.2–1 (continued)
AM100A

Effect of Testing Temperature on Typical Mechanical Properties

	Room temperature			200°F			300°F		
Temper	TS, ksi	TYS, ksi	Elong., %	TS, ksi	TYS, ksi	Elong., %	TS, ksi	TYS, ksi	Elong., %
T4				34		1.5	23		9
T6				–			24	9	4

	400°F			500°F			600°F		
T4				12		22			
T6	17	6.5	25	12	4	45	8.5	2.5	60

	700°F			–108°F		
F				22	18	1
T4				38	18	7
T6	5.5	1.5	100	39	26	2

	Temperature, °F	Hardness Bhn	RE	Charpy, ft-lb
F	–108	63	75	0.8
T4	–108	60	73	2.5
T6	–108	85	90	0.8

1.2–2
AZ63A

AZ63A is a general sand casting alloy having good strength, ductility, and toughness. AZ63A castings have more tendency to porosity and shrinkage problems than AZ91C alloy castings, but otherwise their properties are similar.

Typical chemical composition, percent by weight: Al, 6.0; Mn, 0.15 minimum; Zn, 3.0; Mg, balance

Physical constants and thermal properties
Density at 68°F, lb/cu in.$^{-1}$: 0.0656
Coefficient of thermal expansion, °F^{-1} × 10^{-6}, 65–212°F: 14.5
Modulus of elasticity, psi: 6.5 × 10^6
Poisson's ratio: 0.35
Melting range, °F: 850–1,130
Specific heat, Btu/lb^{-1}/°F^{-1}, 70°F: 0.32
Electrical resistivity, microhm-cm, 70°F: 12.1; 300°F: 14.1; 500°F: 15.8
Thermal conductivity, Btu/hr^{-1}/ft^{-2}/°F^{-1}/ft, 70°F, 35.5; 300°F: 43.0; 500°F: 49.5

Heat Treatments

Treatment	Temperature, °F	Time, hours	Cooling
Solution[a]	740	10	Air
Aging	450	5	Air or furnace
Stabilizing[b]	300	4	Air

[a]An atmosphere with at least 0.5% SO$_2$ is required.
[b]Used to minimize growth in castings to be used at elevated temperatures.

1.2–2 (continued)
AZ63A

Specifications

	Ingot	Sand castings	Permeable mold castings	Welding rod
AMS		4420[a]		
ASTM	B93	B80		
SAE		50		
Federal		QQ-M-56	QQ-M-55	
Military				MIL-R-6944

[a]AMS 4420, as cast; 4422, heat treated; 4424 heat treated and aged.

Typical Room Temperature Mechanical Properties

Temper	TS, ksi	YS, ksi	CYS, ksi	Elong., %	Hardness Bhn	RE	Shear strength, ksi	Charpy impact strength, ft-lb
F	26/29	11/14	14	6	50	59	16	1.0
T4	34/40	11/14	14	12	55	66	17	2.5
T5	26/30	12/15	14	4	55	66	17	2.6
T6	34/40	16/19	19	5	73	83	19	1.1
T7	40	17	17	6	64	76	20	

	Bearing, ksi		Fatigue strength, ksi				
Temper	Ultimate	Yield	10^5 cycles	10^6 cycles	10^7 cycles	10^8 cycles	5×10^8 cycles
F	60	40					11
T4	60	44					12
T5	60	40					11
T6	75	52					11
T7	75	47					13

Effect of Testing Temperature on Typical Mechanical Properties

	Room temperature			200°F			300°F		
Temper	TS, ksi	TYS, ksi	Elong., %	TS, ksi	TYS, ksi	Elong., %	TS, ksi	TYS, ksi	Elong., %
F	29	14	4.5	30		4.5	24		20
T4	40	14	12	35	13	14	29	12	15
T5	30	15	4	29	15	8	26	13	33
T6	40	19	5	36	17	11	24	15	15

	400°F			500°F			600°F		
	TS	TYS	Elong.	TS	TYS	Elong.	TS	TYS	Elong.
F	15		50	10		38			
T4	15	11	20						
T5	21	10	45	17	7	56	8	4	65
T6	18	12	18	12	9	15	8	6	20

1.2–2 (continued)
AZ63A

Effect of Exposure at a Given Temperature on Tensile Properties of AZ63A-T6 –
Separately Cast Test Bars

Property	Temperature, °F		Exposure time, hours						
	Exposure	Testing	0	25	100	500	1,000	2,500	5,000
TS, ksi	200	70	36.0	39.9	39.6	39.7	37.3	39.8	–
TYS, ksi	200	70	18.9	22.3	20.2	20.6	20.1	20.5	–
E, %	200	70	3.5	4.8	5.5	5.4	4.2	5.6	–
TS, ksi	200	200	38.1	36.0	36.4	–	38.0	36.6	–
TYS, ksi	200	200	17.0	18.3	18.0	–	18.1	17.2	–
E, %	200	200	15.8	10.7	17.2	–	16.1	15.5	–
TS, ksi	300	70	36.0	39.1	37.9	39.7	39.1	37.3	38.0
TYS, ksi	300	70	18.9	21.1	20.9	21.0	20.8	21.1	21.4
E, %	300	70	3.5	4.3	3.8	4.0	3.7	3.7	3.7
TS, ksi	300	300	27.1	24.8	24.6	–	25.7	24.0	24.6
TYS, ksi	300	300	15.7	14.9	15.0	–	14.9	15.0	14.6
E, %	300	300	35.7	39.0	35.1	–	34.3	33.8	39.1
TS, ksi	400	70	36.0	39.9	38.5	37.0	37.3	37.0	36.2
TYS, ksi	400	70	18.9	18.7	19.9	17.8	20.1	17.9	17.8
E, %	400	70	3.5	6.1	5.2	4.8	4.2	5.2	5.2
TS, ksi	400	400	17.9	16.2	16.2	–	16.4	16.4	17.4
TYS, ksi	400	400	11.1	10.7	10.6	–	10.3	10.2	10.1
E, %	400	400	34.5	36.2	35.9	–	38.8	36.7	37.0

1.2–2 (continued)
AZ63A

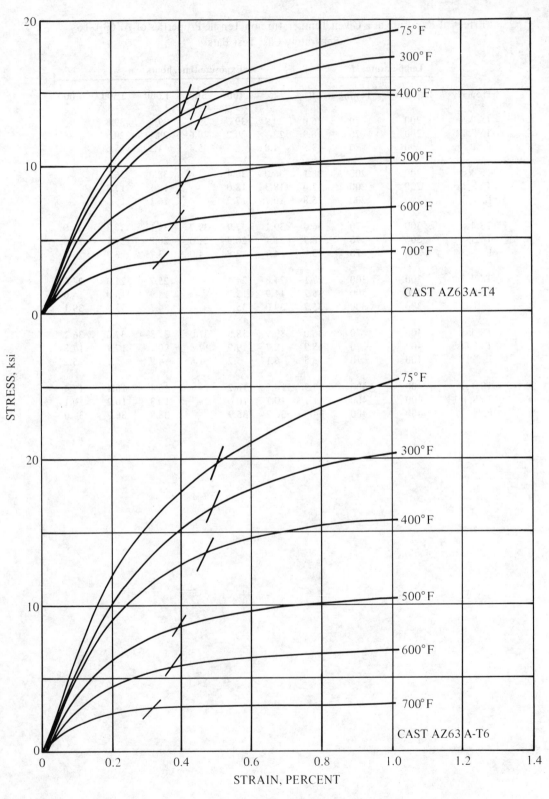

FIGURE A.   Tension stress-strain relationships for cast AZ63A-T6 and AZ63A-T4.

1.2–2 (continued)
AZ63A

Typical Creep Strengths

Stress, ksi

Temper	Temperature, °F	0.1% Total extension					0.2% Total extension				
		Minutes	Hours				Minutes	Hours			
		10	1	10	100	1,000	10	1	10	100	1,000
T6	200	7.9	7.1	5.9	4.6	3.5	11.7	10.7	9.2	7.4	5.7
	300	5.8	5.0	3.7	2.4	1.4	8.8	7.7	6.1	4.1	2.6
	400	3.8	3.0	1.7			5.9	4.9	3.0		
	500	2.4	1.5				3.4	2.4	0.9		

Temper	Temperature, °F	0.5% Total extension					1.0% Total extension				
		Minutes	Hours				Minutes	Hours			
		10	1	10	100	1,000	10	1	10	100	1,000
T6	200	18.8	17.1	14.7	12.3	10.1	22.3	20.5	17.7	14.8	12.3
	300	13.3	11.8	9.6	6.9	4.8	15.9	14.3	11.7	8.7	6.1
	400	8.5	7.1	5.0			10.0	8.5	6.1		
	500	4.7	3.4	1.6			5.6	4.2	2.3		

Table 1.2–3
AZ81A

AZ81A is a sand or permanent mold casting alloy. It is used only as solution treated. It has properties similar to those of AZ91C, but is less susceptible to natural aging.

Typical chemical composition, percent by weight: Al, 7.6; Mn, 0.13 minimum; Zn, 0.7; Mg, balance

Physical constants and thermal properties
Density at 68°F, lb/cu in. $^{-1}$: 0.0649

Coefficient of thermal expansion, $°F^{-1} \times 10^{-6}$, 65–212°F: 14.5

Modulus of elasticity, psi: 6.5×10^6

Modulus of rigidity, psi: 2.4×10^6

Poisson's ratio: 0.35

Melting range, °F: 914–1132

Electrical resistivity, microhom-cm, 70°F: 15.0; 300°F: 16.7; 500°F: 17.8

Thermal conductivity, Btu/hr $^{-1}$ /ft $^{-2}$ /°F $^{-1}$ /ft, 70°F: 28.0; 300°F: 37.0; 500°F: 44.4

Heat Treatments

Treatment	Temperature, °F	Time, hours	Cooling
Solution	775	18	Air or fan
Stabilizing	500	4	Air

Table 1.2–3 (continued)
AZ81A

Specifications

	Sand castings	Permeable mold castings
ASTM	B80	
Federal	QQ-M-56	QQ-M-55

Typical Room Temperature Mechanical Properties

Temper	TS, ksi	YS, ksi	CYS, ksi	Elong., %	Hardness Bhn	Hardness RE	Shear strength, ksi	Charpy impact strength, ft-lb
T4	34/40	11/12	12	15	55	66	17/21	4.5

Temper	Bearing, ksi Ultimate	Bearing, ksi Yield	Fatigue strength, ksi 10^5 cycles	10^6 cycles	10^7 cycles	10^8 cycles	5×10^8 cycles
T4	58/60	35/44					

Effect of Testing Temperature on Typical Mechanical Properties

Temper	Room temperature TS, ksi	TYS, ksi	Elong., %	200°F TS, ksi	TYS, ksi	Elong., %	300°F TS, ksi	TYS, ksi	Elong., %
T4	40	21	5	37	19	24	27	17	31

Temper	400°F			500°F		
T4	20	11	30	14	10	25

Table 1.2–3 (continued)
AZ81A

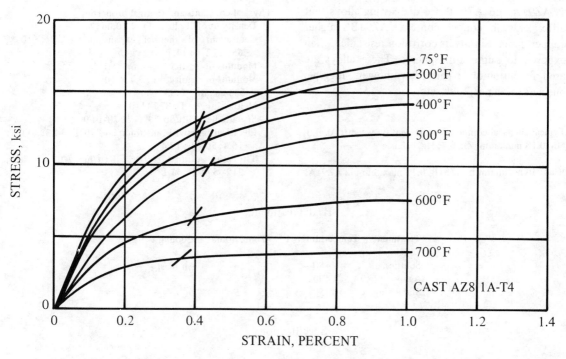

FIGURE B.   Tension stress-strain relationships for cast AZ81A-T4.

Typical Creep Strengths

Stress, ksi

		0.1% Total extension					0.2% Total extension				
			Hours					Hours			
Temper	Temperature, °F	Minutes 10	1	10	100	1000	Minutes 10	1	10	100	1000
T4	200		5.6	5.4	5.2			8.4	8.0	7.4	
	300		5.4	4.0	2.2			7.7	6.5	3.5	
	400		3.4	1.7	1.0			6.0	3.1	1.7	

		0.5% Total extension				
			Hours			
Temper	Temperature °F	Minutes 10	1	10	100	1000
T4	200		12.5	12.0	11.8	
	300			9.0	6.6	
	400				3.0	

Table 1.2–4
AZ91A AND AZ91B[a]

AZ91A and AZ91B are die casting alloys that differ only in copper content. AZ91B's higher copper content limits its corrosion resistance when exposed to saline atmospheres. These alloys are used for automobile parts, lawnmowers, luggage, business machines, tools, etc.

Typical chemical composition, percent by weight: Al, 9.0; Mn, 0.13 minimum; Zn, 0.7; Mg, balance

[a]Cu 0.30 maximum in AZ91B, 0.10 maximum in AZ91A.

Physical constants and thermal properties
Density at 68°F, lb/cu in.$^{-1}$: 0.0652
Coefficient of thermal expansion, °F^{-1} × 10^{-6}, 65–212°F: 14.5
Modulus of elasticity, psi: 6.5 × 10^6
Modulus of rigidity, psi: 2.4 × 10^6
Poisson's ratio. 0.35
Melting range, °F: 875–1105
Specific heat, Btu/lb^{-1}/°F^{-1}, 70°F: 0.25
Electrical resistivity, microhm-cm, 70°F: 14.3; 300°F: 16.4; 500°F: 17.4
Thermal conductivity, Btu/hr^{-1}/ft^{-2}/°F^{-1}/ft, 212–574°F: 41.1

Heat Treatments

Treatment	Temperature, °F	Time, hours	Cooling
Solution	780	16	Air blast
Aging	400	4	—
Stabilizing	450	5	—

Specifications

	Die castings	Ingot
AMS	4490	
ASTM	B94	B93
SAE	501	
Federal	QQ-M-38	

Typical Room Temperature Mechanical Properties

Temper	TS, ksi	YS, ksi	CYS, ksi	Elong., %	Hardness Bhn	RE	Shear strength, ksi	Charpy impact strength, ft-lb
F	33/39	22/24	24	3/6	63	75	20	2.0

	Bearing, ksi		Fatigue strength, ksi				
Temper	Ultimate	Yield	10^5 cycles	10^6 cycles	10^7 cycles	10^8 cycles	5 × 10^8 cycles
F							14

Table 1.2–5
AZ91C

AZ91C is the most commonly employed alloy for sand and permanent mold castings to be used at room temperature. It combines strength and ductility with good foundry characteristics.

Typical chemical composition, percent by weight: Al, 8.7; Mn, 0.13 minimum; Zn, 0.7; Mg, balance

Physical constants and thermal properties
Density at 68°F, lb/cu in^{-1}: 0.0652

Coefficient of thermal expansion, °F^{-1} × 10^{-6}, 65–212°F: 14.5
Modulus of elasticity, psi: 6.5 × 10^6
Modulus of rigidity, psi: 2.4 × 10^6
Poisson's ratio: 0.35
Melting range, °F: 875–1105
Specific heat, Btu/lb^{-1}/°F^{-1}, 70°F: 0.28
Electrical resistivity, microhm-cm, 70°F: 13.6; 300°F: 15.8; 500°F: 16.8
Thermal conductivity, Btu/hr^{-1}/ft^{-2}/°F^{-1}/ft, 70°F: 31.0; 300°F: 39.0; 500°F: 46.5

Heat Treatments

Treatment	Temperature, °F	Time, hours	Cooling
Solution			
T4	780	16	Air blast
Aging			
T6	400	4	
T7	450	5	

Specifications

	Ingot	Sand castings	Permeable mold castings
AMS		4437	
ASTM	B93	B80	B199
SAE		304	
Federal		QQ-M-56	QQ-M-55

Typical Room Temperature Mechanical Properties

Temper	TS, ksi	YS, ksi	CYS, ksi	Elong., %	Hardness Bhn	Hardness RE	Shear strength, ksi	Charpy impact strength, ft-lb
F	23/24	11/14	14	2	52	62	16	0.6
T4	34/40	11/13	13	7/15	53	64	17	3.0
T6	34/40	10/19	19	3/5	66	78	19	1.0

Temper	Bearing, ksi Ultimate	Bearing, ksi Yield
F	60	40
T4	60	44
T6	75	52

Table 1.2—5 (continued)
AZ91C

Effect of Testing Temperature on Typical Mechanical Properties

	Room temperature			200°F			300°F		
Temper	TS, ksi	TYS, ksi	Elong., %	TS, ksi	TYS, ksi	Elong., %	TS, ksi	TYS, ksi	Elong., %
T4	40	15	14	34	14	26	28	14	30
T6	40	21	5	37	19	24	27	17	31

	400°F		
T4	20	13	30
T6	20	14	33

Table 1.2–5 (continued)
AZ91C

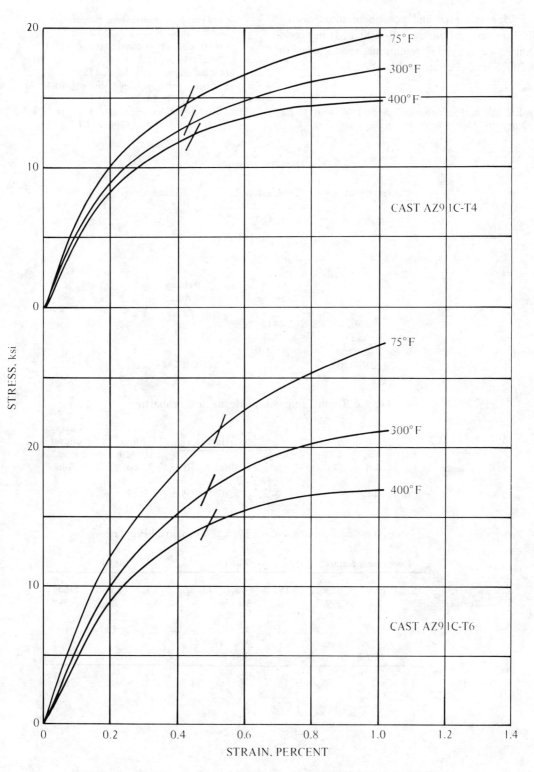

FIGURE C.   Tension stress-strain relationships for cast AZ91C-T4 and AZ91C-T6.

Table 1.2—6
EZ33A

EZ33A is a sand and permanent mold casting alloy for applications at 350–500°F. It has good damping capacity, and retains mechanical strength to 500°F much better than the AZ alloys.

Typical chemical composition, percent by weight: Zn, 2.6; Zr, 0.7; RE, 3.2; Mg, balance

Physical constants and thermal properties

Density at 68°F, lb cu in.$^{-1}$: 0.0659
Coefficient of thermal expansion, °F^{-1} × 10^{-6}, 65–212°F: 14.5
Melting range,°F (liquidus): 1189
Electrical resistivity, microhm-cm, 70°F: 7.0; 300°F: 9.2; 500°F: 11.2
Thermal conductivity, Btu hr^{-1} ft^{-2}/°F/ft, 70°F: 57.6; 300°F: 64.1; 500°F: 67.5

Heat Treatments

Treatment	Temperature, °F	Time, hours
Aging	420	5

Specifications

	Sand castings	Welding rod
AMS	4442	
ASTM	B80	
Federal	QQ-M-56	
Military		MIL-R-6944

Typical Room Temperature Mechanical Properties

Temper	TS, ksi	YS, ksi	CYS, ksi	Elong., %	Hardness Bhn	Hardness RE	Shear strength, ksi	Charpy impact strength, ft-lb
T5	20	14		2	50	59	19	

Effect of Testing Temperature on Typical Mechanical Properties

	Room temperature			200°F			300°F		
Temper	TS, ksi	TYS, ksi	Elong., %	TS, ksi	TYS, ksi	Elong., %	TS, ksi	TYS, ksi	Elong., %
T5	23	16	3	23	15	5	22	14	10

	400°F			500°F			600°F		
T5	21	12	20	18	10	31	12	8	50

Table 1.2–6 (continued)
EZ33A

Effect of Exposure at a Given Temperature on Tensile Properties
of EZ33A-T5 — Separately Cast Test Bars

	Temperature, °F		Exposure time, hours					
Property	Exposure	Testing	0	312	500	1000	2500	5000
TS, ksi	400	400	21.0	–	19.1	18.9	–	–
TYS, ksi	400	400	11.2	–	13.1	14.4	–	–
TS, ksi	500	500	18.3	–	14.1	14.0	–	–
TYS, ksi	500	500	10.0	–	8.6	8.5	–	–
TS, ksi	600	600	11.1	–	11.1	11.1	–	–
TYS, ksi	600	600	7.4	–	7.4	7.4	–	–

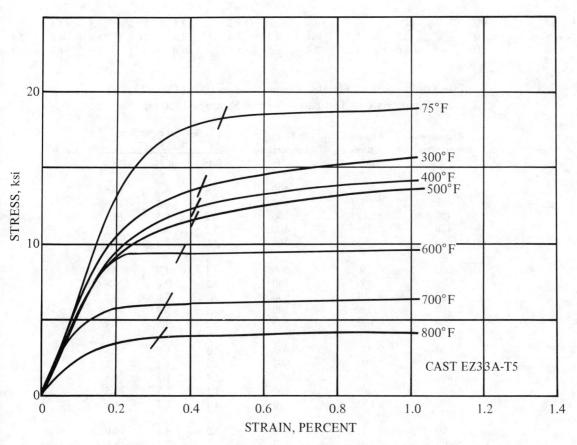

FIGURE D.   Tension stress-strain relationships for cast EZ33A-T5.

Table 1.2–6 (continued)
EZ33A

Typical Creep Strengths

Stress, ksi

		0.1% Total extension					0.2% Total extension				
		Minutes	Hours				Minutes	Hours			
Temper	Temperature, °F	10	1	10	100	1000	10	1	10	100	1000
T5	400	6.0	5.9	5.6	5.1	4.3	9.5	9.3	8.8	8.0	6.2
	500	5.3	4.3	3.2	2.1	1.6	9.2	7.6	5.0	3.0	2.0
	600	4.1	2.6	1.5	0.9	0.7	5.7	3.5	2.0	1.2	0.8

		0.5% Total extension					1.0% Total extension				
		Minutes	Hours				Minutes	Hours			
Temper	Temperature, °F	10	1	10	100	1000	10	1	10	100	1000
T5	400	11.0	11.0	10.9	10.0	8.3	12.8	12.8	12.5	11.6	9.2
	500	10.7	9.2	6.4	4.1	2.6	12.4	10.4	7.4	4.8	3.1
	600	6.6	4.2	2.4	1.5	1.1	8.0	5.1	2.9	1.7	1.2

Effect of Exposure at Elevated Temperature on 100-hr Creep Limits;
Separately Cast Test Bars — Stresses Shown in 1000 psi

Exposure temperature, °F	Exposure time, hours	Testing temperature, °F	% Total extension EZ33A-T5		
			0.1%	0.2%	0.5%
400	0	400	5.4	8.5	10.6
400	1	400	5.3	8.5	10.6
400	10	400	5.2	8.5	10.5
400	100	400	5.2	8.3	10.3
400	1000	400	5.2	7.8	9.7
500	0	500	3.1	3.8	4.8
500	1	500	3.2	3.8	4.8
500	10	500	3.2	3.8	4.8
500	100	500	3.2	3.7	4.6
500	1000	500	3.2	3.5	4.4
600	0	600	1.1	1.3	1.5
600	1	600	1.1	1.3	1.5
600	10	600	1.1	1.3	1.5
600	100	600	1.1	1.3	1.5
600	1000	600	1.1	1.3	1.5
500	1	400	5.5	8.1	10.0
500	10	400	5.7	7.7	9.6
500	20	400	5.8	7.5	9.4
500	100	400	–	–	–
600	1	400	5.3	7.8	9.9
600	10	400	5.2	7.3	9.3
600	20	400	5.2	7.1	9.1
600	1000	400	–	–	–

Table 1.2–7
HK31A

HK31A is a sand casting or wrought alloy with sufficient creep resistance as castings to be used in the 400–700°F temperature range.

As sheet and plate, HK31A has good formability and weldability, and is used for applications at 400–600°F.

Typical chemical composition, percent by weight: Zr, 0.7; Th, 3.2; Mg, balance

Physical constants and thermal properties
Density at 68°F, lb/cu in. $^{-1}$: 0.0647

Coefficient of thermal expansion,°F $^{-1}$ × 10 $^{-6}$, 65–212°F: 14.5
Modulus of elasticity, psi: 6.5 × 10^6
Modulus of rigidity, psi: 2.4 × 10^6
Poisson's ratio: 0.35
Melting range, °F: 1092–1204
Specific heat, Btu/lb $^{-1}$/°F $^{-1}$, 70°F: 0.24; 300°F: 0.26; 500°F: 0.27; 800°F: 0.31
Electrical resistivity, microhm-cm, 70°F: 7.7 (cast, T6); 6.1 (rolled, H24)
Thermal conductivity, Btu/hr $^{-1}$/ft $^{-2}$/°F/ft, 70°F: 52.5 (cast, T6), 660 (rolled, H24); 300°F; 60.5 (cast, T6), 69.7 (rolled, H24); 500°F: 65.5 (cast, T6), 73.4 (rolled, H24)

Heat Treatments

Treatment	Temperature, °F	Time, hours	Cooling
Solution, castings	1050	2	Air or fan
Age	400	16	
H24 sheet may be stress relieved after welding	650	1	
	675	20 min	

Specifications

	Plate and sheet	Sand castings	Welding rod
AMS	4384, 4385	4445	
ASTM	B90	B80	
SAE	507		
Federal		QQ-M-56	
Military	MIL-M-26075		MIL-R-6944

Typical Room Temperature Mechanical Properties

Temper	TS, ksi	YS, ksi	CYS, ksi	Elong., %	Hardness Bhn	Hardness RE	Shear strength, ksi	Charpy impact strength, ft-lb
O	29/33	14/21	15	12/23	47	54	19/22	4
H24	33/37	24/29	23	4/8	57	68	20/23	3
Cast T6	27/32	13/15	15	4/8	55	66	19/21	4

Temper	Bearing, ksi Ultimate	Bearing, ksi Yield	Fatigue strength, ksi 10^5 cycles	10^6 cycles	10^7 cycles	10^8 cycles	5 × 10^8 cycles
H24	61	41	22/25	19/22	18/20		
Cast T6	61	40					

Table 1.2−7 (continued)
HK31A

Effect of Testing Temperature on Typical Mechanical Properties

Temper	Room temperature			200°F			300°F		
	TS, ksi	TYS, ksi	Elong., %	TS, ksi	TYS, ksi	Elong., %	TS, ksi	TYS, ksi	Elong., %
Cast T6	31	16	6	29	16	8	27	15	12
Sheet H24	37	29	8				26	23	20

	400°F			500°F			600°F		
Cast T6	24	14	17	23	13	19	20	12	22
Sheet H24	24	21	21	20	17	19	13	7	70

	700°F			−109°F			−420°F		
Cast T6	13	8	26						
Sheet H24	8	4	>100 (65°F)	45	32	6	57	34	4
Sheet O				43	25	11	55	29	7

Effect of Exposure at a Given Temperature on Tensile Properties
of HK31A-T6 − Separately Cast Test Bars

Property	Temperature, °F		Exposure time, hours					
	Exposure	Testing	0	25	100	500	1000	5000
TS, ksi	400	70	33.0	34.6	34.7	34.8	34.8	−
TYS, ksi	400	70	16.0	18.2	18.3	18.3	18.3	−
TS, ksi	400	400	22.0	24.9	25.3	25.8	26.0	−
TYS, ksi	400	400	14.0	14.1	14.3	14.7	14.8	−
TS, ksi	400	600	19.0	20.3	20.3	20.2	20.2	−
TYS, ksi	400	600	12.0	12.3	12.5	12.8	12.9	−
TS, ksi	500	70	33.0	34.9	35.0	34.0	33.0	−
TYS, ksi	500	70	16.0	18.0	18.0	17.6	17.0	−
TS, ksi	500	500	24.0	23.2	23.0	22.0	21.0	−
TYS, ksi	500	500	14.0	14.0	14.0	12.3	11.0	−
TS, ksi	500	600	19.0	19.7	20.0	19.0	18.0	−
TYS, ksi	500	600	12.0	12.0	12.0	11.5	11.0	−
TS, ksi	600	70	33.0	32.0	30.0	26.0	26.0	−
TYS, ksi	600	70	16.0	15.8	14.0	11.0	11.0	−
TS, ksi	600	600	19.0	17.5	15.0	10.0	9.0	−
TYS, ksi	600	600	12.0	10.5	8.0	4.0	3.0	−

Table 1.2–7 (continued)
HK31A

Mechanical Properties of HK31A-H24 Sheet
After Exposure at Elevated Temperature — Limited Data

Temper	Property	Temperature, °F		Exposure time, hours				
		Exposure	Testing	0	1	100	1000	5000
H24	TS, ksi	400	70	37				37
	TYS, ksi			29	No change in values			29
	Elong., %			8				8
	CYS, ksi			25				25
	TS, ksi	500	70	37	–	–	37	36
	TYS, ksi			29	–	–	29	28
	Elong., %			8	–	–	15	16
	CYS, ksi			25	–	–	22	21
	TS, ksi	600	70	37	36	36	35	34
	TYS, ksi			29	29	27	25	24
	Elong., %			8	11	20	21	20
	CYS, ksi			25	24	19	18	16
	TS, ksi	300	300	26	–	26[a]	–	–
	TYS, ksi			23	–	23[a]	–	–
	Elong., %			20	–	20[a]	–	–
	CYS, ksi			23	–	23[a]	–	–
	TS, ksi	400	400	24	–	21[a]	20	24
	TYS, ksi			21	–	19[a]	18	20
	Elong., %			21	–	31[a]	35	20
	CYS, ksi			22	–	22[a]	20	–

[a]500 hr exposure.

Table 1.2—7 (continued)
HK31A

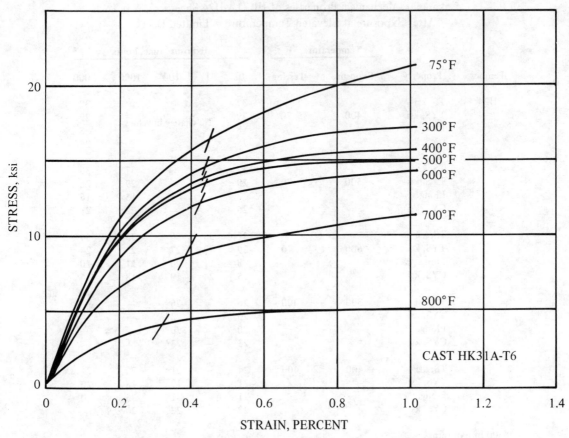

FIGURE E1.   Tension stress-strain relationships for cast HK31A-T6.

Table 1.2–7 (continued)
HK31A

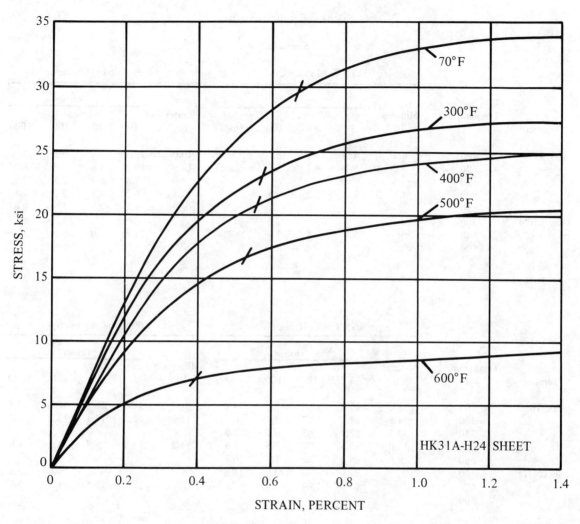

FIGURE E2.   Tensile stress-strain curves for HK31A-H24 sheet at room and elevated temperatures (longitudinal).

Table 1.2–7 (continued)
HK31A

Typical Creep Strengths

Stress, ksi

Temper	Temperature, °F	0.1% Total extension					0.2% Total extension				
		Minutes	Hours				Minutes	Hours			
		10	1	10	100	1000	10	1	10	100	1000
Cast T6	400	6.2	6.0	5.8	5.6	5.4	10.6	10.3	9.8	9.5	9.1
	500	5.8	5.2	4.4	3.5	3.1	10.6	10.0	8.6	6.3	4.2
	550						9.0	7.8	6.4	3.8	2.5
	600	4.8	4.1	3.2	2.1	0.9	7.4	6.2	4.7	2.9	1.1
	660						6.7	4.4	2.3	1.0	0.6
Sheet H24	300				7.6	6.6[a]				12.0	10.5
	400		5.5	5.5	3.9	3.8[a]		11.3	9.0	6.0	5.3
	500				0.6					1.0	

Temper	Temperature, °F	0.5% Total extension					1.0% Total extension				
		Minutes	Hours				Minutes	Hours			
		10	1	10	100	1000	10	1	10	100	1000
Cast T6	400	15.0	15.0	15.0	15.0	14.0	16.1	16.0	16.0	16.0	15.8
	500	14.5	14.0	12.7	9.7	6.8	15.8	15.5	14.4	12.2	7.6
	550	13.5	12.3	9.5	6.3	3.2					
	600	12.0	10.4	7.2	3.5	1.4	14.2	12.3	8.7	4.1	1.6
	660	9.6	6.0	3.2	1.3	0.7					
Sheet H24	300				20.0	18.8[a]				26.0	25.5
	400		15.6	13.4	9.4	7.6[a]		17.4	15.5	11.1	9.0
	500				2.0					3.0	

[a] At 500 hr.

Table 1.2–7 (continued)
HK31A

Effect of Exposure at Elevated Temperature on 100-hr Creep Limits;
Separately Cast Test Bars – Stresses Shown in 1000 psi

Exposure temperature, °F	Exposure time, hours	Testing temperature, °F	% Total extension HK31A-T6		
			0.1%	0.2%	0.5%
400	0	400	8.5	10.6	14.5
400	1	400	8.5	10.6	14.4
400	10	400	8.5	10.6	14.2
400	100	400	8.4	10.6	14.1
400	1000	400	8.4	10.6	14.0
500	0	500	5.1	6.9	10.0
500	1	500	5.1	6.9	10.0
500	10	500	5.1	6.9	10.0
500	100	500	5.0	7.0	10.0
500	1000	500	5.0	7.0	10.0
600	0	600	2.5	3.5	4.2
600	1	600	2.2	2.2	4.2
600	10	600	1.9	1.9	3.8
600	100	600	1.7	1.7	3.3
600	1000	600	1.4	1.4	1.9
500	1	400	7.7	10.4	14.1
500	10	400	7.0	10.2	13.6
500	20	400	6.9	10.1	13.5
500	100	400	6.4	9.7	13.2
600	1	400	6.8	9.5	13.7
600	10	400	5.5	8.2	12.2
600	20	400	5.2	7.8	11.7
600	1000	400	4.5	6.8	10.0
700	1	400	6.3	8.7	12.2
700	4	400	4.9	7.6	10.0
600	1	500	5.1	6.9	9.6
600	10	500	5.1	6.6	8.5
600	20	500	5.0	6.4	8.2
600	100	500	4.4	5.6	7.0
700	0	600	2.4	3.4	4.3
700	1	600	1.9	2.5	3.0
700	4	600	1.6	2.0	2.5

Table 1.2–8
AZ31B

AZ31B is the most commonly used magnesium sheet and plate alloy. Suitable for use up to 275°F, the alloy is both formable and weldable. Weldments must be stress relieved to prevent stress-corrosion cracking. It is also used for general purpose extrusions, and sometimes for forgings.

Typical chemical composition, percent by weight: Al, 3.0; Mn, 0.20 minimum; Zn, 1.0; Mg, balance

Physical constants and thermal properties
Density at 68°F, lb/cu in.$^{-1}$: 0.0640
Coefficient of thermal expansion, °F^{-1} × 10^{-6}, 65–212°F: 14
Modulus of elasticity, psi: 6.5 × 10^6
Modulus of rigidity, psi: 2.4 × 10^6
Poisson's ratio: 0.35
Melting range, °F: 1116–1169
Specific heat, Btu/lb^{-1}/°F^{-1}, 70°F: 0.24; 300°F: 0.26; 500°F: 0.29
Electrical resistivity, microhm-cm, 70°F: 9.2; 300°F: 11.2; 500°F: 12.9
Thermal conductivity, Btu/hr^{-1}/ft^{-2}/°F^{-1}/ft, 70°F: 44.5; 300°F: 53.5; 500°F: 59.0

Heat Treatments

	Temperature, °F	Stress relief
O	650	500°F 15 min
H24		300°F 60 min
H26		300°F 60 min
Extrusions		500°F 15 min

Specifications

	Sheet and plate	Extrusions Bar	Extrusions Tube	Forgings
AMS	4375,4376,4377			
ASTM	B90	B107	B217	B91
SAE	510	510	510	
Federal	QQ-M-44	QQ-M-31	WW-T-825	

Typical Room Temperature Mechanical Properties

Temper	TS, ksi	YS, ksi	CYS, ksi	Elong., %	Hardness Bhn	Hardness RE	Shear strength, ksi	Charpy impact strength, ft-lb
O	32/37	15/22	16	9/21	56	67	21/29	
H24	34/42	18/32	26	6/15	73	83	18/29	
H26	35/40	21/30	24	6/16			18/28	
Extruded bars	38	29	14	15	49	57	19	3.2
Press forged	38	25		15	50	59	19	3.2

Temper	Bearing, ksi Ultimate	Bearing, ksi Yield	Fatigue strength, ksi 10^5 cycles	10^6 cycles	10^7 cycles	10^8 cycles	5×10^8 cycles
O	66/70	37/42	20/24	19/22	18/21		
H24	77	47	22/27	20/24	19/23		
H26	72	46					
Extruded bars	56	33					
Extruded, F			23/29	21/26	19/23	17/21	
Extruded, F			23/28	22/25	20/24		

Table 1.2–8 (continued)
AZ31B

Effect of Testing Temperature on Typical Mechanical Properties

Temper	Room temperature TS, ksi	TYS, ksi	Elong., %	200°F TS, ksi	TYS, ksi	Elong., %	300°F TS, ksi	TYS, ksi	Elong., %
Sheet H24	42	32	17	34	24	35	22	13	57
Extrusions, F	38	28	15	35	21	22	24	14	28

Temper	400°F TS, ksi	TYS, ksi	Elong., %	500°F TS, ksi	TYS, ksi	Elong., %	600°F TS, ksi	TYS, ksi	Elong., %
Sheet H24	13	8	82	8	5	93	6	2	–
Extrusions, F	15	8	42	10	4	47	6	2	66

Temper	–109°F TS, ksi	TYS, ksi	Elong., %	–320°F TS, ksi	TYS, ksi	Elong., %	–420°F TS, ksi	TYS, ksi	Elong., %
Sheet H24	46	33	6	57	37	3	66	40	2

Mechanical Properties of AZ31B-H24 Sheet at Elevated Temperature After Exposure at Temperature – Limited Data

Property	Temperature, °F Exposure	Testing	Exposure, hours 16	48	192	500	1000
TS, ksi			33.7	32.7	33.6	33.2	33.7
TYS, ksi	200	200	24.9	24.8	25.3	24.9	24.4
Elong., %			36.7	40.0	37.2	39.2	34.2
CYS, ksi			23.0	22.8	22.7	22.7	22.3
TS, ksi			28.0	27.7	28.0	27.5	27.8
TYS, ksi	250	250	20.4	19.9	20.8	19.6	21.0
Elong., %			48.2	49.5	49.0	52.5	50.0
CYS, ksi			22.3	21.8	21.4	21.7	21.7
TS, ksi			22.9	22.8	21.6	21.7	21.0
TYS, ksi	300	300	15.8	16.5	16.0	15.7	15.1
Elong., %			52.5	54.7	58.0	62.7	63.7
CYS, ksi			21.8	18.8	19.0	19.5	18.4
TS, ksi			13.3	13.2	13.0	13.4	13.3
TYS, ksi	400	400	10.4	10.2	10.1	10.2	10.8
Elong., %			59.5	83.2	76.2	73.7	71.5
CYS, ksi			12.4	12.1	12.3	10.7	12.3
TS, ksi			8.3	8.7	8.6	8.8	8.9
TYS, ksi	500	500	6.4	7.1	7.2	7.3	6.8
Elong., %			70.5	90.0	91.5	87.0	91.2
CYS, ksi			8.5	8.3	9.0	9.0	8.8
TS, ksi			6.0	–	5.9	6.3	–
TYS, ksi	600	600	–	–	4.7	4.5	–
Elong., %			115.0	–	112.5	112.5	–
CYS, ksi			5.4	–	6.1	5.8	–

Table 1.2–8 (continued)
AZ31B

Typical Creep Strengths

Stress, ksi

Temper	Temperature, °F	0.1% Total extension					0.2% Total extension				
		Minutes 10	1	10	100	500	Minutes 10	1	10	100	500
Sheet, O	200		6.0	5.0	4.0	4.0		10.0	9.0	8.0	7.0
	250		5.0	4.0	2.5	2.0		8.0	7.0	5.0	3.0
	300		4.0	2.5	1.0	0.5		7.0	4.5	2.0	1.0
	350		2.0	1.0				4.0	2.0	0.5	
Sheet, H24	200		6.0	4.0	2.5	2.0		9.0	7.0	4.0	3.0
	250		4.0	2.0	1.0	1.0		6.0	4.0	2.0	1.5
	300		2.0	1.0	0.5			4.0	2.0	1.0	0.5
	350		1.0	0.5				2.0	1.0		
Extruded, F	200		6.0	5.0	4.0	3.5		10.0	9.0	7.0	6.0
	250		5.0	4.0	3.0	2.0		9.0	7.0	5.0	4.0
	300		4.0	3.0	1.5	1.0		7.0	5.0	3.0	2.0
	350		3.0	2.0				5.0	3.0		

Temper	Temperature, °F	0.5% Total extension					1.0% Total extension				
		Minutes 10	1	10	100	500	Minutes 10	1	10	100	500
Sheet, O	200		15.0	14.0	12.0	10.0		16.0	15.0	13.0	11.0
	250		12.0	11.0	9.0	7.0		14.0	12.0	10.0	9.0
	300		10.0	7.0	4.0	3.0		12.0	9.0	5.0	3.5
	350		7.0	4.0				8.0	5.0		
Sheet, H24	200		15.0	11.0	7.0	5.0		19.0	14.0	10.0	7.0
	250		9.0	6.0	4.0	2.0		13.0	8.0	5.0	3.0
	300		6.0	3.0	1.5	1.0		8.0	5.0	2.0	1.0
	350		4.0	1.5				6.0	2.0		
Extruded, F	200		16.0	14.0	12.0	10.0		19.0	17.0	15.0	13.0
	250		13.0	11.0	9.0	7.0		16.0	14.0	11.0	9.0
	300		11.0	8.0	5.0	4.0		12.0	10.0	7.0	5.0
	350		8.0	6.0				10.0	8.0		

Table 1.2–9
HZ32A

HZ32A is a sand casting alloy suitable for use in the 400–700°F temperature range.

Typical chemical composition, percent by weight: Zn, 2.1; Zr, 0.7; Th, 3.2; Mg, balance

Physical constants and thermal properties
Density at 68°F, lb/cu in. $^{-1}$: 0.0659
Coefficient of thermal expansion, $°F^{-1} \times 10^{-6}$,
65–212°F: 14.5
Modulus of elasticity, psi: 6.5×10^6
Modulus of rigidity, psi: 2.5×10^6
Poisson's ratio: 0.3
Melting range, °F: 1026–1198
Specific heat, $Btu/lb^{-1}/°F^{-1}$, 70°F: 0.23
Electrical resistivity, microhm-cm, 70°F: 6.5
Thermal conductivity, $Btu/hr^{-1}/ft^{-2}/°F^{-1}/ft$, 70°F: 62.0; 300°F: 68.0; 500°F: 72.0

Heat Treatments

	Temperature, °F	Time, hours
T5	570–620	16

Specifications

	Sand castings
AMS	4447
ASTM	B80
Federal	QQ-M-56

Typical Room Temperature Mechanical Properties

Temper	TS, ksi	YS, ksi	CYS, ksi	Elong., %	Hardness Bhn	Hardness RE	Shear strength, ksi	Charpy impact strength, ft-lb
T5	27/30	13/14	16	4/7	55/57	66/68		

Effect of Testing Temperature on Typical Mechanical Properties

Temper	Room temperature TS, ksi	TYS, ksi	Elong., %	200°F TS, ksi	TYS, ksi	Elong., %	300°F TS, ksi	TYS, ksi	Elong., %
T5	29	15	6	26	14	15	22	12	23
	400°F			500°F			600°F		
T5	17	10	33	14	9	39	12	8	38
	700°F								
T5	10	7	29						

Table 1.2–9 (continued)
HZ32A

Effect of Exposure at a Given Temperature on Tensile Properties
of HZ32A-T5 — Separately Cast Test Bars

| Property | Temperature, °F | | Exposure time, hours | | | | | |
	Exposure	Testing	0	25	100	500	1000	5000
TS, ksi	400	70	–	–	–	–	31.5	–
TYS, ksi	400	70	–	–	–	–	15.6	–
TS, ksi	400	400	16.5	16.7	16.7	16.8	16.8	–
TYS, ksi	400	400	10.0	10.0	10.0	10.0	10.0	–
TS, ksi	500	70	–	–	–	–	32.6	–
TYS, ksi	500	70	–	–	–	–	15.2	–
TS, ksi	500	500	13.0	13.2	13.3	13.4	13.5	–
TYS, ksi	500	500	7.7	8.3	8.5	8.7	8.8	–
TS, ksi	600	70	–	–	–	–	33.0	–
TYS, ksi	600	70	–	–	–	–	15.0	–
TS, ksi	600	600	11.5	11.5	11.5	11.5	11.5	–
TYS, ksi	600	600	7.4	7.55	7.6	7.7	7.7	–

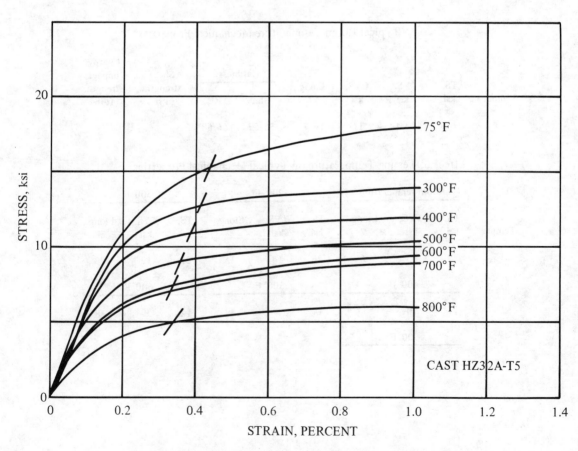

FIGURE F.   Tension stress-strain relationships for cast HZ32A-T5.

Table 1.2–9 (continued)
HZ32A

Typical Creep Strengths

Stress, ksi

Temper	Temperature, °F	0.1% Total extension					0.2% Total extension				
		Minutes	Hours				Minutes	Hours			
		10	1	10	100	1000	10	1	10	100	1000
T5	400	6.2	6.0	5.7	5.3	4.8	8.7	8.4	8.1	7.7	7.4
	500	5.8	5.4	5.0	4.2	3.5	8.2	7.7	6.9	5.9	5.0
	550						7.4	6.8	5.8	4.6	4.0
	600	5.2	4.6	3.5	2.5	1.9	6.5	6.0	5.0	3.6	2.2
	660						6.0	5.0	3.4	2.0	0.7

Temper	Temperature, °F	0.5% Total extension					1.0% Total extension				
		Minutes	Hours				Minutes	Hours			
		10	1	10	100	1000	10	1	10	100	1000
T5	400	10.3	10.2	10.0	9.9	9.8	11.0	11.0	11.0	11.0	10.4
	500	9.3	9.2	9.0	8.0	6.5	10.3	10.3	10.0	9.3	6.9
	550	8.9	8.7	8.0	6.7	4.8					
	600	8.0	7.4	6.3	4.8	2.8	9.2	8.3	6.9	5.4	3.2
	660	7.8	6.1	4.2	2.5	1.4					

Table 1.2–10
AZ92A

AZ92A is a pressure-tight sand and permanent mold casting alloy. It has slightly more strength and less ductility than AZ91C.

Typical chemical composition, percent by weight: Al, 9.0; Mn, 0.10 minimum; Zn, 2.0; Mg, balance

Physical constants and thermal properties
Density at 68°F, lb/cu in.$^{-1}$: 0.0659
Coefficient of thermal expansion, °F^{-1} × 10^{-6}, 65–212°F: 14.5
Modulus of elasticity, psi: 6.5 × 10^6
Modulus of rigidity, psi: 2.4 × 10^6
Poisson's ratio: 0.35
Melting range, °F: 830–1100
Specific heat, Btu/lb^{-1}/°F^{-1}, 70°F: 0.30
Electrical resistivity, microhm-cm, 70°F: 12.5; 300°F: 14.6; 500°F: 16.3
Thermal conductivity, Btu/hr^{-1}/ft^{-2}/°F^{-1}/ft, 70°F: 33.0; 300°F: 41.5; 500°F: 48.0

Heat Treatments

Treatment	Temperature, °F	Time, hours	Cooling
Solution[a]	760	20	Strong air blast
Age	420	14	Air or oven
Stabilizing[b]	500	4	Air

[a]An atmosphere with at least 0.5 SO$_2$ is required.
[b]Used to minimize growth in castings used at elevated temperatures.

Specifications

	Ingot	Sand castings	Investment castings	Permeable mold castings	Welding rod
AMS		4434	4453	4484	
ASTM	B93	B80		B199	B260
SAE		500		503	
Federal		QQ-M-56		QQ-M-55	
Military					MIL-R-6944

Typical Room Temperature Mechanical Properties

Temper	TS, ksi	YS, ksi	CYS, ksi	Elong., %	Hardness Bhn	Hardness RE	Shear strength, ksi	Charpy impact strength, ft-lb
F	23/25	11/14	14	2	60	66	16	0.5
T4	34/40	11/14	14	6/10	55	62	17	2.0
T5	23/25	12/17	20	2	55	66	16	–
T6	34/40	18/22	22	3	70	77	19	0.8
T7	40	21	21	3	78	86	21	–

Temper	Bearing, ksi Ultimate	Bearing, ksi Yield	Fatigue strength, ksi 10^5 cycles	10^6 cycles	10^7 cycles	10^8 cycles	5 × 10^8 cycles
F	50	46					12
T4	68	46					13
T5	50	46					12
T6	80	65					11
T7							13

Table 1.2–9 (continued)
HZ32A

Effect of Exposure at Elevated Temperature on 100-hr Creep Limits;
Separately Cast Test Bars — Stresses Shown in 1000 psi

Exposure temperature, °F	Exposure time, hours	Testing temperature, °F	% Total extension HZ32A-T5		
			0.1%	0.2%	0.5%
400	0	400	5.7	6.8	8.5
400	1	400	5.6	6.8	8.5
400	10	400	5.6	6.7	8.5
400	100	400	5.3	6.7	8.5
400	1000	400	4.8	6.6	8.5
500	0	500	4.6	5.8	7.4
500	1	500	4.8	5.8	7.5
500	10	500	4.9	5.9	7.6
500	100	500	5.1	6.0	7.7
500	1000	600	5.2	6.1	7.8
600	0	600	2.7	3.7	4.8
600	1	600	3.0	3.8	4.9
600	10	600	3.2	4.0	5.0
600	100	600	3.3	4.2	5.2
600	1000	600	3.5	4.4	5.4
500	1	400	5.4	6.9	8.6
500	10	400	5.1	7.0	8.8
500	20	400	5.1	7.0	8.9
500	100	400	4.9	7.1	9.0
600	1	400	5.5	7.0	8.7
600	10	400	5.3	7.2	9.0
600	20	400	5.3	7.2	9.1
600	1000	400	5.2	7.4	9.2
700	1	400	5.4	7.0	8.8
700	4	400	5.3	7.2	9.0
600	1	500	–	–	–
600	10	500	–	–	–
600	20	500	–	–	–
600	100	500	–	–	–
700	0	600	2.8	3.7	4.7
700	1	600	2.7	3.8	5.0
700	4	600	2.6	3.8	5.1

Table 1.2–10
AZ92A

Effect of Testing Temperature on Typical Mechanical Properties

Temper	Room temperature			200°F			300°F		
	TS, ksi	TYS, ksi	Elong., %	TS, ksi	TYS, ksi	Elong., %	TS, ksi	TYS, ksi	Elong., %
T5	25	17	1	24	16	2	23	14	4
T6	40	23	2	37	21	25	28	17	35

	400°F			500°F			600°F		
T5	20	11	15	16	8	32	9	4	61
T6	17	12	36	11	8	33	8	5	49

Effect of Exposure at a Given Temperature on Tensile Properties of AZ92A-T6—Separately Cast Test Bars

Property	Temperature, °F		Exposure time, hours						
	Exposure	Testing	0	25	100	500	1000	2500	5000
TS, ksi	200	70	40.4	40.9	38.7	41.0	40.7	41.5	–
TYS, ksi	200	70	24.0	24.9	23.5	23.6	22.1	23.2	–
E, %	200	70	1.5	2.3	1.7	2.0	3.4	3.8	–
TS, ksi	200	200	40.1	40.5	39.9	–	38.6	40.0	–
TYS, ksi	200	200	20.0	19.3	19.7	–	18.5	20.9	–
E, %	200	200	9.6	7.1	15.0	–	9.0	10.2	–
TS, ksi	300	70	40.4	38.0	39.1	39.1	39.1	39.4	38.7
TYS, ksi	300	70	24.0	25.2	22.8	24.4	20.8	23.7	24.0
E, %	300	70	1.5	1.3	2.0	1.5	3.7	1.7	1.6
TS, ksi	300	300	28.3	27.5	26.3	–	27.2	24.8	27.0
TYS, ksi	300	300	16.9	15.8	15.8	–	15.9	14.6	14.7
E, %	300	300	39.8	41.9	39.6	–	45.6	34.6	41.5
TS, ksi	400	70	40.4	40.0	39.7	40.1	39.2	35.8	35.4
TYS, ksi	400	70	24.0	25.1	23.8	22.3	22.9	19.9	19.7
E, %	400	70	1.5	1.3	1.8	2.4	2.5	2.8	3.1
TS, ksi	400	400	19.6	16.9	17.2	–	17.2	17.9	17.5
TYS, ksi	400	400	12.5	11.4	11.9	–	11.3	11.2	10.2
E, %	400	400	47.5	36.9	30.5	–	36.8	48.5	47.4

Table 1.2–10 (continued)
AZ92A

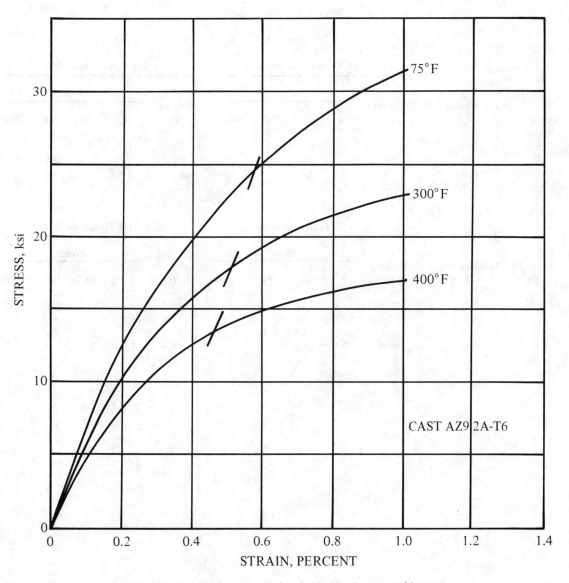

FIGURE G.   Tension stress-strain relationships for cast AZ92A-T6.

Table 1.2–10 (continued)
AZ92A

Typical Creep Strengths

Stress, ksi

Temper	Temperature, °F	0.1% Total extension					0.2% Total extension				
		Minutes	Hours				Minutes	Hours			
		10	1	10	100	1000	10	1	10	100	1000
T6	200	7.9	7.1	6.1	5.1	4.2	13.0	11.7	9.8	8.0	6.5
	300	5.4	4.5	3.3	2.3	1.5	9.2	7.8	5.7	3.8	2.6
	400	3.5	2.5	1.5	0.6		5.7	4.4	2.8	1.3	
	500	2.2	1.2				3.5	2.2	1.0		

Temper	Temperature, °F	0.5% Total extension					1.0% Total extension				
		Minutes	Hours				Minutes	Hours			
		10	1	10	100	1000	10	1	10	100	1000
T6	200	20.5	18.8	15.9	13.0	10.5	25.0	23.2	20.2	17.0	14.2
	300	14.4	12.5	9.8	7.0	4.8	17.4	15.2	12.0	9.0	6.4
	400	9.1	7.2	4.8	2.5	0.9	11.0	8.6	5.8	3.4	1.3
	500	5.0	3.5	1.8			6.4	4.8	2.8		

Table 1.2–11
K1A

K1A is a sand casting alloy having excellent damping capacity.

Typical chemical composition, percent by weight: Zr, 0.6; Mg, balance

Physical constants and thermal properties
Melting range, °F (liquidus): 1205
Electrical resistivity, microhm-cm, 70°F: 5.5; 300°F: 7.7; 500°F: 9.6
Thermal conductivity, Btu/hr^{-1}/ft^{-2}/°F^{-1}/ft, 70°F: 74; 300°F: 76; 500°F: 78

Specifications

	Sand castings
ASTM	B80
Military	MIL-M-45207

Typical Room Temperature Mechanical Properties

Temper	TS, ksi	YS, ksi	CYS, ksi	Elong., %	Hardness Bhn	RE	Shear strength, ksi	Charpy impact strength, ft-lb
F	24	6		14			8.1	

Table 1.2–12
QE22A

QE22A is the only magnesium alloy that contains silver. It is a sand casting alloy.

Typical chemical composition, percent by weight: Zr, 0.7; RE, 2.2; Ag, 2.5; Mg, balance

Physical constants and thermal properties
Melting range, °F (liquidus): 1190
Electrical resistivity, microhm-cm, 70°F: 7.1; 300°F: 9.1; 500°F: 10.9
Thermal conductivity, Btu/hr^{-1}/ft^{-2}/°F^{-1}/ft, 70°F: 57; 300°F: 64; 500°F: 69

Specifications

	Sand castings
AMS	4418

Typical Room Temperature Mechanical Properties

Temper	TS, ksi	YS, ksi	CYS, ksi	Elong., %	Hardness Bhn	Hardness RE	Shear strength, ksi	Charpy impact strength, ft-lb
T6	35	25		2	78	86		

Table 1.2–13
ZE41A

ZE41A is a sand casting alloy with improved castability over ZK alloys.

Typical chemical composition, percent by weight: Zn, 4.2; Zr, 0.7; RE, 1.2; Mg, balance

Physical constants and thermal properties
Density at 68°F, lb/cu in.$^{-1}$: 0.0657

Coefficient of thermal expansion, °F^{-1} × 10^{-6}, 65–212°F: 14.5
Modulus of elasticity, psi: 6.5 × 10^6
Modulus of rigidity, psi: 2.4 × 10^6
Poisson's ratio: 0.35
Melting range, °F (liquidus): 1185
Electrical resistivity, microhm-cm, 70°F: 5.6; 300°F: 7.8; 500°F: 9.7
Thermal conductivity, Btu/hr^{-1}/ft^{-2}/°F^{-1}/ft, 70°F: 71.5; 300°F: 74.5; 500°F: 76.4

Heat Treatments

Temperature, °F	Time, hours	Cooling
625	2	Air
350	16	Air

Specifications

	Sand castings
AMS	4439
ASTM	B80

Typical Room Temperature Mechanical Properties

Temper	TS, ksi	YS, ksi	CYS, ksi	Elong., %	Hardness Bhn	Hardness RE	Shear strength, ksi	Charpy impact strength, ft-lb
T5	29	19.5		2	62	72	22	

Effect of Testing Temperature on Typical Mechanical Properties

Temper	Room temperature TS, ksi	Room temperature TYS, ksi	Room temperature Elong., %	200°F TS, ksi	200°F TYS, ksi	200°F Elong., %	300°F TS, ksi	300°F TYS, ksi	300°F Elong., %
T5	30	20	4	27	19	8	24	17	15

Temper	400°F TS, ksi	400°F TYS, ksi	400°F Elong., %	500°F TS, ksi	500°F TYS, ksi	500°F Elong., %	600°F TS, ksi	600°F TYS, ksi	600°F Elong., %
T5	19	14	29	14	10	40	11	8	43

Table 1.2–13 (continued)
ZE41A

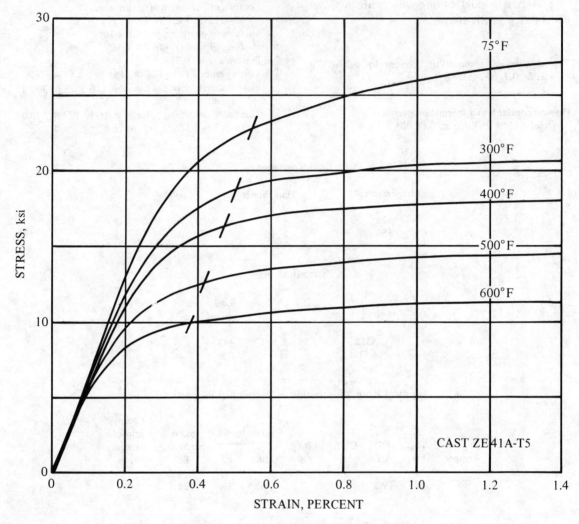

FIGURE H.   Tension stress-strain relationship for cast ZE41A-T5.

Table 1.2–13 (continued)
ZE41A

Typical Creep Strengths

Stress, ksi

Temper	Temperature, °F	Minutes 10	0.1% Total extension Hours				Minutes 10	0.2% Total extension Hours			
			1	10	100	1,000		1	10	100	1,000
T5	200		6.8	6.6	6.1	5.4		12.3	12.0	11.0	9.8
	300		6.3	6.2	6.0	5.0		10.7	10.3	9.9	9.1
	400		5.5	4.8	3.4	2.1		9.7	8.1	6.0	3.3
	500		4.1	2.3	1.0	0.8		5.6	3.4	1.8	1.0

Temper	Temperature, °F	Minutes 10	0.5% Total extension Hours				Minutes 10	1.0% Total extension Hours			
			1	10	100	1,000		1	10	100	1,000
T5	200		20.0	19.0	18.1	16.5					
	300		16.3	15.2	14.3	12.5					
	400		15.1	13.2	10.7	5.4					
	500		8.0	5.1	3.0	1.4		9.5	6.2	3.6	1.7

Table 1.2–14
ZH62A

ZH62A is a sand casting alloy with improved castability over ZK alloys.

Typical chemical composition, percent by weight: Zn, 5.7; Zr, 0.7; Th, 1.8; Mg, balance

Physical constants and thermal properties
Density at 68°F, lb/cu in.$^{-1}$: 0.0670
Coefficient of thermal expansion, °F^{-1} × 10^{-6}, 65–212°F: 14.5

Modulus of elasticity, psi: 6.5 × 10^6
Modulus of rigidity, psi: 2.5 × 10^6
Poisson's ratio: 0.3
Melting range, °F (liquidus): 1169
Specific heat, Btu/lb^{-1}/°F^{-1}, 70°F: 0.23
Electrical resistivity, microhm-cm, 70°F: 6.4; 300°F: 8.6; 500°F: 10.5
Thermal conductivity, Btu/hr^{-1}/ft^{-2}/°F^{-1}/ft, 70°F: 62.4; 300°F: 68.0; 500°F: 71.4

Heat Treatments

Treatment	Temperature, °F	Time, hours
Aging	480	12

Specifications

	Sand castings
AMS	4438
ASTM	B80
Federal	QQ-M-56

Typical Room Temperature Mechanical Properties

Temper	TS, ksi	YS, ksi	CYS, ksi	Elong., %	Hardness Bhn	RE	Shear strength, ksi	Charpy impact strength, ft-lb
T5	35/40	22/25	15	4/6	70	77		

Effect of Testing Temperature on Typical Mechanical Properties

Temper	Room temperature TS, ksi	TYS, ksi	Elong., %	200°F TS, ksi	TYS, ksi	Elong., %	300°F TS, ksi	TYS, ksi	Elong., %
T5	40	27	8	33	23	20	26	20	24

Temper	400°F TS, ksi	TYS, ksi	Elong., %	500°F TS, ksi	TYS, ksi	Elong., %
T5	19	15	28	14	10	30

Table 1.2–14 (continued)
ZH62A

Typical Creep Strengths

Stress, ksi

| | | 0.1% Total extension | | | | | 0.2% Total extension | | | | |
| | | | Hours | | | | | Hours | | | |
Temper	Temperature, °F	Minutes 10	1	10	100	1,000	Minutes 10	1	10	100	1,000
T5	200	6.5	6.4	6.1	5.5	4.3	12.0	12.0	11.7	10.8	8.7
	300	5.9	5.7	5.3	4.6	3.2	10.6	9.9	8.9	7.6	5.5
	400	5.0	4.3	3.4	2.6	1.7	8.8	7.0	5.2	3.6	2.2
	500	3.8	2.7	1.9	1.4	0.9	5.9	3.7	2.4	1.7	1.2

| | | 0.5% Total extension | | | | | 1.0% Total extension | | | | |
| | | | Hours | | | | | Hours | | | |
Temper	Temperature, °F	Minutes 10	1	10	100	1,000	Minutes 10	1	10	100	1,000
T5	200	21.0	19.9	18.2	16.3	13.6	26.3	24.8	22.3	19.9	16.9
	300	17.7	16.3	14.5	12.6	10.3	20.6	19.2	17.5	15.6	13.0
	400	13.3	10.7	7.8	5.5	3.3	17.0	13.8	10.3	7.2	4.0
	500	8.6	5.4	3.2	2.0	1.5	9.9	6.2	3.6	2.2	1.6

Table 1.2—15
ZK51A

ZK51A is a sand casting alloy with good strength and ductility at room temperature. Complex sections are difficult to cast.

Typical chemical composition, percent by weight: Zn, 4.6; Zr, 0.7; Mg, balance

Physical constants and thermal properties
Density at 68°F, lb/cu in.$^{-1}$: 0.0653

Coefficient of thermal expansion, $°F^{-1} \times 10^{-6}$, 65–212°F: 14.5
Modulus of elasticity, psi: 6.5×10^6
Melting range, °F: 1020–1185
Specific heat, $Btu/lb^{-1}/°F^{-1}$, 70°F: 0.24
Electrical resistivity, microhm-cm, 70°F: 5.7; 300°F: 8.0; 500°F: 9.9
Thermal conductivity, $Btu/hr^{-1}/ft^{-2}/°F^{-1}/ft$, 70°F: 63.1; 300°F: 69.7; 500°F: 73.8

Heat Treatments

Treatment	Temperature, °F	Time, hours
Aging	350	12

Specifications

	Sand castings
AMS	4443
ASTM	B80
Federal	QQ-M-56

Typical Room Temperature Mechanical Properties

Temper	TS, ksi	YS, ksi	CYS, ksi	Elong., %	Hardness Bhn	RE	Shear strength, ksi	Charpy impact strength, ft-lb
T5	34/40	20/24	24	5/8	65	77	22	

Temper	Bearing, ksi Ultimate	Yield
T5	72	47

Effect of Testing Temperature on Typical Mechanical Properties

Temper	Room temperature TS, ksi	TYS, ksi	Elong., %	200°F TS, ksi	TYS, ksi	Elong., %	300°F TS, ksi	TYS, ksi	Elong., %
T5	40	26	8	30	21	12	23	17	14

Temper	400°F TS, ksi	TYS, ksi	Elong., %	500°F TS, ksi	TYS, ksi	Elong., %	600°F TS, ksi	TYS, ksi	Elong., %
T5	17	13	17	12	9	16	8	6	16

Table 1.2–15 (continued)
ZK51A

Typical Creep Strengths

Stress, ksi

		0.1% Total extension					0.2% Total extension				
		Minutes	Hours				Minutes	Hours			
Temper	Temperature, °F	10	1	10	100	1,000	10	1	10	100	1,000
T5	200	6.3	5.9	5.3	4.5	3.5	11.8	10.7	9.5	8.0	6.5
	300	5.4	4.7	3.8	3.0	2.1	9.0	7.8	6.4	5.0	3.7
	400	4.8	3.9	2.9	2.0	1.2	7.2	5.8	4.4	3.0	1.7

		0.5% Total extension					1.0% Total extension				
		Minutes	Hours				Minutes	Hours			
Temper	Temperature °F	10	1	10	100	1,000	10	1	10	100	1,000
T5	200	19.5	18.0	16.3	14.3	12.0	22.2	21.0	19.0	17.2	14.5
	300	14.3	12.8	11.0	9.2	7.1	17.0	15.7	13.7	11.5	9.0
	400	10.5	9.0	7.2	5.4	3.5	12.3	11.0	9.2	7.2	4.9

Table 1.2–16
ZK61A

ZK61A is a sand casting alloy with foundry characteristics similar to those of ZK51A. This alloy has high strength and ductility at room temperature.

Typical chemical composition, percent by weight: Zn, 6.0; Zr, 0.8; Mg, balance

Physical constants and thermal properties
Modulus of elasticity, psi: 6.5×10^6
Melting range, °F (liquidus): 1175

Specifications

	Sand castings
AMS	4444
ASTM	B80
Federal	QQ-M-56

Table 1.2–16 (continued)
ZK61A

Typical Room Temperature Mechanical Properties

Temper	TS, ksi	YS, ksi	CYS, ksi	Elong., %	Hardness Bhn	RE	Shear strength, ksi	Charpy impact strength, ft-lb
T6	40	26		5				

Table 1.2–17
AM60A

AM60A is a die casting alloy with improved toughness. It is used for automobile wheels.

Typical chemical composition, percent by weight: Al, 6.0; Mn, 0.13 minimum; Mg, balance

Physical constants and thermal properties
Modulus of elasticity, psi: 6.5×10^6
Melting range, °F (liquidus): 1140

Typical Room Temperature Mechanical Properties

Temper	TS, ksi	YS, ksi	CYS, ksi	Elong., %	Hardness Bhn	RE	Shear strength, ksi	Charpy impact strength, ft-lb
F	38	20		12				

Table 1.2–18
AS41A

AS41A is a die casting alloy. It has higher creep strength at moderately elevated temperatures than AZ91B, and also has higher conductivity. It is used in automotive engines and housings.

Typical chemical composition, percent by weight: Al, 4.2; Mn, 0.35; Si, 1.0; Mg, balance

Physical constants and thermal properties
Modulus of elasticity, psi: 6.5×10^6
Melting range, °F (liquidus): 1150

Typical Room Temperature Mechanical Properties

Temper	TS, ksi	YS, ksi	CYS, ksi	Elong., %	Hardness Bhn	RE	Shear strength, ksi	Charpy impact strength, ft-lb
F	35	20		10				

Table 1.2−19
HM21A

HM21A is a sheet, plate, and forging alloy usable to 800°F. It is used in missile and aircraft applications.

Typical chemical composition, percent by weight: Mn, 0.80 minimum; Th, 2.0; Mg, balance

Physical constants and thermal properties
Density at 68°F, lb/cu in.$^{-1}$: 0.0643; 500°F, 0.0629

Modulus of elasticity, psi: 6.5×10^6
Modulus of rigidity, psi: 2.4×10^6
Poisson's ratio: 0.35
Melting range, °F: 1121−1202
Specific heat, Btu/lb^{-1}/°F^{-1}, 70°F: 0.24; 300°F: 0.26; 500°F: 0.27
Electrical resistivity, microhm-cm, 70°F: 5.0; 300°F: 7.2; 500°F: 9.1
Thermal conductivity, Btu/hr^{-1}/ft^{-2}/°F^{-1}/ft, 70°F: 79.1; 300°F: 80.3; 500°F: 81.3

Heat Treatments

Treatments	Temperature, °F	Time, hours
Solution	850	
Aging (T5, forgings)	450	16

Specifications

	Sheet and plate
AMS	4390
ASTM	B90
Military	MIL-M-8917

Typical Room Temperature Mechanical Properties

Temper	TS, ksi	YS, ksi	CYS, ksi	Elong., %	Hardness Bhn	Hardness RE	Shear strength, ksi	Charpy impact strength, ft-lb	Bearing, ksi Ultimate	Bearing, ksi Yield
T8	30/37	18/33	15/24	8/12			18/20		60/67	36/41

Effect of Testing Temperature on Typical Mechanical Properties

Temper	Room temperature TS, ksi	Room temperature TYS, ksi	Room temperature Elong., %	200°F TS, ksi	200°F TYS, ksi	200°F Elong., %	300°F TS, ksi	300°F TYS, ksi	300°F Elong., %
Sheet, T8	35	27	12	27	24	15	22	21	20
Forging, T5	38	23	10				23	19	29

Temper	400°F TS, ksi	400°F TYS, ksi	400°F Elong., %	500°F TS, ksi	500°F TYS, ksi	500°F Elong., %	600°F TS, ksi	600°F TYS, ksi	600°F Elong., %
Sheet, T8	19	18	30	17	15	25	15	13	17
Forging, T5	20	16	29	18	14	26	17	13	25

Table 1.2–19 (continued)
HM21A

Effect of Testing Temperature on Typical Mechanical Properties (continued)

	700° F			800° F			900° F		
Sheet, T8	11	8	50	5	3	100	2	1	–
Forging, T5	12	9	45						

	–109° F			–320° F			–420° F		
Sheet, T8	43	26	6	50	26	5	52	27	5

Mechanical Properties of HM21A-T8 Sheet After Exposure
at Elevated Temperature – Limited Data

		Temperature, °F		Exposure time, hours				
Temper	Property	Exposure	Testing	0	1	100	1000	5000
T8	TS, ksi			33	33	33	33	33
	TYS, ksi	600	70	25	25	25	25	25
	Elong., %			7	7	7	7	7
	TS, ksi			33	–	32	–	–
	TYS, ksi	700	70	25	–	24	–	–
	Elong., %			7	–	9	–	–
	TS, ksi			33	–	30	–	–
	TYS, ksi	800	70	25	–	18	–	–
	Elong., %			7	–	12	–	–
	TS, ksi	600	600	14	No change in values			14
	TYS, ksi			12				12
	Elong., %			15				15
	CYS, ksi			13				13
	TS, ksi	700	700	11	–	11	7	–
	TYS, ksi			8	–	8	5	–
	Elong., %			50	–	50	112	–
	CYS, ksi			9	–	9	6	–

Table 1.2–19 (continued)
HM21A

Typical Creep Strengths

Stress, ksi

Temper	Temperature, °F	0.1% Total extension					0.2% Total extension				
		Minutes	Hours				Minutes	Hours			
		10	1	10	100	1000	10	1	10	100	1000
Sheet, T8	400		5.8	5.8	5.8			11.7	11.6	11.4	
	500		5.6	5.6	5.6			9.8	9.0	7.0	
	600		5.2	4.4	4.3			7.2	5.5	5.0	
	700		3.0	2.6	2.3			3.7	3.2	2.6	
Forging, T5	400				15.4[a]					10.7	
	500				10.8[a]					9.0	
	600				6.9[a]					6.5	
	700				3.3[a]					3.2	

Temper	Temperature, °F	0.5% Total extension					1.0% Total extension				
		Minutes	Hours				Minutes	Hours			
		10	1	10	100	1000	10	1	10	100	1000
Sheet, T8	400		16.9	16.6	13.5			17.7	17.5	13.9	
	500		14.4	12.0	9.0			15.3	13.3	9.7	
	600		8.7	6.9	6.0			9.6	8.1	6.8	
	700		5.1	4.2	3.3			5.7	4.5	3.6	
Forging, T5	400				14.6						
	500				11.0						
	600				7.6						
	700				4.8						

[a] Creep extension

Table 1.2−20
LA141A

LA141A is the lightest alloy available for aerospace applications. It is 25% lighter than conventional magnesium alloys and 27% lighter than beryllium. It is easily formed and machined. It has been produced as sheet, plate, foil, extrusions, forgings, wire, and castings.

Typical chemical composition, percent by weight: Al, 1.2; Mn, 0.15 minimum; Li, 14.0; Mg, balance

Physical constants and thermal properties
Density at 68°F, lb/cu in.$^{-1}$: 0.049

Coefficient of thermal expansion,°F^{-1} × 10^{-6}, 65−212°F: 21.8
Modulus of elasticity, psi: 7.2 × 10^6 at −105°F, 6.2 × 10^6 at + 70°F, 5.0 × 10^6 at 150°F, 3.4 × 10^6 at 250°F, 2.6 × 10^6 at 300°F
Melting range, °F (liquidus): 1075 ± 10
Specific heat Btu/lb^{-1}/°F^{-1}, 70°F: 0.346; 300°F: 0.348; 500°F: 0.350
Electrical resistivity, microhm-cm, 70°F: 15.2
Thermal conductivity, Btu/hr^{-1}/ft^{-2}/°F^{-1}/ft, 70°F: 301; 300°F: 287; 500°F: 280

Heat treatments
Stabilized at 350°F ± 25 for 3−6 hr.

Specifications

	Sheet and plate
AMS	4386

Typical Room Temperature Mechanical Properties

Temper	TS, ksi	YS, ksi	CYS, ksi	Elong., %	Hardness Bhn	Hardness RE	Shear strength, ksi	Charpy impact strength, ft-lb
T7	18/21	13/19	19	10/27	54	65	13	15 longitudinal
Extrusion	20.2	15.7		22				32 transverse
Casting	17.7	12.3		17				

Temper	Bearing, ksi Ultimate	Bearing, ksi Yield	Fatigue strength, ksi 10^5 cycles	10^6 cycles	10^7 cycles	10^8 cycles	5 × 10^8 cycles
T7	46.5	31				8.0	

Effect of Testing Temperature on Typical Mechanical Properties

Temper	Room temperature TS, ksi	TYS, ksi	Elong., %	200°F TS, ksi	TYS, ksi	Elong., %	300°F TS, ksi	TYS, ksi	Elong., %
T7	20	18	18	11		33	5		47[a]

Temper	−100°F TS, ksi	TYS, ksi	Elong., %	−300°F TS, ksi	TYS, ksi	Elong., %	−420°F TS, ksi	TYS, ksi	Elong., %
T7	28	21	12	32	27	11	42	39	11

[a]At 250°F

Table 1.2–20 (continued)
LA141A

LA141A Alloy Transverse and Longitudinal Bearing Strength

| | Bearing strength, psi | | | |
| | Ultimate | | Yield | |
Temperature, °F	Design	Typical	Design	Typical
Room temperature	43,000	–	27,000	–
Longitudinal	–	46,500	–	31,000
Transverse	–	46,400	–	33,000
150°	28,000	–	22,000	–
Longitudinal	–	30,600	–	26,000
Transverse	–	30,100	–	24,700

Charpy Impact (V-notch) Toughness of LA141A Alloy

| | Surface notch | | Short transverse notch | |
Temperature, °F	Longitudinal, ft-lb	Transverse, ft-lb	Longitudinal, ft-lb	Transverse, ft-lb
–320	15	31	11	15
–110	15	30	9	14
–20	15	30	9	15
+75	15	32	10	14
+150	14	28	9	15

Typical Creep Strengths

Stress, ksi

| | | 0.1% Creep extension | | | | | 0.2% Total extension | | | | |
| | | Minutes | Hours | | | | Minutes | Hours | | | |
Temper	Temperature, °F	10	1	10	100	1000	10	1	10	100	1000
T7	70			6.8	6.0				6.6	6.1	
	100			5.8	4.8				5.8	5.0	
	200			2.0	1.5				2.4	1.4	
	250			1.3	1.1				1.5	1.2	

| | | 0.5% Total extension | | | | |
| | | Minutes | Hours | | | |
Temper	Temperature, °F	10	1	10	100	1000
T7	70			8.3	8.1	
	100			7.7	6.7	
	200			3.5	2.1	
	250			1.9	1.4	

Table 1.2–21
ZM21

ZM21 is a general purpose alloy for extrusions and sheet. Stress relief after welding is not required, and it can be used for parts requiring good damping capacity.

Typical chemical composition, percent by weight: Mn, 0.95; Zn, 1.9; Mg, balance

Typical Room Temperature Mechanical Properties

Temper	TS, ksi	YS, ksi	CYS, ksi	Elong., %	Hardness Bhn	Hardness RE	Shear strength, ksi	Charpy impact strength, ft-lb
Sheet and plate	40	26		8				
Extrusions	37	27		17			18	

Table 1.2–22
AZ21A

AZ21A is an extrusion alloy developed exclusively as an anode material for primary batteries.

Typical chemical composition, percent by weight: Al, 2.0; Mn, 0.15 maximum; Zn, 1.0; Ca, 0.2; Mg, balance

Table 1.2–23
AZ61A

AZ61A is a general purpose extrusion and forging alloy with good mechanical properties and intermediate cost. It has higher strength than AZ31B, and is rarely used in sheet form.

Typical chemical composition, percent by weight: Al, 6.5; Mn, 0.15 minimum; Zn, 1.0; Mg, balance

Physical constants and thermal properties
Density at 68°F, lb/cu in.$^{-1}$:0.065

Coefficient of thermal expansion,°F^{-1} × 10^{-6}, 65–212°F: 14; 65–750°F: 16
Modulus of elasticity, psi: $6.5 × 10^6$
Modulus of rigidity, psi: $2.4 × 10^6$
Poisson's ratio: 0.35
Melting range, °F: 977–1145
Specific heat, Btu/lb^{-1}/°F^{-1}, 70°F: 0.25
Electrical resistivity, microhm-cm, 70°F: 12.5
Thermal conductivity, Btu/hr^{-1}/ft^{-2}/°F^{-1}/ft, 212–650°F: 46.0

Specifications

	Extrusions Bars	Extrusions Tubes	Forgings
AMS	4350	4350	4358
ASTM	B107	B217	B91
SAE	520	520	531
Federal	QQ-M-31	WW-T-825	QQ-M-40

Table 1.2–23 (continued)
AZ61A

Typical Room Temperature Mechanical Properties

Temper	TS, ksi	YS, ksi	CYS, ksi	Elong., %	Hardness Bhn	Hardness RE	Shear strength, ksi	Charpy impact strength, ft-lb
Extruded bar, F	36/45	16/33	19	7/16	55/60	66/72	19/20	3.0
Extruded tubes	41	24	16	14	50	60		
Forgings, F	38/43	22/26	18	6/12	55	66	19/21	2.2
Sheet	44	32	22	8				

Temper	Bearing, ksi Ultimate	Bearing, ksi Yield	10^5 cycles	10^6 cycles	10^7 cycles	10^8 cycles	5×10^8 cycles
Extruded bar, F	68	41	24/30	22/27	20/25	18/23	
F			17/19	12/15	11/14		
F			23/26	19/22	18/20		
Forging, F			24/28	21/25	19/23	17/22	
F			15/18	11/13	10/12		

Effect of Testing Temperature on Typical Mechanical Properties

Temper	Room temperature TS, ksi	TYS, ksi	Elong., %	200°F TS, ksi	TYS, ksi	Elong., %	300°F TS, ksi	TYS, ksi	Elong., %
Extrusions, F	45	33	16	41	26	23	31	19	32

Temper	400°F TS, ksi	TYS, ksi	Elong., %	600°F TS, ksi	TYS, ksi	Elong., %	-100°F TS, ksi	TYS, ksi	Elong., %
Extrusions, F	21	14	48	7.5	5	70	48	38	9

Temper	-200°F TS, ksi	TYS, ksi	Elong., %	-300°F TS, ksi	TYS, ksi	Elong., %
Extrusions, F	51	43	6	55	46	4

Table 1.2—23 (continued)
AZ61A

Typical Creep Strengths

Stress, ksi

Temper	Temperature, °F	0.1% Total extension Minutes 10	1	10	100	500	0.2% Total extension Minutes 10	1	10	100	500
Extrusions, F	200		6.0	5.0	4.0	3.0		11.0	10.0	7.0	6.0
	250		5.0	4.0	2.0	1.0		9.0	7.0	4.0	3.0
	300		3.5	2.0				7.0	4.0	1.0	
	350		2.0					4.5	1.5		

Temper	Temperature, °F	0.5% Total extension Minutes 10	1	10	100	500	1.0% Total extension Minutes 10	1	10	100	500
Extrusions, F	200		20.0	18.0	14.0	12.0		25.0	22.0	19.0	16.0
	250		16.0	13.0	8.0	6.0		20.0	16.0	11.0	8.0
	300		12.0	8.0	3.0	1.0		16.0	11.0	5.0	2.0
	350		9.0	4.0				11.0	6.0		

Table 1.2–24
AZ80A

AZ80A is available as extruded products and forgings. It has higher strength than the other AZ alloys, and like them is limited to use below 300°F.

Typical chemical composition, percent by weight: Al, 8.5; Mn, 0.12 minimum; Zn, 0.5; Mg, balance

Physical constants and thermal properties
Density at 68°F, lb/cu in. $^{-1}$: 0.065
Modulus of elasticity, psi: 6.5 × 10^6
Modulus of rigidity, psi: 2.4 × 10^6

Poisson's ratio: 0.35
Melting range, °F: 914–1130
Specific heat, Btu/lb $^{-1}$/°F $^{-1}$, 70°F: 0.25
Electrical resistivity, microhm-cm, 70°F: 15.6 (F), 12.2 (T5); 300°F: 17.3 (F), 14.4 (T5); 500°F: 18.3 (F), 16.3 (T5)
Thermal conductivity, Btu/hr $^{-1}$/ft $^{-2}$/°F $^{-1}$/ft, 70°F: 27.6 (F), 34.4 (T5); 300°F: 35.8 (F), 41.9 (T5); 500°F: 43.1 (F), 47.9 (T5)

Heat treatments
Annealing temperature, 725°F; stress relief 500°F, 15 min; aging for extrusions, 400°F, 60 min (F5).

Specifications

	Extruded bars	Forgings
AMS	—	4360
ASTM	B107	B91
SAE	523	532
Federal	QQ-M-31	QQ-M-40

Typical Room Temperature Mechanical Properties

Temper		TS, ksi	YS, ksi	CYS, ksi	Elong., %	Hardness Bhn	RE	Shear strength, ksi	Charpy impact strength, ft-lb
Extruded bar,	F	42/49	27/36		4/11	66	77	19/22	
	T5	45/55	30/40	35	2/7	80	88	20/24	
Forgings,	F	42/48	26/33	25	5/11	69	80	20/22	
	T5	34/50	22/36	28	2/6	72	82	20/23	
	T6	50	36	27	5	72			

Temper		Bearing, ksi Ultimate	Yield	Fatigue strength, ksi 10^5 cycles	10^6 cycles	10^7 cycles	10^8 cycles	5 × 10^8 cycles
Extruded bar,	F	80	51	25/30	23/28	21/26	20/24	
	F			19/21	13/16	12/14		
Forgings,	F			28/30	24/26	20/22	18/20	
	T5			26/30	22/25	19/21	16/19	
	T6			16/20	13/16	12/15		
	T6			23/27	19/22	16/19	14/16	

Table 1.2–24 (continued)
AZ80A

Effect of Testing Temperature on Typical Mechanical Properties

		Room temperature			200°F			300°F		
	Temper	TS, ksi	TYS, ksi	Elong., %	TS, ksi	TYS, ksi	Elong., %	TS, ksi	TYS, ksi	Elong., %
	F	49	36	11	44	32	18	35	25	25
Forging,	T5	47	36	8				22	17	50
Forging,	T6	53	40	4				30	21	30
Extrusion,	T5	55	43	5	48	33	20	35	24	26

		400°F			500°F			600°F		
	F	28	17	35	16	11	57			
Forging,	T5	12	10	70						
Forging,	T6	22	15	49	14	8	83	9	5	123
Extrusion,	T5	22	15	46	12	8	82	10	5	120

		0°F			−100°F		
	F	51	36	10	56	39	8

Typical Creep Strengths

Stress, ksi

		0.1% Total extension					0.2% Total extension				
			Hours					Hours			
Temper	Temperature, °F	Minutes 10	1	10	100	1000	Minutes 10	1	10	100	1000
Forging, T5	300				2.8					3.5	
	400				0.5					0.8	
T6	300				2.0					2.8	
	400				0.6					0.9	

		0.5% Total extension				
			Hours			
Temper	Temperature, °F	Minutes 10	1	10	100	1000
Forging, T5	300				5.6	
	400				2.0	
T6	300				5.9	
	400				1.8	

Table 1.2–25
HM31A

HM31A is an extrusion alloy primarily used for elevated-temperature structural applications. It can be used to 800°F. It has superior modulus of elasticity at elevated temperatures.

Typical chemical composition, percent by weight: Mn, 1.20 minimum; Th, 3.0; Mg, balance

Physical constants and thermal properties
  Density at 68°F, lb/cu in.$^{-1}$: 0.0654; at 900°F, 0.0625

Coefficient of thermal expansion, °F^{-1} × 10^{-6}, 65–212°F: 14.5; 68–600°F: 15.6; 68–1000°F: 16.8

Modulus of elasticity, psi: 6.5 × 10^6 at 70°F, 6.1 × 10^6 at 300°F, 5.9 × 10^6 at 400°F, 5.6 × 10^6 at 600°F

Poisson's ratio: 0.35

Melting range, °F: 1121–1202

Specific heat, Btu/lb^{-1}/°F^{-1}, 70°F: 0.24; 300°F: 0.26; 600°F: 0.28; 900°F: 0.33

Electrical resistivity, microhm-cm, 70°F: 6.6; 300°F: 8.8; 500°F: 10.6

Thermal conductivity, Btu/hr^{-1}/ft^{-2}/°F^{-1}/ft, 70°F: 60.9; 300°F: 66.8; 500°F: 70.9

Specifications

	Extrusions
AMS	4388, 4389
Military	MIL-M-8916

Typical Room Temperature Mechanical Properties

Temper	TS, ksi	YS, ksi	CYS, ksi	Elong., %	Hardness Bhn	RE	Shear strength, ksi	Charpy impact strength, ft-lb	Bearing, ksi Ultimate	Yield
T5	37/42	26/33	27	4/10			22		70	50

Effect of Testing Temperature on Typical Mechanical Properties

Temper	Room temperature TS, ksi	TYS, ksi	Elong., %	200°F TS, ksi	TYS, ksi	Elong., %	300°F TS, ksi	TYS, ksi	Elong., %
Extrusions, T5	45	38	10	34	30	17	27	25	22
	400°F			500°F			600°F		
Extrusions, T5	24	21	26	21	18	25	17	14	24
	700°F			800°F			900°F		
Extrusions, T5	13	11	27	8	7	46	4	3	66

Table 1.2–25 (continued)
HM31A

Typical Creep Strengths

Stress, ksi

Temper	Temperature, °F	0.1% Total extension Minutes 10	Hours 1	10	100	1000	0.2% Total extension Minutes 10	Hours 1	10	100	1000
Extrusion, T5	400		5.8	5.8	5.8/16[a]					10.9	
	500		5.6	5.6	5.6/11[a]					9.8	
	600		5.4	5.4	5.4/6[a]					7.6	
	700									2.0	

Temper	Temperature, °F	0.5% Total extension Minutes 10	Hours 1	10	100	1000
Extrusions, T5	400				16.5	
	500				13.9	
	600				9.7	
	700				2.3	

[a]Creep extension

Table 1.2–26
ZK60A

ZK60A is an extrusion and forging alloy having high strength and ductility.

Typical chemical composition, percent by weight: Zn, 5.5; Zr, 0.45; Mg, balance

Physical constants and thermal properties
Density at 68°F, lb/cu in. $^{-1}$: 0.066
Coefficient of thermal expansion, $°F^{-1} \times 10^{-6}$, 65–212°F: 26

Modulus of elasticity, psi: 6.5×10^6
Modulus of rigidity, psi: 2.4×10^6
Poisson's ratio: 0.35
Melting range, °F: 968–1175
Thermal conductivity, $Btu/hr^{-1}/ft^{-2}/°F^{-1}/ft$, 70°F: 67.8 (F), 70.2 (T5)

Heat treatments
T5: age at 300°F for 24 hr, air cool.

Specifications

	Extrusions Bars	Tubes	Forgings
AMS	4352	4352	4362
ASTM	B107	B217	B91
SAE	524	524	
Federal	QQ-M-31	WW-T-825	QQ-M-40

Table 1.2–26 (continued)
ZK60A

Typical Room Temperature Mechanical Properties

Temper		TS, ksi	YS, ksi	CYS, ksi	Elong., %	Hardness Bhn	RE	Shear strength, ksi	Charpy impact strength, ft-lb
Extruded bar,	F	40/49	28/38	33	5/14	75	84	22/27	
	T5	43/53	31/44	36	4/11	82	88	22/26	
Extruded tube,	F	46	34	25	12	75	84		
	T5	50	40	29	11	82	88		
Forgings,	T5	38/44	20/31	23	7/16	65	77	24	

Temper		Bearing, ksi Ultimate	Yield	Fatigue strength, ksi 10^5 cycles	10^6 cycles	10^7 cycles	10^8 cycles	5×10^8 cycles
Extruded bar,	F	80	55	19/32	26/29	23/26	20/23	
	T5			26/35	21/29	18/26	17/23	
	F			41/45	38/42	38/42		
	T5			42/48	41/45	40/44		
Forgings,	T5	61	41	17/21	13/16	11/14		

Effect of Testing Temperature on Typical Mechanical Properties

Temper		Room temperature TS, ksi	TYS, ksi	Elong., %	200°F TS, ksi	TYS, ksi	Elong., %	300°F TS, ksi	TYS, ksi	Elong., %
Extrusions,	T5	52	39	13	37	32	40	25	22	46
Forgings,	T5	45	31	15				23	17	78
	T6	49	40	8				27	23	33

Temper		400°F TS	TYS	Elong.	500°F TS	TYS
Extrusions,	T5	15	12	81	6	3
Forgings,	T5	14	10	88		
	T6	18	15	37		

Table 1.2–26 (continued)
ZK60A

Typical Creep Strengths

Stress, ksi

| | | 0.1% Total extension | | | | | | 0.2% Total extension | | | | |
| | | | Hours | | | | | | Hours | | | |
Temper	Temperature, °F	Minutes 10	1	10	100	1000	Minutes 10	1	10	100	1000
Forging, T5	300				0.9[a]					1.3	
	400				0.2[a]					0.3	
T6	300				3.3[a]					3.8	
	400				1.2[a]					1.5	

| | | 0.5% Total extension | | | | |
| | | | Hours | | | |
Temper	Temperature, °F	Minutes 10	1	10	100	1000
Forging, T5	300				2.3	
	400				0.5	
T6	300				6.2	
	400				2.3	

[a]Creep extension

Table 1.2–27
TA54A

TA54A is a general purpose hammer forging alloy.

Typical chemical composition, percent by weight: Al, 3.5; Mn, 0.20 minimum; Zn, 0.30 maximum; Sn, 5.0; Mg, balance

Specifications

	Forgings
ASTM	B91
Federal	QQ-M-40

Typical Room Temperature Mechanical Properties

TS, ksi	YS, ksi	CYS, ksi	Elong., %	Hardness Bhn	RE	Shear strength, ksi	Charpy impact strength, ft-lb
36	22		7	52	62	16	

1.3 ALUMINUM-BASE ALLOYS

Aluminum base alloys are divided into two categories, wrought and cast. Wrought alloys have a systematic identification according to the alloying elements. For the Aluminum Association designation, lxxx is 99.00% aluminum or greater. The second digit indicates special purity controls and the last two digits indicate the minimum aluminum beyond 99.00%. Thus, a 1030 Al has 99.30% aluminum and no special control of individual impurities. The other series designations are

Al plus Cu as principal alloying element — 2xxx

Al plus Mn as principal alloying element — 3xxx

Al plus Si as principal alloying element — 4xxx

Al plus Mg as principal alloying element — 5xxx

Al plus Mg and Si as principal alloying elements — 6xxx

Al plus Zn as prinicipal alloying element — 7xxx

Al plus other major alloying element — 8xxx

The second digit refers to alloying modifications, and the last two digits identify the alloy, usually from its former commercial designation; thus, 75S is now 7075. The system of temper designations used for all forms of wrought and cast aluminum is briefly reviewed here. The individual company data sheets should be consulted for specific heat treatments.

F — as fabricated
O — annealed
H — strain hardened

H1 — strain hardened only
H2 — strain hardened and partially annealed
H3 — strain hardened and stabilized by low temperature treatment. A second digit is used to indicate tempers between O (annealed) and 8 (full hard, final degree of strain hardening, although 9 is used for extra-hard tempers). A third digit indicates variations from these tempers.
W — solution heat treated, spontaneous aging at room temperature
T — thermally treated to produce stable tempers
T1 — partially solution heat treated and naturally aged to stable condition
T2 — annealed (for improved castings)
T3 — solution heat treated and cold worked
T4 — solution heat treated and naturally aged to stable condition
T5 — partially solution heat treated and artificially aged
T6 — solution heat treated and artificially aged
T7 — solution heat treated and stabilized, beyond point of maximum hardness
T8 — solution heat treated, cold worked, and artificially aged
T9 — solution heat treated, artificially aged, and cold worked
T10 — partially solution heat treated, artificially aged, and cold worked
Additional digits may be added for variation in treatments.

1.3.1 WROUGHT ALUMINUM ALLOYS

Table 1.3.1—1
WROUGHT ALUMINUM EC

Typical chemical composition, percent by weight: Al, 99.45 minimum

Physical constants and thermal properties
  Density, lb/in.^{3}: 0.098
  Coefficient of thermal expansion, (70–200°F) in./in./°F × 10^{-6}: 13.2
  Modulus of elasticity, psi: tension, 10 × 10^6

Poisson's ratio: 0.345
Melting range, °F: 1,195–1,215
Thermal conductivity, Btu/ft^2/hr/in./°F, 70°F: 1,620
Electrical resistivity, ohms/cmil/ft, 70°F: 16.8

Heat treatments
Annealing temperature 650°F.

Table 1.3.1–1 (continued)
WROUGHT ALUMINUM EC

Tensile Properties

Annealed

Temperature, °F	T.S., psi	Y.S., psi	Elong., in 2 in., %	Hardness, Brinell
75	12,000 (O)	4,000 (O)	23 (O)	–
75	27,000 (H19)	24,000 (H19)	1.5 (H19)	–

Fatigue Strength

Hard (H19)

Test temperature, °F	Stress, psi	Cycles to failure
75	7,000	5×10^8

Specifications

ASTM: EC

Table 1.3.1–2
WROUGHT ALUMINUM 1060

Typical chemical compostion, percent by weight: Mn, 0.03; Fe, 0.35; Si, 0.25; Ti, 0.03; Al, 99.60 minimum; Cu, 0.05; Mg, 0.03; Zn, 0.05

Physical constants and thermal properties
Density, lb/in.^{3}: 0.098
Coefficient of thermal expansion, (70–200°F) in./in./°F $\times 10^{-6}$: 13.1

Modulus of elasticity, psi: tension, 10×10^6
Melting range, °F: 1,195–1,215
Thermal conductivity, Btu/ft^2/hr/in./°F, 70°F: 1,536
Electrical resistivity, ohms/cmil/ft, 70°F: 16.8

Heat treatments
Annealing temperature 650°F.

Tensile Properties

Annealed

Temperature, °F	T.S., psi	Y.S., psi	Elong., in 2 in., %	Hardness, Brinell
75	10,000	4,000	43	19
75	19,000 (H18)	18,000 (H18)	6 (H18)	35 (H18)

Fatigue Strength

Test temperature, °F	Stress, psi	Cycles to failure
75	3,000	5×10^8 (annealed)
75	6,500	5×10^8 (H18)

Specifications

ASTM: 996A; BD1S

Table 1.3.1−3
WROUGHT ALUMINUM 1100

Typical chemical composition, percent by weight: Mn, 0.05; Si + Fe, 1.0; Al, 99.0; Cu, 0.20; Zn, 0.10

Physical constants and thermal properties
Density, lb/in.^{3}: 0.098
Coefficient of thermal expansion, (70−200°F) in./in./°F × 10^{-6}: 13.1
Modulus of elasticity, psi: tension, 10 × 10^6

Melting range, °F: 1,190−1,215
Specific heat, Btu/lb/°F, 70°F: 0.22
Thermal conductivity, Btu/ft^2/hr/in./°F, 70°F: 1,536
Electrical resistivity, ohms/cmil/ft, 70°F: 17.5

Heat treatments
Annealing temperature 650°F; hot working temperature 500−950°F.

Tensile Properties

Temperature, °F	T. S., psi	Y. S., psi	Elong., in 2 in., %	Hardness, Brinell
Annealed				
75	13,000	5,000	45	23
212	11,000	5,000	45	
300	8,500	4,500	55	
400	6,000	3,500	65	
500	4,000	2,000	75	
600	2,500	1,500	80	
700	2,000	1,000	85	
H14 Temper				
75	18,000	17,000	20	32
212	16,000	15,000	20	
300	13,000	12,000	22	
400	9,500	7,000	25	
500	4,000	2,500	75	
600	2,500	1,500	80	
700	2,000	1,000	85	

Fatigue Strength

Test temperature, °F	Stress, psi	Cycles to failure
75	5,000	5 × 10^8 (annealed)
75	7,000	5 × 10^8 (H 14)

Specifications

ASTM: 990A; SAE-AMS: 25; 2S

Table 1.3.1–4
WROUGHT ALUMINUM 2011

Typical chemical composition, percent be weight: Fe, 0.70; Si, 0.40; Al, balance; Zn, 0.30; Cu, 5.0–6.0; Pb, 0.2–0.6; Bi, 0.2–0.6

Physical constants and thermal properties
  Density, lb/in.^{3}: 0.102
  Coefficient of thermal expansion, (70–200°F) in./in./°F × 10^{-6}: 12.8
  Modulus of elasticity, psi: tension, 10.2 × 10^6

Melting range, °F: 995–1190
Specific heat, Btu/lb/°F, 70°F: 0.25
Thermal conductivity, Btu/ft^2/hr/in./°F, 70°F: 990
Electrical resistivity, ohms/cmil/ft, 70°F: 28.8

Heat treatments
  Annealing temperature (start) 750°F; solution temperature 950°F; aging temperature 320°F, 12–16 hr for T6.

Tensile Properties

Heat Treated

Temperature, °F	T. S., psi	Y. S., psi, 0.2% offset	Elong., in 2 in., %	Hardness, Brinell
T3 Temper				
75	55,000	43,000	15	95
212	47,000	34,000	16	
300	28,000	19,000	25	
400	16,000	11,000	35	
500	6,500	4,000	45	
600	3,500	2,000	90	
700	2,500	1,500	125	
T8 Temper				
75	59,000	45,000	12	100
300	28,000	20,000	24	
400	16,000	11,000	35	
500	6,500	4,000	45	
600	3,500	2,000	90	
700	2,500	1,500	125	

Fatigue Strength

Test temperature, °F	Stress, psi	Cycles to failure
75	18,000	5 × 10^8 (T3)
75	18,000	5 × 10^8 (T8)

Specifications

ASTM: CB60A; SAE-AMS: 202; 11S

Table 1.3.1–5
WROUGHT ALUMINUM 2014

Typical chemical composition, percent by weight: Mn, 0.4–1.2; Fe, 1.0; Si, 0.5–1.2; Cr, 0.10; Ti, 0.15; Al, balance; Cu, 3.9–5.0; Mg, 0.2–0.8; Zn, 0.25

Physical constants and thermal properties
Density, lb/in.^{3}: 0.101
Coefficient of thermal expansion, (70–200°F) in./in./°F $\times$ 10^{-6}: 12.8
Modulus of elasticity, psi: tension, 10.6 $\times$ 10^6

Melting range, °F: 950–1,180
Specific heat, Btu/lb/°F, 70°F: 0.22
Thermal conductivity, Btu/ft^2/hr/in./°F, 70°F; 1,332
Electrical resistivity, ohms/cmil/ft, 70°F: 20.70 (O); 25.86 (T6)

Heat treatments
Annealing temperature (start) 775°F; solution temperature 940°F; aging temperature 340°F, 8–12 hr for T6.

Tensile Properties

Temperature, °F	T.S., psi	Y.S., psi 0.2% offset	Elong., in 2 in., %	Hardness, Brinell
Annealed				
75	27,000	14,000	18	45
T4 Temper				
75	62,000	42,000	20	105
300	46,000	33,000		
400	18,000	12,000		
500	11,000	8,500		
600	6,500	5,000		
700	4,500	3,500		
T6 Temper				
75	70,000	60,000	13	135
212	66,000	56,000	14	
300	47,000	40,000	15	
400	18,000	12,000	35	
500	11,000	8,500	45	
600	6,500	5,000	65	
700	4,500	3,500	70	

Fatigue Strength

Test temperature, °F	Stress, psi	Cycles to failure
75	13,000	5 $\times$ 10^8 (annealed)
75	20,000	5 $\times$ 10^8 (T4)
75	18,000	5 $\times$ 10^8 (T6)

Specifications

ASTM: CS41; SAE-AMS: 260; 14S

Table 1.3.1–6
WROUGHT ALUMINUM 2017

Typical chemical composition, percent by weight: Mn,
0.4–1.0; Fe, 1.0; Si, 0.8; Cr, 0.10; Al, balance; Cu,
3.5–4.5; Mg, 0.2–0.8; Zn, 0.25

Physical constants and thermal properties
Density, lb/in.^{3}: 0.101
Coefficient of thermal expansion, (70–200°F)
   in./in./°F × 10^{-6}: 13.1
Modulus of elasticity, psi: tension, 10.5 × 10^6

Melting range, °F: 955–1,185
Specific heat, Btu/lb/°F, 70°F: 0.22
Thermal conductivity, Btu/ft^2/hr/in./°F, 70°F: 1,193
Electrical resistivity, ohms/cmil/ft, 70°F: 23.0 (O);
   34.5 (T4)

Heat treatments
Annealing temperature (start) 775°F; solution temperature 940°F.

Tensile Properties

Temperature, °F	T.S., psi	Y.S., psi, 0.2% offset	Elong., in 2 in., %	Hardness, Brinell
		Annealed		
75	26,000	10,000	22	45
		T4 Temper		
75	62,000	40,000	22	105
212	56,000	37,000	18	
300	40,000	30,000	16	
400	22,000	17,000	28	
500	12,000	9,500	45	
600	6,500	5,000	95	
700	4,500	3,500	100	

Fatigue Strength

Test temperature, °F	Stress, psi	Cycles to failure
75	13,000	5 × 10^8 (annealed)
75	18,000	5 × 10^8 (T4)

Specifications

ASTM: CM41A; SAE-AMS: 26; 17S

Table 1.3.1−7
WROUGHT ALUMINUM 2021

Typical chemical composition, percent by weight: Mn, 0.20−0.40; Fe, 0.40; Si, 0.40; Al, balance; Zr, 0.10−0.25; Cu, 5.8−6.8; Cd, 0.05−0.20; Zn, 0.25

Modulus of elasticity, psi: tension, 10.7×10^6
Melting range, °F: 997−1,195

Physical constants and thermal properties
Density, lb/in.³: 0.103
Coefficient of thermal expansion, (70−200°F) in./in./°F × 10^{-6}: 12.6

Heat treatments
Annealing temperature (start) 775°F; solution temperature 985°F; aging temperature 325°F.

Tensile Properties

Temperature, °F	T.S., psi	Y.S., psi, 0.2% offset	Elong., in 2 in., %	Hardness, Brinell
		Annealed		
75	24,000	10,000	23	44
		T81 Temper		
75	73,000	63,000	9	137

Table 1.3.1−8
WROUGHT ALUMINUM 2024

Typical chemical composition, percent by weight: Mn, 0.30−0.90; Fe, 0.5; Si, 0.5; Cr, 0.10; Al, balance; Zr, 0.10−0.25; Cu, 3.8−4.9; Cd, 0.05−0.20; Mg, 1.2−1.8; Zn, 0.25

Melting range, °F: 935−1,180
Specific heat, Btu/lb/°F, 70°F: 0.22
Thermal conductivity, Btu/ft²/hr/in./°F, 70°F: 1,310
Electrical resistivity, ohms/cmil/ft, 70°F: 20.7 (O); 34.50 (T4)

Physical constants and thermal properties
Density, lb/in.³: 0.100
Coefficient of thermal expansion, (70−200°F) in./in./°F × 10^{-6}: 12.9
Modulus of elasticity, psi: tension, 10.6×10^6

Heat treatments
Annealing temperature (start) 775°F; solution temperature 930°F; aging temperature 375°F, 11−13 hr for T6.

Tensile Properties

Temperature, °F	T.S., psi	Y.S., psi, 0.2% offset	Elong., in 2 in., %	Hardness, Brinell
		Annealed		
75	27,000	11,000	20	47
		T3 Temper		
75	65,000	45,000	18	120
300	55,000	50,000	11	
400	29,000	22,000	23	
500	12,000	9,000	55	
600	8,000	6,000	75	
700	5,500	4,000	100	

Table 1.3.1–8 (continued)
WROUGHT ALUMINUM 2024

Fatigue Strength

Test temperature, °F	Stress, psi	Cycles to failure
75	13,000	5×10^8 (annealed)
75	20,000	5×10^8 (T3)

Specifications

ASTM: CG42A; SAE-AMS: 24; 24S

Table 1.3.1–9
WROUGHT ALUMINUM 2219

Typical chemical composition, percent by weight: Mn, 0.20–0.40; Fe, 0.30; Si, 0.20; Ti, 0.02–0.10; Al, balance; Zr, 0.10–0.25; Cu, 5.8–6.8; V, 0.05–0.15; Mg, 0.02; Zn, 0.10

Physical constants and thermal properties
Density, lb/in.^{3}: 0.103
Coefficient of thermal expansion, (70–200°F) in./in./°F $\times 10^{-6}$: 12.4

Modulus of elasticity, psi: tension, 10.6×10^6
Melting range, °F: 1,010–1,190
Thermal conductivity, Btu/ft^2/hr/in./°F, 70°F: 1,200
Electrical resistivity, ohms/cmil/ft, 70°F: 23.4 (O); 34.2 (T6)

Heat treatments
Annealing temperature (start) 775°F; solution temperature 960°F; aging temperature 340°F.

Tensile Properties

Temperature, °F	T.S., psi	Y.S., psi, 0.2% offset	Elong., in 2 in., %	Hardness, Brinell
		Annealed		
75	25,000	11,000	18	130 (T87 temper)
		T6 Temper		
212	54,000	35,000	16	
300	46,000	31,000	20	
400	38,000	27,000	23	
500	31,000	22,000	25	
600	7,000	6,000	55	
700	4,500	4,000	60	

Fatigue Strength

Heat Treated T87

Test temperature, °F	Stress, psi	Cycles to failure
70	15,000	5×10^8

Table 1.3.1—10
WROUGHT ALUMINUM 3003

Typical chemical composition, percent by weight: Mn, 1.0–1.5; Fe, 0.7; Si, 0.6; Al, balance; Cu, 0.20; Zn, 0.10

Physical constants and thermal properties
Density, lb/in.^{3}: 0.099
Coefficient of thermal expansion, (70–200°F) in./in./°F × 10^{-6}: 12.9

Modulus of elasticity, psi: tension, 10 × 10^6
Melting range, °F: 1,190–1,210
Specific heat, Btu/lb/°F, 70°F: 0.22
Thermal conductivity, Btu/ft^2/hr/in./°F, 70°F: 1,332
Electrical resistivity, ohms/cmil/ft, 70°F: 20.7

Heat treatments
Annealing temperature 775°F; hot working temperature 500–950°F.

Tensile Properties

Temperature, °F	T.S., psi	Y.S., psi	Elong., in 2 in., %	Hardness, Brinell
		Annealed		
75	16,000	6,000	40	28
212	13,000	5,500	43	
300	11,000	5,000	47	
400	8,500	4,500	60	
500	6,000	3,500	65	
600	4,000	2,500	70	
700	3,000	2,000	70	
		H14 Temper		
75	22,000	21,000	16	40
212	21,000	19,000	16	
300	18,000	16,000	16	
400	14,000	9,000	20	
500	7,500	4,000	60	
600	4,000	2,500	70	
700	3,000	2,000	70	

Fatigue Strength

Test temperature, °F	Stress, psi	Cycles to failure
75	7,000	5 × 10^8 (annealed)
75	9,000	5 × 10^8 (H14)

Specifications

ASTM: M1A; SAE-AMS: 29; 3S

Table 1.3.1–11
WROUGHT ALUMINUM 3004

Typical chemical composition, percent by weight: Mn, 1.0–1.5; Fe, 0.7; Si, 0.30; Al, balance; Mg, 0.8–1.3; Cu, 0.25; Zn, 0.25

Melting range, °F 1,165–1,205
Specific heat, Btu/lb/°F, 70°F: 0.22
Thermal conductivity, Btu/ft²/hr/in./°F, 70°F: 1,126
Electrical resistivity, ohms/cmil/ft, 70°F: 24.6

Physical constants and thermal properties
Density, lb/in.³: 0.098
Coefficient of thermal expansion, (70–200°F) in./in./ °F × 10⁻⁶: 13.3
Modulus of elasticity, psi: tension, 10 × 10⁶

Heat treatments
Annealing temperature 650°F; hot working temperature 500–950°F.

Tensile Properties

Temperature, °F	T.S., psi	Y.S., psi	Elong., in 2 in., %	Hardness, Brinell
		Annealed		
75	26,000	10,000	25	45
212	26,000	10,000	25	
300	22,000	10,000	35	
400	14,000	9,500	55	
500	10,000	7,500	70	
600	7,500	5,000	80	
700	5,000	3,000	90	
		H34 Temper		
75	35,000	29,000	12	63
212	34,000	29,000	13	
300	28,000	25,000	22	
400	21,000	15,000	35	
500	14,000	7,500	55	
600	7,500	5,000	80	
700	5,000	3,000	90	

Fatigue Strength

Test temperature, °F	Stress, psi	Cycles to failure
75	14,000	5 × 10⁸ (annealed)
75	15,000	5 × 10⁸ (H34)

Specifications

ASTM: MG11A; SAE-AMS: 20; 4S

Table 1.3.1—12
WROUGHT ALUMINUM 4032

Typical chemical composition, percent by weight: Fe, 1.0; Si, 11.0–13.5; Cr, 0.10; Ni, 0.50–1.3; Al, balance; Cu, 0.50–1.3; Mg, 0.8–1.3; Zn, 0.25

Physical constants and thermal properties
Density, lb/in.^{3}: 0.097
Coefficient of thermal expansion, (70–200°F) in./in./°F × 10^{-6}: 10.8

Modulus of elasticity, psi: tension, 10.0 × 10^6
Melting range, °F: 990–1,060
Thermal conductivity, Btu/ft^2/hr/in./°F, 70°F: 1,070
Electrical resistivity, ohms/cmil/ft, 70°F: 26

Heat treatments
Anneal 775°F; solution temperature 940–960°F; age 335–345°F for 8–12 hr for T6.

Tensile Properties

T6 Temper

Temperature, °F	T.S., psi	Y.S., psi	Elong., in 2 in., %	Hardness, Brinell
75	55,000	46,000	9	120
212	50,000	44,000	9	
300	37,000	33,000	9	
400	13,000	9,000	30	
500	8,000	5,500	50	
600	5,000	3,000	70	
700	3,500	2,000	90	

Fatigue Strength

T6 Temper

Test temperature, °F	Stress, psi	Cycles to failure
75	16,000	5 × 10^8

Table 1.3.1–13
WROUGHT ALUMINUM 5005

Typical chemical composition, percent by weight: Mn, 0.20; Fe, 0.70; Si, 0.40; Cr, 0.10; Al, balance; Mg, 0.5–1.1; Cu, 0.20; Zn, 0.25

Physical constants and thermal properties
Density, lb/in.^{3} : 0.097
Coefficient of thermal expansion, (70–200°F) in./in./°F × 10^{-6} : 13.3

Modulus of elasticity, psi: tension, 10 × 10^6
Melting range, °F: 1,170–1,205
Specific heat, Btu/lb/°F, 70°F: 0.23
Thermal conductivity, Btu/ft^2/hr/in./°F, 70°F: 1,392
Electrical resistivity, ohms/cmil/ft, 70°F: 19.8

Heat treatments
Annealing temperature 650°F.

Tensile Properties

Temperature, °F	T.S., psi	Y.S., psi	Elong., in 2 in., %	Hardness, Brinell
		Annealed		
75	18,000	6,000	30	28
		H14 Temper		
75	23,000	22,000	6	

Specifications

ASTM: G1B; K155

Table 1.3.1—14
WROUGHT ALUMINUM 5050

Typical chemical composition, percent by weight: Mn, 0.10; Fe, 0.7; Si, 0.40; Cr, 0.10; Al, balance; Mg, 1.0–1.8; Cu, 0.20; Zn, 0.25

Physical constants and thermal properties
Density, lb/in.^{3} : 0.097
Coefficient of thermal expansion, (70–200°F) in./in./ °F × 10^{-6} : 13.2

Modulus of elasticity, psi: tension, 10×10^6
Melting range, °F: 1,160–1,205
Specific heat, Btu/lb/°F, 70°F: 0.22
Thermal conductivity, Btu/ft^2/hr/in./°F, 70°F: 1,332
Electrical resistivity, ohms/cmil/ft, 70°F: 20.4

Heat treatments
Annealing temperature 650°F.

Tensile Properties

Temperature, °F	T.S., psi	Y.S., psi	Elong., in 2 in., %	Hardness, Brinell
Annealed				
75	21,000	8,000	24	36
212	21,000	8,000	28	
300	19,000	8,000	38	
400	14,000	7,500	58	
500	9,000	6,000	67	
600	6,000	4,000	80	
700	4,000	3,000	95	
H34 Temper				
75	28,000	24,000	8	53
212	32,000	27,000	18	
300	26,000	23,000	25	
400	14,000	7,000	58	
500	8,000	7,000	59	
600	6,000	4,000	–	
700	4,000	3,000	–	

Fatigue Strength

Test temperature, °F	Stress, psi	Cycles to failure
75	12,000	5×10^8 (annealed)
75	13,000	5×10^8 (H34)

Specifications

ASTM: 50S; R305; SAE-AMS: 207

Table 1.3.1–15
WROUGHT ALUMINUM 5052

Typical chemical composition, percent by weight: Mn, 0.10; Si + Fe, 0.45; Cr, 0.15–0.35; Al, balance; Mg, 2.2–2.8; Cu, 0.10; Zn, 0.10

Physical constants and thermal properties
Density, lb/in.^{3} : 0.097
Coefficient of thermal expansion, (70–200°F) in./in./°F × 10^{-6} : 13.2

Modulus of elasticity, psi: tension, 10.2 × 10^6
Melting range, °F: 1,100–1,200
Specific heat, Btu/lb/°F, 70°F: 0.22
Thermal conductivity, Btu/ft^2/hr/in./°F, 70°F: 960
Electrical resistivity, ohms/cmil/ft, 70°F: 29.4

Heat treatments
Annealing temperature 650°F.

Tensile Properties

Temperature, °F	T.S., psi	Y.S., psi	Elong., in 2 in., %	Hardness, Brinell
		Annealed		
75	28,000	13,000	30	47
212	28,000	13,000	35	
300	24,000	13,000	50	
400	18,000	11,000	65	
500	12,000	7,500	80	
600	7,500	5,000	100	
700	5,000	3,000	130	
		H34 Temper		
75	38,000	31,000	14	68
212	38,000	30,000	16	
300	31,000	27,000	25	
400	23,000	14,000	40	
500	12,000	7,000	80	
600	7,500	5,000	100	
700	5,000	3,000	130	

Fatigue Strength

Test temperature, °F	Stress, psi	Cycles to failure
75	16,000	5 × 10^8 (annealed)
75	18,000	5 × 10^8 (H34)

Specifications

ASTM: GR20A; SAE-AMS: 201; 52S

Table 1.3.1–16
WROUGHT ALUMINUM 5056

Typical chemical composition, percent by weight: Mn, 0.05–0.20; Fe, 0.40; Si, 0.30; Cr, 0.05–0.20; Al, balance; Mg, 4.5–5.6; Cu, 0.10; Zn, 0.10

Physical constants and thermal properties
Density, lb/in.³ : 0.095
Coefficient of thermal expansion, (70–200°F) in./in./ °F × 10⁻⁶ : 13.4

Modulus of elasticity, psi: tension, 10.3×10^6
Melting range, °F: 1,055–1,180
Specific heat, Btu/lb/°F, 70°F: 0.22
Thermal conductivity, Btu/ft/hr/in./°F, 70°F: 809
Electrical resistivity, ohms/cmil/ft, 70°F: 35.4

Heat treatments
Annealing temperature 650°F.

Tensile Properties

Temperature, °F	T.S., psi	Y.S., psi	Elong., in 2 in., %	Hardness, Brinell
		Annealed		
75	42,000	22,000	35	65
300	31,000	17,000	55	
400	22,000	13,000	65	
500	16,000	10,000	80	
600	11,000	7,000	100	
700	6,000	4,000	130	
		H38 Temper		
75	60,000	50,000	15	100
300	38,000	31,000	30	
400	26,000	18,000	50	
500	16,000	10,000	80	
600	11,000	7,000	100	
700	6,000	4,000	130	

Fatigue Strength

Test temperature, °F	Stress, psi	Cycles to failure
75	20,000	5×10^8 (annealed)
75	22,000	5×10^8 (H38)

Specifications

ASTM: GM50A; 56S

Table 1.3.1–17
WROUGHT ALUMINUM 5083

Typical chemical composition, percent by weight: Mn, 0.3–1.0; Fe, 0.40; Si, 0.40; Cr, 0.05–0.25; Ti, 0.15; Al, balance; Mg, 4.0–4.9; Cu, 0.10; Zn, 0.25

Physical constants and thermal properties
  Density, lb/in.^{3}: 0.096
  Coefficient of thermal expansion, (70–200°F) in./in./ °F × 10^{-6}: 13.2

Modulus of elasticity, psi: tension, 10.3×10^6
Melting range, °F: 1,060–1,180
Specific heat, Btu/lb/°F, 70°F: 0.23
Thermal conductivity, Btu/ft^2/hr/in./°F, 70°F: 816
Electrical resistivity, ohms/cmil/ft, 70°F: 35.4

Heat treatments
  Annealing temperature 650°F

Tensile Properties

Temperature, °F	T.S., psi	Y.S., psi	Elong., in 2 in., %	Hardness, Brinell
		Annealed		
75	42,000	21,000	22	–
212	44,000	22,000	33	
300	30,000	19,000	50	
400	23,000	17,000	62	
500	18,000	12,000	74	
600	10,000	6,000	111	
		H113 Temper		
75	46,000	33,000	16	–
212	46,000	31,000	28	
300	32,000	23,000	43	
400	23,000	18,000	60	
500	18,000	9,000	63	
600	10,000	4,000	76	

Fatigue Strength

H113 Temper

Test temperature, °F	Stress, psi	Cycles to failure
75	23,000	5×10^8

Specifications

ASTM: GM41A; LK183

Table 1.3.1–18
WROUGHT ALUMINUM 5086

Typical chemical composition, percent by weight: Mn, 0.20–0.7; Fe, 0.50; Si, 0.40; Cr, 0.05–0.25; Ti, 0.15; Al, balance; Mg, 3.5–4.5; Cu, 0.10; Zn, 0.25

Physical constants and thermal properties
Density, lb/in.³ : 0.096
Coefficient of thermal expansion, (70–200°F) in./in./°F × 10⁻⁶ : 13.2

Modulus of elasticity, psi: tension, 10.3×10^6
Melting range, °F: 1,084–1,184
Specific heat, Btu/lb/°F, 70°F: 0.23
Thermal conductivity, Btu/ft² /hr/in./°F, 70°F: 876
Electrical resistivity, ohms/cmil/ft, 70°F: 33.0

Heat treatments
Annealing temperature 650°F.

Tensile Properties

Temperature, °F	T.S., psi	Y.S., psi	Elong., in 2 in., %	Hardness, Brinell
		Annealed		
75	38,000	17,000	22	–
200	40,000	18,000	27	
300	29,000	15,000	38	
400	24,000	15,000	51	
500	17,000	13,000	73	
600	11,000	8,000	86	
		H34 Temper		
75	–	–	–	
200	45,000	33,000	14	
300	33,000	26,000	28	
400	28,000	21,000	34	
500	18,000	13,000	62	
600	12,000	9,000	86	

Fatigue Strength

H34 Temper

Test temperature, °F	Stress, psi	Cycles to failure
75	16,000	5×10^8

Specifications

ASTM: GM40A; K186

Table 1.3.1–19
WROUGHT ALUMINUM 5090

Typical chemical composition, percent by weight: Fe, 0.50; Si, 0.40; Cr, 0.15–0.25; Al, balance; B, 0.005–0.015; Mg, 6.6–7.4

Physical constants and thermal properties
Density, $lb/in.^3$: 0.095
Coefficient of thermal expansion, (70–200°F) in./in./°F × 10^{-6}: 13.2

Modulus of elasticity, psi: tension, 10.3×10^6
Melting range, °F: 1,040–1,157
Specific heat, Btu/lb/°F, 70°F: 0.23
Thermal conductivity, $Btu/ft^2/hr/in./°F$, 70°F: 780
Electrical resistivity, ohms/cmil/ft, 70°F: 41

Heat treatments
Annealing temperature 650°F.

Tensile Properties

Temperature, °F	T.S., psi	Y.S., psi	Elong., in 2 in., %	Hardness, Brinell
		Annealed		
75	50,000	22,000	25	95
		H34 Temper		
75	59,000	40,000	15	121

Fatigue Strength

H32 Temper

Test temperature, °F	Stress, psi	Cycles to failure
75	19,000	5×10^8

Table 1.3.1–20
WROUGHT ALUMINUM 5154

Typical chemical composition, percent by weight: Mn, 0.10; Si + Fe, 0.45; Cr, 0.15–0.25; Ti, 0.20; Al, balance; B, 0.005–0.015; Mg, 3.1–3.9; Cu, 0.10; Zn, 0.20

Physical constants and thermal properties
Density lb/in.^{3} : 0.095
Coefficient of thermal expansion, (70–200°F) in./in./ °F × 10^{-6} : 13.3

Modulus of elasticity, psi: tension, 10.3 × 10^6
Melting range, °F: 1,100–1,190
Specific heat, Btu/lb/°F, 70°F: 0.23
Thermal conductivity, Btu/ft^2/hr/in./°F, 70°F: 876
Electrical resistivity, ohms/cmil/ft, 70°F: 31.8

Heat treatments
Annealing temperature 650°F.

Tensile Properties

Temperature, °F	T.S., psi	Y.S., psi	Elong., in 2 in., %	Hardness, Brinell
		Annealed		
75	35,000	17,000	27	58
212	35,000	17,000	30	
300	29,000	17,000	40	
400	22,000	14,000	55	
500	16,000	9,000	70	
600	10,000	6,000	100	
700	6,000	4,000	130	
		H34 Temper		
75	42,000	33,000	13	73
300	34,000	28,000	25	
400	25,000	16,000	35	
500	16,000	9,000	70	
600	10,000	6,000	100	
700	6,000	4,000	130	

Fatigue Strength

Test temperature, °F	Stress, psi	Cycles to failure
75	17,000	5 × 10^8 (annealed)
75	19,000	5 × 10^8 (H34)

Specifications

ASTM: GR40A; SAE-AMS: 208; A54S

Table 1.3.1−21
WROUGHT ALUMINUM 5252

Typical chemical composition, percent by weight: Mn,
0.10; Fe, 0.10; Si, 0.08; Al, balance; Mg, 2.2−2.8; Cu,
0.10

Physical constants and thermal properties
  Density, lb/in.³ : 0.097
  Coefficient of thermal expansion, (70−200°F) in./in./
    °F × 10⁻⁶ : 13.2
  Modulus of elasticity, psi: tension, 10 × 10⁶
  Melting range, °F: 1,100−1,200
  Thermal conductivity, Btu/ft² /hr/in./°F, 70°F: 960
  Electrical resistivity, ohms/cmil/ft, 70°F: 29.4

Tensile Properties

H25 Temper

Temperature, °F	T.S., psi	Y.S., psi	Elong., in 2 in., %	Hardness, Brinell
75	34,000	25,000	11	68

Table 1.3.1−22
WROUGHT ALUMINUM 5454

Typical chemical composition, percent by weight: Mn,
0.50−1,0; Si + Fe, 0.40; Cr, 0.05−0.20; Ti, 0.20; Al,
balance; Mg, 2.4−3.0; Cu, 0.10; Zn, 0.25

Physical constants and thermal properties
  Density, lb/in.³ : 0.097
  Coefficient of thermal expansion, (70−200°F) in./in./
    °F × 10⁻⁶ : 13.1

Modulus of elasticity, psi: tension, 10.2 × 10⁶
Melting range, °F: 1,115−1,195
Thermal conductivity, Btu/ft² /hr/in./°F, 70°F: 936
Electrical resistivity, ohms/cmil/ft, 70°F: 30.6

Heat treatments
  Annealing temperature 650°F.

Tensile Properties

Annealed

Temperature, °F	T.S., psi	Y.S., psi	Elong., in 2 in., %	Hardness, Brinell
75	36,000	17,000	22	62
212	40,000	24,000	45	
300	30,000	23,000	46	
400	24,000	20,000	44	
500	16,000	11,000	63	
600	10,000	6,500	100	

Specifications

ASTM: GM31A

Table 1.3.1–23
WROUGHT ALUMINUM 5456

Typical chemical composition, percent by weight: Mn, 0.5–1.0; Si + Fe, 0.40; Cr, 0.05–0.20; Ti, 0.20; Al, balance; Mg, 4.7–5.5; Cu, 0.10; Zn, 0.25

Physical constants and thermal properties
Density, lb/in.^{3} : 0.096
Coefficient of thermal expansion, (70–200°F) in./in./°F × 10^{-6} : 13.3

Modulus of elasticity, psi: tension, 10.3 × 10^6
Melting range, °F: 1,060–1,180
Specific heat, Btu/lb/°F, 70°F: 0.23
Thermal conductivity, Btu/ft^2/hr/in./°F, 70°F: 816
Electrical resistivity, ohms/cmil/ft, 70°F: 35.4

Heat treatments
Annealing temperature 775°F.

Tensile Properties

Annealed

Temperature, °F	T.S., psi	Y.S., psi	Elong., in 2 in., %	Hardness, Brinell
75	45,000	23,000	24	75

Specifications

ASTM: GM51A

Table 1.3.1–24
WROUGHT ALUMINUM 5457

Typical chemical composition, percent by weight: Mn, 0.15–0.45; Fe, 0.10; Si, 0.08; Al, balance; Mg, 0.8–1.2; Cu, 0.20

Physical constants and thermal properties
Density, lb/in.^{3} : 0.098
Coefficient of thermal expansion, (70–200°F) in./in./°F × 10^{-6} : 13.2

Modulus of elasticity, psi: tension, 10 × 10^6
Melting range, °F: 1,165–1,210
Thermal conductivity, Btu/ft^2/hr/in./°F, 70°F: 1,224
Electrical resistivity, ohms/cmil/ft, 70°F: 23.4

Heat treatments
Annealing temperature 650°F.

Tensile Properties

Annealed

Temperature, °F	T.S., psi	Y.S., psi	Elong., in 2 in., %	Hardness, Brinell
75	13,000	5,000	27	23

Specifications

SAE-AMS: 252

Table 1.3.1—25
WROUGHT ALUMINUM 6061

Typical chemical composition, percent by weight: Mn, 0.15; Fe, 0.7; Si, 0.4–0.8; Cr, 0.15–0.35; Ti, 0.15; Al, balance; Mg, 0.8–1.2; Cu, 0.15–0.40; Zn, 0.25

Physical constants and thermal properties
Density, lb/in.^{3} : 0.098
Coefficient of thermal expansion, (70–200°F) in./in./ °F × 10^{-6} : 13.0
Modulus of elasticity, psi: tension, 10.0 × 10^6

Melting range, °F: 1,080–1,200
Specific heat, Btu/lb/°F, 70°F: 0.23
Thermal conductivity, Btu/ft^2 /hr/in./°F, 70°F: 1,188
Electrical resistivity, ohms/cmil/ft, 70°F: 22.8

Heat treatments
Annealing temperature 775°F; solution temperature 970°F; aging temperature 320–350°F, 6–10 hr for T6.

Tensile Properties

Temperature, °F	T.S., psi	Y.S., psi, 0.2% offset	Elong., in 2 in., %	Hardness, Brinell
		Annealed		
75	18,000	8,000	30	30
300	16,000	8,000	30	
400	9,000	6,500	55	
500	5,500	4,000	70	
600	4,000	2,500	85	
700	3,000	2,000	95	
		T6 Temper		
75	45,000	40,000	17	95
212	41,000	40,000	16	
300	34,000	31,000	20	
400	18,000	15,000	28	
500	10,000	7,000	50	
600	5,000	4,000	100	
700	3,000	2,000	95	

Fatigue Strength

Test temperature, °F	Stress, psi	Cycles to failure
75	9,000	5 × 10^8 (annealed)
75	14,000	5 × 10^8 (T6)

Specifications

ASTM: GS11A; SAE-AMS: 281; 61S

Table 1.3.1−26
WROUGHT ALUMINUM 6063

Typical chemical composition, percent by weight: Mn, 0.10; Fe, 0.35; Si, 0.2−0.6; Cr, 0.10; Ti, 0.10; Al, balance; Mg, 0.45−0.9; Cu 0.10; Zn, 0.10

Modulus of elasticity, psi: tension, 10.0×10^6
Melting range, °F: 1,140−1,205
Specific heat, Btu/lb/°F, 70°F: 0.23
Thermal conductivity, Btu/ft^2/hr/in./°F, 70°F: 1,188
Electrical resistivity, ohms/cmil/ft, 70°F, 19.8

Physical constants and thermal properties
Density, lb/in.^{3}: 0.098
Coefficient of thermal expansion, (70−200°F) in./in./°F $\times 10^{-6}$: 13.0

Heat treatments
Annealing temperature 750°F; aging temperature 350−450°F, 6−8 hr for T6.

Tensile Properties

Temperature, °F	T.S., psi	Y.S., psi, 0.2% offset	Elong., in 2 in., %	Hardness, Brinell
Annealed				
75	13,000	7,000	−	−
T6 Temper				
75	35,000	31,000	12	73
212	33,000	31,000	15	
300	22,000	20,000	20	
400	9,000	6,000	49	
500	5,000	3,500	68	
600	2,500	2,000	107	
700	2,000	1,500	110	

Fatigue Strength

Test temperature, °F	Stress, psi	Cycles to failure
75	8,000	5×10^8 (annealed)
75	10,000	5×10^8 (T6)

Specifications

ASTM: GS10A; SAE-AMS: 212; 63S

Table 1.3.1–27
WROUGHT ALUMINUM 6066

Typical chemical composition, percent by weight: Mn, 0.6–1.1; Fe, 0.50; Si, 0.9–1.8; Cr, 0.40; Ti, 0.20; Al, balance; Cu, 0.7–1.2; Mg, 0.8–1.4; Zn, 0.25

Melting range, °F: 1,050–1,200
Thermal conductivity, Btu/ft²/hr/in./°F, 70°F: 1,008
Electrical resistivity, ohms/cmil/ft, 70°F: 28

Physical constants and thermal properties
Density, lb/in.³: 0.098
Coefficient of thermal expansion, (70–200°F) in./in./°F × 10⁻⁶: 12.9
Modulus of elasticity, psi: tension, 10.0 × 10⁶

Heat treatments
Annealing temperature 775°F; solution temperature 970°F; aging temperature 320–350°F, 6–10 hr for T6.

Tensile Properties

Temperature, °F	T.S., psi	Y.S., psi, 0.2% offset	Elong., in 2 in., %	Hardness, Brinell
		Annealed		
75	22,000	12,000	18	43
		T6 Temper		
75	57,000	52,000	12	120

Fatigue Strength

Heat Treated T6 Temper

Test temperature, °F	Stress, psi	Cycles to failure
75	16,000	5 × 10⁸

Specifications

ASTM: SG11B; 66S

Table 1.3.1–28
WROUGHT ALUMINUM 6101

Typical chemical composition, percent by weight: Mn, 0.03; Fe, 0.50; Si, 0.30–0.7; Cr, 0.03; Al, balance; Mg, 0.35–0.8; Cu, 0.10; Zn, 0.10

Modulus of elasticity, psi: tension, 10.0×10^6
Melting range, °F: 1,140–1,205
Thermal conductivity, Btu/ft²/hr/in./°F, 70°F: 1,500
Electrical resistivity, ohms/cmil/ft, 70°F: 18

Physical constants and thermal properties
Density, lb/in.³: 0.098
Coefficient of thermal expansion, (70–200°F) in./in./°F $\times 10^{-6}$: 13

Heat treatments
Annealing temperature 775 °F.

Tensile Properties

Heat Treated T6 Temper

Temperature, °F	T.S., psi	Y.S., psi, 0.2% offset	Elong., in 2 in., %	Hardness, Brinell
75	32,000	28,000	15	71

Specifications

ASTM: GS10B

Table 1.3.1–29
WROUGHT ALUMINUM 6262

Typical chemical composition, percent by weight: Mn, 0.15; Fe, 0.7; Si, 0.40–0.8; Cr, 0.04–0.14; Ti, 0.15; Al, balance; Cu, 0.15–0.40; Mg, 0.8–1.2; Pb, 0.6; Bi, 0.6; Zn, 0.25

Modulus of elasticity, psi: tension, 10.0×10^6
Melting range, °F: 1,140–1,205
Thermal conductivity, Btu/ft²/hr/in./°F, 70°F: 1,188
Electrical resistivity, ohms/cmil/ft, 70°F: 23.4

Physical constants and thermal properties
Density, lb/in.³: 0.098
Coefficient of thermal expansion, (70–200°F) in./in./°F $\times 10^{-6}$: 13

Heat treatments
Annealing temperature 775°F.

Tensile Properties

Heat Treated T9 Temper

Temperature, °F	T.S., psi	Y.S., psi, 0.2% offset	Elong., in 2 in., %	Hardness, Brinell
75	58,000	55,000	10	120

Fatigue Strength

Test temperature, °F	Stress, psi	Cycles to failure
75°F	13,000	5×10^8

Table 1.3.1–30
WROUGHT ALUMINUM 7039

Typical chemical composition, percent by weight: Mn,
0.10–0.40; Fe, 0.40; Si, 0.30; Cr, 0.15–0.25; Ti, 0.10;
Al, balance; Mg, 2.3–3.3; Zn, 3.5–4.5; Cu, 0.10

Physical constants and thermal properties
Density, lb/in.³: 0.099
Electrical resistivity, ohms/cmil/ft, 70°F: 29.4

Tensile Properties

Temperature, °F	T.S., psi	Y.S., psi, 0.2% offset	Elong., in 2 in., %	Hardness, Brinell
		Annealed		
75	32,000	–	–	–
		T61 Temper		
75	60,000	50,000	14	123

Table 1.3.1–31
WROUGHT ALUMINUM 7049

Typical chemical composition, percent by weight: Fe,
0.30; Si, 0.40; Cr, 0.15; Al, balance; Zn, 7.6; Mg, 2.5;
Cu, 1.5

Physical constants and thermal properties
Density, lb/in.³: 0.102
Coefficient of thermal expansion, (70–200°F)
in./in./°F × 10⁻⁶: 13.0

Modulus of elasticity, psi: tension, 10.2×10^6
Melting range, °F: 890–1,160
Specific heat, Btu/lb/°F, 70°F: 0.23
Electrical resistivity, ohms/cmil/ft, 70°F: 25.8

Heat treatments
Solution temperature 875°F; age at room
temperature, 250°F, and 325°F.

Tensile Properties

Heat Treated T73 Temper

Temperature, °F	T.S., psi	Y.S., psi, 0.2% offset	Elong., in 2 in., %	Hardness, Brinell
75	66,000–72,000	55,000–62,000	2–9	145

Fatigue Strength

Heat Treated T73 Temper

Test temperature, °F	Stress, psi	Cycles to failure
75	23,000	5×10^8

Table 1.3.1−32
WROUGHT ALUMINUM 7075

Typical chemical composition, percent by weight: Mn, 0.30; Fe, 0.70; Si, 0.50; Cr, 0.18−0.40; Ti, 0.20; Al, balance; Zn, 5.1−6.1; Mg, 2.1−2.9; Cu, 1.2−2.0

Physical constants and thermal properties
Density, lb/in.^{3}: 0.101
Coefficient of thermal expansion, (70−200°F) in./in./°F × 10^{-6}: 13.1

Modulus of elasticity, psi: tension, 10.4 × 10^{-6}
Melting range, °F: 890−1,180
Specific heat, Btu/lb/°F, 70°F: 0.23
Thermal conductivity, Btu/ft^2/hr/in./°F, 70°F: 840
Electrical resistivity, ohms/cmil/ft, 70°F: 34.2

Heat treatments
Annealing temperature 775°F; solution temperature 870°F; aging temperature 250°F, 24−28 hr for T6.

Tensile Properties

Temperature, °F	T.S., psi	Y.S., psi, 0.2% offset	Elong., in 2 in., %	Hardness, Brinell
		Annealed		
75	33,000	15,000	16	60
300	19,000	13,000	40	
400	14,000	11,000	60	
500	11,000	9,000	65	
600	8,500	6,500	75	
700	6,500	5,000	70	
		T6 Temper		
75	83,000	73,000	11	150
212	74,000	68,000	11	
300	30,000	26,000	30	
400	18,000	16,000	45	
500	13,000	11,000	52	
600	8,500	6,500	75	
700	6,500	5,000	70	

Fatigue Strength

Heat Treated T6 Temper

Test temperature, °F	Stress, psi	Cycles to failure
75	23,000	5 × 10^8

Specifications

ASTM: 2G62A; SAE-AMS: 215; 75S

Table 1.3.1–33
WROUGHT ALUMINUM 7079

Typical chemical composition, percent by weight: Mn, 0.10–0.30; Fe, 0.40; Si, 0.30; Cr, 0.10–0.25; Ti, 0.10; Al, balance; Zn, 3.8–4.8; Mg, 2.9–3.7; Cu, 0.40–0.8

Physical constants and thermal properties
Density, lb/in.^{3}: 0.099
Coefficient of thermal expansion, (70–200°F) in./in./°F × 10^{-6}: 13.1

Modulus of elasticity, psi: tension, 10.3 × 10^6
Melting range, °F: 900–1,180
Thermal conductivity, Btu/ft^2/hr/in./°F, 70°F: 840
Electrical resistivity, ohms/cmil/ft, 70°F: 33

Heat treatments
Annealing temperature 775°F; solution temperature 830°F; aging temperature 230–250°F, 46–50 hr for T6.

Tensile Properties

Heat Treated T6 Temper

Temperature, °F	T.S., psi	Y.S., psi, 0.2% offset	Elong., in 2 in., %	Hardness, Brinell
75	78,000	68,000	14	145
212	65,000	60,000	15	
300	34,000	30,000	38	
400	16,000	13,000	60	
500	11,000	8,500	100	
600	8,000	6,000	175	

Fatigue Strength

Heat Treated T6 Temper

Test temperature, °F	Stress, psi	Cycles to failure
75	23,000	5 × 10^8

Specifications

X79S

Table 1.3.1–34
WROUGHT ALUMINUM 7178

Typical chemical composition, percent by weight: Mn, 0.30; Fe, 0.70; Si, 0.50; Cr, 0.18–0.40; Al, balance; Zn, 6.3–7.3; Mg, 2.4–3.1; Cu, 1.6–2.4; Ti, 0.20

Modulus of elasticity, psi: tension, 10.4×10^6
Melting range, °F: 890–1,165
Thermal conductivity, $Btu/ft^2/hr/in./°F$, 70°F: 840
Electrical resistivity, ohms/cmil/ft, 70°F: 33

Physical constants and thermal properties
Density, $lb/in.^3$: 0.102
Coefficient of thermal expansion, (70–200°F) $in./in./°F \times 10^{-6}$: 13.0

Heat treatments
Annealing temperature 775°F; solution temperature 870°F; aging temperature 250–310°F, 24–28 hr for T6.

Tensile Properties

Temperature, °F	T.S., psi	Y.S., psi, 0.2% offset	Elong., in 2 in., %	Hardness, Brinell
		Annealed		
75	33,000	15,000	16	60
		T6 Temper		
75	88,000	78,000	10	160
300	32,000	29,000	36	

Specifications

XA78S

1.3.2 CAST ALUMINUM ALLOYS

Table 1.3.2–1
CAST ALUMINUM 201.0

Typical chemical composition, percent by weight: Cu, 4.7; Ag, 0.7; Al, balance

Physical constants and thermal properties
Density, lb/in.^{3} : 0.101
Coefficient of thermal expansion, (70–200°F) in./in./°F × 10^{-6} : 10.7
Melting range, °F: 1,200–1,060

Electrical conductivity (as cast), % IACS[a]: 27–32

[a]% IACS = percent of International Annealed Copper Standard, based on equal volume.

Heat treatments
Solution temperature 940–970 °F; aging temperature 305–315 °F.

Tensile Properties

Solution Treated and Aged

Temperature, °F	T.S., psi	Y.S., psi, 0.2% offset	Elong., in 2 in., %	Hardness, Brinell
75	65,000	55,000	9	110

Table 1.3.2–2
CAST ALUMINUM 208.0

Typical chemical composition, percent by weight: Si, 3.0; Al, balance; Cu, 4.0

Physical constants and thermal properties
Density, lb/in.^{3} : 0.101
Coefficient of thermal expansion, (70–200°F) in./in./°F × 10^{-6} : 12.2
Melting range, °F: 1,160–970
Specific Heat Btu/lb./°F, 70°F

Thermal conductivity, Btu/ft^2 /hr/in./°F, 70°F: 840
Electrical conductivity (as cast) % IACS[a]: 31

[a]% IACS = percent of International Annealed Copper Standard, based on equal volume.

Heat treatments
Annealing temperature 650°F.

Tensile Properties

As Cast

Temperature, °F	T.S., psi	Y.S., psi, 0.2% offset	Elong., in 2 in., %	Hardness, Brinell
75	21,000	14,000	2.5	55

Table 1.3.2–3
CAST ALUMINUM 222.0

Typical chemical composition, percent by weight: Cu, 4.0; Mg, 0.25; Al, balance

Physical constants and thermal properties
Density, lb/in.³: 0.107
Coefficient of thermal expansion, (70–200°F) in./in./°F × 10⁻⁶: 12.2
Melting range, °F: 1,155–965

Thermal conductivity, Btu/ft²/hr/in./°F, 70°F: 924
Electrical conductivity (as cast), % IACS[a]: 33

[a]% IACS = percent of International Annealed Copper Standard, based on equal volume.

Heat treatments
Annealing temperature 650°F; solution temperature 950°F; aging temperature 310°F.

Tensile Properties

Solution Treated and Aged

Form	Temperature, °F	T.S., psi	Y.S., psi 0.2% offset	Elong. in 2 in., %	Hardness, Brinell
Sand casting	75	40,000	20,000	0.5	115
Permanent mold casting	75	48,000	36,000	<0.5	140

Fatigue Strength

Solution Treated and Aged

Form	Test temperature, °F	Stress, psi	Cycles to failure
Sand	75	8,500	5×10^8
Mold	75	9,000	5×10^8

Table 1.3.2–4
CAST ALUMINUM 242.0

Typical chemical composition, percent by weight: Ni, 2.0; Al, balance; Cu, 4.0; Mg, 1.5

Physical constants and thermal properties
Density, lb/in.³: 0.102
Coefficient of thermal expansion, (70–200°F) in./in./°F × 10⁻⁶: 12.5
Melting range, °F: 1,175–990
Thermal conductivity, Btu/ft²/hr/in./°F, 70°F: 1,044

Electrical conductivity (as cast), % IACS[a]: 38.5

[a]% IACS = percent of International Annealed Copper Standard, based on equal volume.

Heat treatments
Annealing temperature 650°F; solution temperature 950°F; aging temperature 450°F.

Tensile Properties

Solution Treated and Aged

Form	Temperature, °F	T.S., psi	Y.S., psi, 0.2% offset	Elong., in 2 in., %	Hardness, Brinell
Sand casting	75	28,000	25,000	2.0	75
Permanent mold casting	75	47,000	42,000	0.5	110

Table 1.3.2—4 (continued)
CAST ALUMINUM 242.0

Fatigue Strength

Solution Treated and Aged

Test temperature, °F	Stress, psi	Cycles to failure
75	9,500	5×10^8

Table 1.3.2—5
CAST ALUMINUM 295.0

Typical chemical composition, percent by weight: Cu, 4.5; Al, balance

Physical constants and thermal properties
Density, lb/in.³: 0.101
Coefficient of thermal expansion, (70–200°F) in./in./°F × 10⁻⁶: 12.7
Melting range, °F: 1,190–970

Thermal conductivity, Btu/ft²/hr/in./°F, 70°F: 990
Electrical conductivity (as cast), % IACS[a]: 36

[a]% IACS = percent of International Annealed Copper Standard, based on equal volume.

Heat treatments
Annealing temperature 650°F; solution temperature 960°F; aging temperature 310°F.

Tensile Properties

Solution Treated and Aged

Temperature, °F	T.S., psi	Y.S., psi, 0.2% offset	Elong., in 2 in., %	Hardness, Brinell
75	36,000	24,000	5.0	75

Fatigue Strength

Solution Treated and Aged

Test temperature, °F	Stress, psi	Cycles to failure
75	8,000	5×10^8

Table 1.3.2—6
CAST ALUMINUM B295.0

Typical chemical composition, percent by weight: Si, 2.5; Al, balance; Cu, 4.5

Physical constants and thermal properties
Density, lb/in.³: 0.100
Coefficient of thermal expansion, (70–200°F) in./in./°F × 10⁻⁶: 12.2
Melting range, °F: 1,190–970
Thermal conductivity, Btu/ft²/hr/in./°F, 70°F: 1,110

Electrical conductivity (as cast), % IACS[a]: 42.5

[a]% IACS = percent of International Annealed Copper Standard, based on equal volume.

Heat treatments
Annealing temperature 650°F; solution temperature 950.°F; aging temperature 310°F.

Table 1.3.2–6 (continued)
CAST ALUMINUM B295.0

Tensile Properties

Solution Treated and Aged

Temperature, °F	T.S., psi	Y.S., psi, 0.2% offset	Elong., in 2 in., %	Hardness, Brinell
75	40,000	26,000	5.0	90

Fatigue Strength

Solution Treated and Aged

Test temperature, °F	Stress, psi	Cycles to failure
75	10,000	5×10^8

Table 1.3.2–7
CAST ALUMINUM 308.0

Typical chemical composition, percent by weight: Si, 5.5; Al, balance; Cu, 4.5

Physical constants and thermal properties
Density, lb/in.^{3}: 0.101
Coefficient of thermal expansion, (70–200°F) in./in./°F $\times 10^{-6}$: 11.9
Melting range, °F: 1,135–970

Thermal conductivity, Btu/ft^2/hr/in./°F, 70°F: 984
Electrical conductivity (as cast), % IACS[a]: 37

[a]% IACS = percent of International Annealed Copper Standard, based on equal volume.

Heat treatments
Annealing temperature 650°F.

Tensile Properties

As Cast

Temperature, °F	T.S., psi	Y.S., psi, 0.2% offset	Elong., in 2 in., %	Hardness, Brinell
75	28,000	16,000	2.0	70

Table 1.3.2—8
CAST ALUMINUM 319.0

Typical chemical composition, percent by weight: Mn, 0.10; Fe, 0.6; Si, 5.5—7.0; Ni, 0.10; Ti, 0.20; Al, balance; Cu, 3.0—4.5; Mg, 0.10; Zn, 0.10

Physical constants and thermal properties
Density, lb/in.^{3}: 0.100
Coefficient of thermal expansion, (70—200°F) in./in./°F × 10^{-6}: 12.0
Melting range, °F: 1,120—950

Thermal conductivity, Btu/ft^2/hr/in./°F, 70°F: 792
Electrical conductivity (as cast), % IACS[a]: 28

[a]% IACS = percent of International Annealed Copper Standard, based on equal volume.

Heat treatments
Annealing temperature 650°F; solution temperature 940°F; aging temperature 310°F.

Tensile Properties

Form	Temperature, °F	T.S., psi	Y.S., psi, 0.2% offset	Elong., in 2 in., %	Hardness, Brinell
As Cast					
Sand casting	75	27,000	18,000	2.0	70
Permanent mold casting	75	34,000	19,000	2.5	85
Solution Treated and Aged					
Sand casting	75	36,000	24,000	2.0	80
Permanent mold casting	75	40,000	27,000	3.0	95

Table 1.3.2—9
CAST ALUMINUM 355.0

Typical chemical composition, percent by weight: Mn, 0.05 Fe, 0.16—0.30; Si, 4.5—5.5; Ti, 0.20; Al, balance; Cu, 1.0—1.5; Mg, 0.45—0.6; Zn, 0.05

Physical constants and thermal properties
Density, lb/in.^{3}: 0.098
Coefficient of thermal expansion, (70—200°F) in./in./°F × 10^{-6}: 12.4
Melting range, °F: 1,150—1,015

Thermal conductivity, Btu/ft^2/hr/in./°F, 70°F: 1,044
Electrical conductivity (as cast), % IACS[a]: 37

[a]% IACS = percent of International Annealed Copper Standard, based on equal volume.

Heat treatments
Annealing temperature 650°F; solution temperature 980°F; aging temperature 310°F.

Table 1.3.2—9 (continued)
CAST ALUMINUM 355.0

Tensile Properties

Form	Temperature, °F	T.S., psi	Y.S., psi, 0.2% offset	Elong., in 2 in., %	Hardness, Brinell
Solution Treated and Aged					
Sand casting	75	35,000	25,000	3.0	80
Permanent mold casting	75	42,000	27,000	4.0	90
355 T6 Temper					
Sand casting	75	35,000	25,000	3	
	300	30,000	25,000	1.5	
	400	13,000	9,000	12	
	500	8,000	5,000	22	
	600	6,000	3,500	30	
355 T6 Temper					
Permanent mold casting	75	43,000	27,000	4	
	300	32,000	24,000	2	
	400	12,000	9,000	20	
	500	8,000	6,000	25	
	600	4,500	3,000	50	

Fatigue Strength

Solution Treated and Aged

Form	Test temperature, °F	Stress, psi	Cycles to failure
Sand casting	75	9,000	5×10^8
Permanent mold casting	75	10,000	5×10^8

Table 1.3.2−10
CAST ALUMINUM C355.0

Typical chemical composition, percent by weight: Mn, 0.05; Fe, 0.15; Si, 4.5−5.5; Ti, 0.20; Al, balance; Cu, 1.0−1.5; Mg, 0.45−0.6; Zn, 0.05

Physical constants and thermal properties
Density, lb/in.^{3}: 0.097
Coefficient of thermal expansion, (70−200°F) in./in./°F × 10^{-6}: 12.4
Melting range, °F: 1,150−1,015

Thermal conductivity, Btu/ft^2/hr/in./°F, 70°F: 984
Electrical conductivity (as cast), % IACS[a]: 37

[a]% IACS = percent of International Annealed Copper Standard, based on equal volume.

Heat treatments
Annealing temperature 650°F; solution temperature 980°F; aging temperature 310°F.

Tensile Properties

Solution Treated and Aged

Temperature, °F	T.S., psi	Y.S., psi, 0.2% offset	Elong., in 2 in., %	Hardness, Brinell
75	46,000	34,000	6.0	100

Table 1.3.2−11
CAST ALUMINUM 356.0

Typical chemical composition, percent by weight: Mn, 0.05; Fe, 0.13−0.30; Si, 6.5−7.5; Ti, 0.20; Al, balance; Mg, 0.25−0.40; Cn, 0.10; Zn, 0.05

Physical constants and thermal properties
Density, lb/in.^{3}: 0.097
Coefficient of thermal expansion, (70−200°F) in./in.°F × 10^{-6}: 11.9
Melting range, °F: 1,135−1,035

Thermal conductivity, Btu/ft^2/hr/in./°F, 70°F: 1,104
Electrical conductivity (as cast), % IACS[a]: 41

[a]% IACS = percent of International Annealed Copper Standard, based on equal volume.

Heat treatments
Annealing temperature 650°F; solution temperature 1,000°F; aging temperature 310°F.

Tensile Properties

Form	Temperature, °F	T.S., psi	Y.S., psi, 0.2% offset	Elong., in 2 in., %	Hardness, Brinell
Solution Treated and Aged					
Sand casting	75	25,000	20,000	2.0	60
Permanent mold casting	75	38,000	27,000	5.0	80
356 T6 Temper					
Sand casting	75	33,000	24,000	3.5	
	300	21,000	16,000	5	
	400	13,000	9,000	8	
	500	8,000	5,500	20	
	600	6,000	3,000	45	
356 T6 Temper					
Permanent mold casting	75	40,000	27,000	5	
	300	19,000	17,000	5	
	400	11,000	9,000	10	
	500	6,500	5,000	40	
	600	4,500	3,500	50	

Table 1.3.2–12
CAST ALUMINUM A356.0

Typical chemical composition, percent by weight: Mn, 0.05; Fe, 0.12; Si, 6.5–7.5; Ti, 0.20; Al, balance; Mg, 0.25–0.40; Cu, 0.10; Zn, 0.05

Physical constants and thermal properties
Density, lb/in.^{3}: 0.097
Coefficient of thermal expansion, (70–200°F) in./in./°F × 10^{-6}: 11.9
Melting range, °F: 1,135–1,035

Thermal conductivity, Btu/ft^2/hr/in./°F, 70°F: 1,104
Electrical conductivity (as cast), % IACS[a]: 41

[a]% IACS= percent of International Annealed Copper Standard, based on equal volume.

Heat treatments
Annealing temperature 650°F; solution temperature 1,000°F; aging temperature 310°F.

Tensile Properties

Solution Treated and Aged

Temperature, °F	T.S., psi	Y.S., psi, 0.2% offset	Elong., in 2 in., %	Hardness, Brinell
75	41,000	30,000	10.0	90

Table 1.3.2–13
CAST ALUMINUM A380.0

Typical chemical composition, percent by weight: Mn, 0.10; Fe, 0.6; Si, 7.5–9.5; Ni, 0.10; Al, balance; Cu, 3.0–4.0; Mg, 0.05; Zn, 0.10

Physical constants and thermal properties
Density, lb/in.^{3}: 0.097
Coefficient of thermal expansion, (70–200°F) in./in./°F × 10^{-6}: 11.7

Melting range, °F: 1,100–1,000
Thermal conductivity, Btu/ft^2/hr/in./°F, 70°F: 696
Electrical conductivity (as cast), % IACS[a]: 25

[a]% IACS = percent of International Annealed Copper Standard, based on equal volume.

Tensile Properties

As Cast

Temperature, °F	T.S., psi	Y.S., psi, 0.2% offset	Elong., in 2 in., %	Hardness, Brinell
75	47,000	23,000	4	80

Table 1.3.2–14
CAST ALUMINUM A390.0

Typical chemical composition, percent by weight: Si, 17.0; Al, balance; Cu, 4.5; Mg, 0.55

Physical constants and thermal properties
Density, lb/in.^{3}: 0.099
Coefficient of thermal expansion, (70–200°F) in./in./°F × 10^{-6}: 10.3
Melting range, °F: 1,200–945

Electrical conductivity (as cast), % IACS[a]: 20

[a]% IACS = percent of International Annealed Copper Standard, based on equal volume.

Heat treatments
Annealing temperature 450°F (T5 temper); solution temperature 935°F.

Table 1.3.2—14 (continued)
CAST ALUMINUM A390.0

Tensile Properties

Form	Temperature, °F	T.S., psi	Y.S., psi, 0.2% offset	Elong., in 2 in., %	Hardness, Brinell
As Cast					
Sand cast	75	26,000	26,000	<1	100
Permanent mold casting	75	29,000	29,000	<1	110
Die casting	75	40,500	35,000	1	120
Solution Treated and Aged					
Sand cast		40,000	40,000	<1	140
Permanent mold casting		45,000	45,000	<1	145

Table 1.3.2—15
CAST ALUMINUM A393.0

Typical chemical composition, percent by weight: Fe, 1.3; Si, 22.0; Al, balance; Cu, 0.9; Mg, 1.0

Melting range, °F: 1,290—1,040

Physical constants and thermal properties
Density, lb/in.^{3}: 0.096
Coefficient of thermal expansion, (70—200°F) in./in./°F × 10^{-6}: 9.0

Heat treatments
Annealing temperature 650°F; solution temperature 960°F (T6 temper); aging temperature 340°F (T5 temper).

Tensile Properties

Solution Treated and Aged

Temperature, °F	T.S., psi	Y.S., psi, 0.2% offset	Elong., in 2 in., %	Hardness, Brinell
75	40,000	Low ductility	<0.50	130

Table 1.3.2—16
CAST ALUMINUM A413.0

Typical chemical composition, percent by weight: Fe, 1.3; Si, 12.0; Al, balance; Cu, 0.6; Mg, 0.10

Melting range, °F: 1,085—1,065
Thermal conductivity, Btu/ft^2/hr/in./°F, 70°F: 840
Electrical conductivity (as cast), % IACS[a]: 31

Physical constants and thermal properties
Density, lb/in.^{3}: 0.096
Coefficient of thermal expansion, (70—200°F) in./in./°F × 10^{-6}: 11.4

[a]% IACS = percent of International Annealed Copper Standard, based on equal volume.

Tensile Properties

As Cast

Temperature, °F	T.S., psi	Y.S., psi, 0.2% offset	Elong., in 2 in., %	Hardness, Brinell
75	35,000	16,000	3.5	80

Table 1.3.2–17
CAST ALUMINUM 443.0

Typical chemical composition, percent by weight: Fe, 0.8; Si, 5.2; Al, balance; Cu, 0.6

Physical constants and thermal properties
Density, lb/in.^{3}: 0.097
Coefficient of thermal expansion, (70–200°F) in./in./°F $\times 10^{-6}$: 12.2
Melting range, °F: 1,175–1,070

Thermal conductivity, Btu/ft^2/hr/in./°F, 70°F: 1,008
Electrical conductivity (as cast), % IACS[a]: 38

[a] % IACS = percent of International Annealed Copper Standard, based on equal volume.

Heat treatments
Annealing temperature 650°F.

Tensile Properties

As Cast

Form	Temperature, °F	T.S., psi	Y.S., psi, 0.2% offset	Elong., in 2 in., %	Hardness, Brinell
Sand casting	75	19,000	8,000	8	40
Permanent mold casting	75	23,000	9,000	10	45

Table 1.3.2–18
CAST ALUMINUM B443.0

Typical chemical composition, percent by weight: Si, 5.25; Al, balance; Cu, 0.15 maximum

Physical constants and thermal properties
Density, lb/in.^{3}: 0.097
Coefficient of thermal expansion, (70–200°F) in./in./°F $\times 10^{-6}$: 12.2
Melting range, °F: 1,170–1,070

Thermal conductivity, Btu/ft^2/hr/in./°F, 70°F: 1,020
Electrical conductivity (as cast), % IACS[a]: 38

[a] % IACS = percent of International Annealed Copper Standard, based on equal volume.

Heat treatments
Annealing temperature 650°F.

Tensile Properties

As Cast

Temperature, °F	T.S., psi	Y.S., psi, 0.2% offset	Elong., in 2 in., %	Hardness, Brinell
75	21,000	–	5	–

Table 1.3.2—19
CAST ALUMINUM 514.0

Typical chemical composition, percent by weight: Mg, 4.0; Al, balance

Physical constants and thermal properties
Density, lb/in.^{3} : 0.096
Coefficient of thermal expansion, (70—200°F) in./in./°F × 10^{-6} : 13.3
Melting range, °F: 1,185—1,110

Thermal conductivity, Btu/ft^2/hr/in./°F, 70°F: 960
Electrical conductivity (as cast), % IACS[a]: 35

[a]% IACS = percent of International Annealed Copper Standard, based on equal volume.

Heat treatments
Annealing temperature 650°F.

Tensile Properties

As Cast

Temperature, °F	T.S., psi	Y.S., psi, 0.2% offset	Elong., in 2 in., %	Hardness, Brinell
75	25,000	12,000	9	50

Table 1.3.2—20
CAST ALUMINUM 518.0

Typical chemical composition, percent by weight: Fe, 1.8; Al, balance; Mg, 8.0

Physical constants and thermal properties
Density, lb/in.^{3} : 0.093
Coefficient of thermal expansion, (70—200°F) in./in./°F × 10^{-6} : 13.3

Melting range, °F: 1,150—995
Thermal conductivity, Btu/ft^2/hr/in./°F, 70°F: 672
Electrical conductivity (as cast), % IACS[a]: 24

[a]% IACS = percent of International Annealed Copper Standard, based on equal volume.

Tensile Properties

As Cast

Temperature, °F	T.S., psi	Y.S., psi, 0.2% offset	Elong., in 2 in., %	Hardness, Brinell
75	45,000	27,000	8	80

Table 1.3.2–21
CAST ALUMINUM 520.0

Typical chemical composition, percent by weight: Mg, 10.0; Al, balance

Physical constants and thermal properties
Density, lb/in.^{3}: 0.093
Coefficient of thermal expansion, (70–200°F) in./in./°F × 10^{-6}: 13.6
Melting range, °F: 1,120–840
Thermal conductivity, Btu/ft^2/hr/in./°F, 70°F: 612

Electrical conductivity (as cast), % IACS[a]: 21

[a]% IACS = percent of International Annealed Copper Standard, based on equal volume.

Heat treatments
Annealing temperature 650°F; solution temperature 810°F.

Tensile Properties

Solution Treated and Aged

Temperature, °F	T.S., psi	Y.S., psi, 0.2% offset	Elong., in 2 in., %	Hardness, Brinell
75	48,000	26,000	16	75

Table 1.3.2–22
CAST ALUMINUM D712.0

Typical chemical composition, percent by weight: Cr, 0.5; Ti, 0.15; Al, balance; Zn, 5.5; Mg, 0.6

Physical constants and thermal properties
Density, lb/in.^{3}: 0.100
Coefficient of thermal expansion, (70–200°F) in./in./°F × 10^{-6}: 13.7
Melting range, °F: 1,140–1,060

Thermal conductivity, Btu/ft^2/hr/in./°F, 70°F: 960
Electrical conductivity (as cast), % IACS[a]: 25

[a]% IACS = percent of International Annealed Copper Standard, based on equal volume.

Heat treatments
Aging temperature 350°F.

Tensile Properties

As Cast

Temperature, °F	T.S., psi	Y.S., psi, 0.2% offset	Elong., in 2 in., %	Hardness, Brinell
75	34,000	25,000	4	–

1.4 NICKEL-BASE ALLOYS

E. B. Fernsler
Huntington Alloy Products Division
The International Nickel Company

Nickel, in substantially pure form, is widely used in industry for its corrosion resistance combined with moderate mechanical strength and good thermal and electrical conductivity. The principal high nickel alloy contains approximately 99.5% nickel and small amounts of manganese, magnesium, and carbon. A wide variety of nickel alloys with widely differing properties are commercially available. These have been listed in the following six categories:

Nickel alloys
Nickel-copper alloys
Nickel-chromium alloys
Nickel-molybdenum alloys

Nickel-iron-chromium alloys
High temperature-high strength alloys

Alloys containing a high concentration of nickel in a matrix containing a larger amount of some other constituent are listed under the major element. An example of this is an austenitic stainless steel containing a higher concentration of iron than of nickel, and thus listed in Section 1.1. TD-nickel and TD-NiCr were not included because they are not currently commercially available. The new mechanically alloyed Inconel®MA753 alloy which is both dispersion and precipitation hardened has been included.

1.4.1 NICKEL ALLOYS

Table 1.4.1−1
NICKEL 200

Chemical composition, percent by weight: C, 0.08; Mn, 0.18; Fe, 0.20; S, 0.005; Si, 0.18; Ni, 99.5; Cu, 0.13

Physical constants and thermal properties
Density, lb/in.^{3}: 0.321
Coefficient of thermal expansion, (70−200°F) in./in./°F × 10^{-6}: 7.4
Modulus of elasticity, psi (dynamic): tension, 29.6 × 10^6; torsion, 11.7 × 10^6
Poisson's ratio: 0.264
Melting range, °F: 2,615−2,635

Specific heat, Btu/lb/°F, 70°F: 0.109
Thermal conductivity, Btu/ft²/hr/in./°F, 70°F: 520
Electrical resistivity, ohms/cmil/ft, 70°F: 57
Curie temperature, °F: annealed, 680
Permeability (70°F, 200 Oe): annealed − ferromagnetic

Heat treatments
Generally used in annealed (1,300−1,700°F) condition.

Tensile Properties

Annealed

Temperature, °F	T.S., psi	Y.S., psi, 0.2% offset	Elong., in 2 in., %	Hardness, Brinell
70	67,000	21,500	47.0	—
200	66,500	22,300	46.0	—
400	66,500	20,200	44.0	—
600	66,200	20,200	47.0	—
800	44,000	16,500	65.0	—
1,000	31,500	13,500	69.0	—
1,200	21,500	10,000	76.0	—
1,400	14,000	7,000	89.0	—
1,600	8,200	3,600	110.0	—
1,800	5,400	2,300	198.0	—
2,000	3,500	1,400	205.0	—

Table 1.4.1−1 (continued)
NICKEL 200

Creep Strength
(Stress, psi, to Produce 1% Creep)

Annealed

Test temperature, °F	10,000 hr
600	40,000
700	13,000
800	6,000

Fatigue Strength

Annealed

Test temperature, °F	Stress, psi	Cycles to failure
70	33,000	10^8

Impact Strength

Annealed

Test temperature, °F	Type test	Strength, ft-lb
+70	Charpy − V-notched	228

Specifications

ASTM: B160, B161, B162, B163

Producers and Tradenames

Huntington Alloy Products Division, INCO: Nickel 200
G. O. Carlson: Carlson 200
Industrial Stainless: N-200

Table 1.4.1−2
NICKEL 201

Chemical composition, percent by weight: C, 0.01; Mn, 0.18; Fe, 0.20; S, 0.005; Si, 0.18; Ni, 99.5; Cu, 0.13

Physical constants and thermal properties
 Density, lb/in.^{3}: 0.321
 Coefficient of thermal expansion, (70−200°F) in./in./°F × 10^{-6}: 7.4
 Modulus of elasticity, psi: tension, 30 × 10^6
 Specific heat, Btu/lb/°F, 70°F: 0.109

Thermal conductivity, Btu/ft^2/hr/in./°F, 70°F: 549
Electrical resistivity, ohms/cmil/ft, 70°F: 51
Curie temperature, °F: annealed, 680
Permeability (70°F, 200 Oe): annealed − ferromagnetic

Heat treatments
 Generally used in annealed (1,200−1,600°F) condition.

Tensile Properties

Annealed

Temperature, °F	T.S., psi	Y.S., psi, 0.2% offset	Elong., in 2 in., %	Hardness, Brinell
70	58,500	15,000	50	−
200	56,100	15,400	45	−
400	54,000	14,800	44	−
600	52,500	15,300	42	−
800	41,200	13,500	58	−
1,000	33,100	12,100	60	−
1,200	22,200	10,200	74	−

Rupture Strength, 1,000 hr

Annealed

Test temperature, °F	Strength, psi	Elong., in 2 in., %	Reduction of area, %
1,000	7,900	−	−
1,100	11,000	−	−
1,200	15,000	−	−

Creep Strength (Stress, psi, to Produce 1% Creep)

Annealed

Test temperature, °F	10,000 hr
1,000	5,000
1,100	3,800
1,200	2,050
1,300	1,000

Specifications

ASTM: B160, B161, B162, B163; SAE-AMS: AMS 5553

Producers and Tradenames

Huntington Alloy Products Division, INCO: Nickel 201

Table 1.4.1—3
NICKEL 205

Chemical composition, percent by weight: C. 0.08; Mn, 0.18; Fe, 0.10; S, 0.004; Si, 0.08; Ni, 99.5; Cu, 0.08; Ti, 0.03; Mg, 0.05

Physical constants and thermal properties
See Nickel 200

Heat treatments
All other properties same as Nickel 200.

Creep Strength

See Nickel 200

Fatigue Strength

See Nickel 200

Impact Strength

See Nickel 200

Specifications

ASTM: B162, F9

Producers and Tradenames

Huntington Alloy Products Division, INCO: Nickel 205

Table 1.4.1—4
NICKEL 270

Chemical composition, percent by weight: C, 0.01; Fe, 0.003; Ni, 99.98

Physical constants and thermal properties
Density, lb/in.^{3}: 0.321
Coefficient of thermal expansion, (70—200°F) in./in./°F $\times$ 10^{-6}: 7.4
Modulus of elasticity, psi: tension, 30 $\times$ 10^6
Melting range, °F: 2,650

Specific heat, Btu/lb/°F, 70°F: 0.11
Thermal conductivity, Btu/ft^2/hr/in./°F, 70°F: 599
Electrical resistivity, ohms/cmil/ft, 70°F: 45
Curie temperature, °F: annealed, 676
Permeability (70°F, 200 Oe): annealed — ferromagnetic

Heat treatments
Usually used in annealed condition (800—1,000°F).

Tensile Properties

Annealed

Temperature, °F	T.S., psi	Y.S., psi 0.2% offset	Elong., in 2 in., %	Hardness, Brinell
70	50,000	16,000	50	80

Producers and Tradenames

Huntington Alloy Products Division, INCO: Nickel 270

Table 1.4.1–5
PERMANICKEL® ALLOY 300

Chemical composition, percent by weight: C, 0.20; Mn, 0.25; Fe, 0.30; S, 0.005; Si, 0.18; Ni, 98.5; Cu, 0.13; Ti, 0.40; Mg, 0.35

Physical constants and thermal properties
  Density, lb/in.^{3}: 0.316
  Coefficient of thermal expansion, (70–200°F) in./in./°F × 10^{-6}: 6.8
  Modulus of elasticity, psi: tension, 30 × 10^5; torsion, 11 × 10^6
  Specific heat, Btu/lb/°F, 70°F: 0.106

Thermal conductivity, Btu/ft^2/hr ft, in./°F, 70°F: 400
Electrical resistivity, ohms/cmil/ft, 70°F: 95
Curie temperature, °F: annealed, 600; age hardened, 563
Permeability (70°F, 200 Oe): annealed − 26; age hardened − 26

Heat treatments
Solution anneal 2,000°F, water quench. Age harden 16 hr at 910–930°F.

Tensile Properties

Annealed and Aged

Temperature, °F	T.S., psi	Y.S., psi, 0.2% offset	Elong., in 2 in., %	Hardness, Brinell
70	176,000	140,000	16	–
200	170,000	136,000	18	–
400	172,000	135,000	20	–
600	166,000	136,000	17	–

Fatigue Strength

Full Hard Plus Aged

Test temperature, °F	Stress, psi	Cycles to failure
70	37,000	10^8

Specifications

ASTM: F-290

Producers and Tradenames

Huntington Alloy Products Division, INCO: Permanickel alloy 300

Table 1.4.1−6
DURANICKEL® ALLOY 301

Chemical composition, percent by weight: C, 0.15; Mn, 0.25; Fe, 0.30; S, 0.005; Si, 0.50; Ni, 96.5; Cu, 0.13; Ti, 0.63; Al, 4.38

Physical constants and thermal properties
Density, lb/in.³: 0.298
Coefficient of thermal expansion, (70−200°F) in./in./°F × 10⁻⁶: 7.2
Modulus of elasticity, psi: tension, 30 × 10⁶; torsion, 11 × 10⁶
Poisson's ratio: 0.31
Melting range, °F: 2,550−2,620

Specific heat, Btu/lb/°F, 70°F: 0.104
Thermal conductivity, Btu/ft²/hr/in./°F, 70°F: 165
Electrical resistivity, ohms/cmil/ft, 70°F: 255
Curie temperature, °F: annealed, 60−120; age hardened, 200
Permeability (70°F, 200 Oe): annealed, 4.28; age hardened, 10.58

Heat treatments
Anneal 1,600−1,800°F. Age 16 hr at 1,080−1,100° plus furnace cool to 900° at rate of not more than 15° per hour. Air cool or quench.

Tensile Properties

Hot Rolled and Aged

Temperature, °F	T.S., psi	Y.S., psi, 0.2% offset	Elong., in 2 in., %	Hardness, Brinell
70	185,000	132,000	28	−
600	168,000	120,000	29	−
800	155,000	114,000	24	−
1,000	118,000	99,000	7	−
1,200	69,000	54,000	4	−
1,400	25,000	14,000	60	−

Fatigue Strength

Cold Drawn and Aged

Test temperature, °F	Stress, psi	Cycles to failure
70	55,400	10^8

Producers and Tradenames

Huntington Alloy Products Division, INCO: Duranickel alloy 301

Table 1.4.1—7
BERYLCO® NICKEL 440

Chemical composition, percent by weight: Ni, balance; Ti, 0.50; Be, 1.95

Physical constants and thermal properties
Density, lb/in.³: 0.302
Coefficient of thermal expansion, (70–200°F) in./ in./°F × 10⁻⁶: 8.0

Modulus of elasticity, psi: tension, 30 × 10⁶
Poisson's ratio: 0.295
Thermal conductivity, Btu/ft²/hr/in./°F, 70°F: 18.3
Electrical resistivity, ohms/cmil/ft, 70°F: 144

Heat treatments
Anneal at 1,825°F. Age 1½ hr at 950°F.

Tensile Properties

Annealed and Aged

Temperature, °F	T.S., psi	Y.S., psi, 0.2% offset	Elong., in 2 in., %	Hardness, Brinell
70	220,000	156,000	19	–
200	225,000	154,000	18	–
400	225,000	152,000	17	–
600	215,000	150,000	15	–
800	190,000	145,000	5	–
1,000	135,000	113,000	2	–
1,200	55,000	35,000	22	–

Producers and Tradenames

Beryllium Corporation: Berylco Nickel 440
Cast beryllium-nickel alloys are also available from
Brush Beryllium Co., Cleveland, Ohio.

Table 1.4.1—8
NICKEL 210 (ACI CZ-100)

Chemical composition, percent by weight: C, 0.80; Mn, 0.90; Fe, 0.50; S, 0.008; Si, 1.6; Ni, 95.6

Physical constants and thermal properties
Density, lb/in.³: 0.301
Coefficient of thermal expansion, (70–200°F) in./ in./°F × 10⁻⁶: 7.5

Modulus of elasticity, psi: tension, 21.5 × 10⁶
Melting range, °F: 2,450–2,600
Specific heat, Btu/lb/°F, 70°F: 0.13
Thermal conductivity, Btu/ft²/hr/in./°F, 70°F: 410
Electrical resistivity, ohms/cmil/ft, 70°F: 125

Heat treatments
As cast.

Tensile Properties

As Cast

Temperature, °F	T.S., psi	Y.S., psi, 0.2% offset	Elong., in 2 in., %	Hardness, Brinell
70	58,000	–	22	110

Table 1.4.1—8 (continued)
NICKEL 210 (ACI CZ-100)

Specifications
ASTM: A-296 Grade CZ-100

Producers and Tradenames

Aloyco, Inc.: CZ-100
Esco Corp.: Esco® 21
The Hica Corp.: Alloy 5
Lebanon Steel Foundry (L)Nickel
Quaker Alloy Casting Co.: Q Ni

Table 1.4.1—9
HASTELLOY® D

Chemical composition, percent by weight: C, 0.12 maximum; Mn, 0.87; Fe, 2.0 maximum; Si, 9.25; Cr, 1.00 maximum; Ni, balance; Cu, 3.00; Co, 1.50 maximum

Physical constants and thermal properties
Density, lb/in.^{3}: 0.282
Coefficient of thermal expansion, (70–200°F) in./in./°F $\times$ 10^{-6}: 6.1

Modulus of elasticity, psi: tension, 28.9 $\times$ 10^6
Melting range, °F: 2,065–2,220
Specific heat, Btu/lb/°F, 70°F: 0.109
Thermal conductivity, Btu/ft^2/hr/in./°F, 70°F: 145
Electrical resistivity, ohms/cmil/ft, 70°F: 679

Heat treatments
Solution heat treat 1,800–1,850°F, furnace cool.

Tensile Properties

Cast

Temperature, °F	T.S., psi	Y.S., psi 0.2% offset	Elong., in 2 in., %	Hardness, Brinell
70	115,000	–	1	–

Impact Strength

Cast

Test temperature, °F	Type test	Strength, ft-lb
+70	Izod	1–2

Producers and Tradenames

Stellite Division of Cabot Corp.: Hastelloy alloy D

1.4.2 NICKEL-COPPER ALLOYS

Table 1.4.2-1
MONEL® ALLOY 400

Chemical composition, percent by weight: C, 0.15; Mn, 1.00; Fe, 1.25; S, 0.012; Si, 0.25; Ni, 66.5; Cu, 31.5

Physical constants and thermal properties
Density, lb/in.³ :0.319
Coefficient of thermal expansion, (70–200°F) in./in./°F × 10⁻⁶ : 7.7
Modulus of elasticity, psi: tension, 26.0 × 10⁶; torsion, 9.5 × 10⁶

Poisson's ratio: 0.32
Melting range, °F: 2,370–2,460
Specific heat, Btu/lb/°F, 70°F: 0.105
Thermal conductivity, Btu/ft² /hr/in./°F, 70°F: 151
Electrical resistivity, ohms/cmil/ft,70°F: 307
Curie temperature, °F: annealed, 20–50

Heat treatments
Generally used in annealed condition.

Tensile Properties

Annealed

Temperature, °F	T.S., psi	Y.S., psi, 0.2% offset	Elong., in 2 in., %	Hardness, Brinell
70	79,000	30,000	48	–
600	75,000	21,500	50	–
800	63,500	21,000	50	–
1,000	45,500	20,000	26	–
1,200	26,500	14,500	36	–
1,400	17,500	11,000	44	–
1,600	9,000	6,500	52	–
1,800	5,000	2,500	60	–

Rupture Strength, 1,000 hr

Cold Drawn, Annealed

Test temperature, °F	Strength, psi	Elong., in 2 in., %	Reduction of area, %
700	70,000	–	–
900	42,000	–	–
1,100	17,000	–	–

Creep Strength (Stress, psi, to Produce 1% Creep)

Cold Drawn, Annealed

Test temperature, °F	10,000 hr	100,000 hr
750	30,000	–
800	24,000	–
900	16,000	–
1,000	9,500	–

Table 1.4.2–1 (continued)
MONEL® ALLOY 400

Fatigue Strength

Annealed

Test temperature, °F	Stress, psi	Cycles to failure
70	33,500	10^8

Impact Strength

Cold Drawn, Annealed

Test temperature, °F	Type test	Strength, ft-lb
–310	Charpy–V-notched	212
–112	Charpy–V-notched	219
75	Charpy–V-notched	216

Specifications

ASTM: B127, B163, B164, B165, B395; Federal QQ-N-281 Class A

Producers and Tradenames

Huntington Alloy Products Division, INCO: Monel alloy 400
G. O. Carlson Co.: Carlson 400
Bridgeport Brass: Bronel® 400
H. M. Harper: HMH 400

Table 1.4.2–2
MONEL® ALLOY 404

Chemical composition, percent by weight: C, 0.08; Mn, 0.05; Fe, 0.25; S, 0.012; Si, 0.05; Cu, 44.0; Ni, 54.5; Al, 0.03

Physical constants and thermal properties
Density, lb./in.³ : 0.321
Coefficient of thermal expansion, (70–200°F) in./in./°F × 10^{-6} : 7.4
Modulus of elasticity, psi: tension, 24.5×10^6; torsion, 9.4×10^6

Poisson's ratio: 0.295
Melting range,°F: 2,370–2,460
Specific heat, Btu/lb/°F, 70°F: 0.099
Thermal conductivity, Btu/ft²/hr/in./°F, 70°F: 146
Electrical resistivity, ohms/cmil/ft,70°F: 300
Curie temperature, °F: annealed, –110
Permeability (70°F, 200 Oe): annealed, 1.002

Heat treatments
Usually used in annealed condition.

Table 1.4.2–2 (continued)
MONEL® ALLOY 404

Tensile Properties

Annealed

Temperature, °F	T.S., psi	Y.S., psi, 0.2% offset	Elong., in 2 in., %	Hardness, Brinell
70	65,700	22,300	–	–
200	61,300	19,800	–	–
400	57,700	17,700	–	–
600	57,300	17,400	–	–
800	52,000	17,000	–	–
1,000	39,600	15,500	–	–
1,200	27,400	13,600	–	–
1,400	17,000	11,400	–	–
1,600	10,400	7,500	–	–
1,800	6,300	4,300	–	–
2,000	–	–	–	–

Producers and Tradenames

Huntington Alloy Products Division, INCO: Monel alloy 404

Table 1.4.2–3
MONEL® ALLOY R-405

Chemical composition, percent by weight: C, 0.15; Mn, 1.00; Fe, 1.25; S, 0.043; Si, 0.25; Cu, 31.5; Ni, 66.5

Physical constants and thermal properties
Density, lb/in.³: 0.319
Coefficient of thermal expansion, (70–200°F) in./in./°F × 10⁻⁶: 7.6
Modulus of elasticity, psi: tension, 26.0 × 10⁶; torsion, 9.5 × 10⁶
Poisson's ratio: 0.32

Melting range, °F: 2,370–2,460
Specific heat, Btu/lb /°F, 70°F: 0.105
Thermal conductivity, Btu/ft²/hr/in./°F, 70°F: 151
Electrical resistivity, ohms/cmil/ft,70°F: 307
Curie temperature,°F: annealed, 20–50

Heat treatments
Generally used in cold drawn stress relieved (1,000–1,200°F) condition.

Tensile Properties

Cold Drawn – Stress Relieved

Temperature, °F	T.S., psi	Y.S., psi, 0.2% offset	Elong., in 2 in., %	Hardness, Brinell
70	83,000	74,000	28	180

Rupture Strength, 1,000 hr

Not used in this service

Table 1.4.2−3 (continued)
MONEL® ALLOY R-405

Creep Strength

Not used in this service

Fatigue Strength

Cold Drawn − Stress Relieved

Test temperature, °F	Stress, psi	Cycles to failure
70	40,000	10^8

Impact Strength

Cold Drawn

Test temperature, °F	Type test	Strength, ft-lb
+70	Charpy−V-notched or Charpy − ʊ -notched	140

Specifications

ASTM: B164; SAE-AMS: 4674, 7234; Federal QQ-N-281 Class B

Producers and Tradenames

Huntington Alloy Products Division, INCO: Monel alloy R-405
H. M. Harper: HMH 405

Table 1.4.2−4
MONEL® ALLOY R-500

Chemical composition, percent by weight: C, 0.13; Mn, 0.75; Fe, 1.00; S, 0.005; Si, 0.25; Cu, 29.5; Ni, 66.5; Ti, 0.60; Al, 2.73

Physical constants and thermal properties
  Density, lb/in.³ :0.306
  Coefficient of thermal expansion, (70−200°F) in./in./°F × 10^{-6} : 7.6
  Modulus of elasticity, psi: tension, 26×10^6 ; torsion 9.5×10^6
  Poisson's ratio: 0.32
  Melting range, °F: 2,400−2,460

Specific heat, Btu/lb/°F, 70°F: 0.100
Thermal conductivity, Btu/ft² /hr/in./°F, 70°F: 121
Electrical resistivity, ohms/cmil/ft,70°F:370
Curie temperature, °F: annealed, <−210; age hardened, <−150
Permeability (70°F, 200 Oe): annealed, 1.001; age hardened, 1.002

Heat treatments
  Solution treat 1,800°F − water quench or fast air cool. Age 1,100°F/16 hr, furnace cool ar rate of 15−25°F per hour to 900°F. Air cool.

Table 1.4.2–4 (continued)
MONEL® ALLOY R-500

Tensile Properties

Hot Rolled and Aged

Temperature, °F	T.S., psi	Y.S., psi, 0.2% offset	Elong., in 2 in., %	Hardness, Brinell
70	160,000	111,000	23.5	–
200	150,000	108,000	23.5	–
400	149,000	103,000	24.0	–
600	146,000	105,000	23.0	–
800	124,000	105,000	8.5	–
1,000	95,000	92,000	3.0	–
1,200	80,000	80,000	1.5	–
1,400	45,000	30,000	8.0	–
1,600	21,000	–	23.0	–
1,800	6,000	–	47.5	–
2,000	3,000	–	81.5	–

Rupture Strength, 1,000 hr

Hot Rolled and Aged

Test temperature, °F	Strength, psi	Elong., in 2 in., %	Reduction of area, %
900	46,500	–	–
1,000	30,000	–	–
1,100	27,000	–	–

Creep Strength (Stress, psi, to Produce 1% Creep)

Age Hardened

Test temperature, °F	10,000 hr	100,000 hr
800	88,000	–
900	48,000	–
1,000	21,000	–
1,100	9,100	–

Fatigue Strength

Cold Drawn, Aged

Test temperature, °F	Stress, psi	Cycles to failure
70	52,000	10^8
1,000	48,000	10^8

Table 1.4.2–4 (continued)
MONEL® ALLOY R-500

Impact Strength

Aged

Test temperature, °F	Type test	Strength, ft-lb
–320	Charpy–V-notched	31
–110	Charpy–V-notched	34
+ 70	Charpy–V-notched	37

Specifications

SAE-AMS: 4676; Federal QQ-N-286

Producers and Tradenames

Huntington Alloy Products Division, INCO: Monel alloy R-500

Table 1.4.2–5
MONEL® ALLOY 502

Chemical composition, percent by weight: C, 0.05; Mn, 0.75; Fe, 1.00; S, 0.005; Si, 0.25; Cu, 28.0; Ni, 66.5; Ti, 0.25; Al, 3.00

Physical constants and thermal properties
Density, lb/in.³ : 0.305
Coefficient of thermal expansion, (70–200°F) in./in./°F $\times 10^{-6}$: 6.2
Modulus of elasticity, psi: tension, 26×10^6 ; torsion, 9.5×10^6
Poisson's ratio: 0.32
Melting range, °F: 2,400–2,460
Specific heat, Btu/lb/°F, 70°F: 0.100

Thermal conductivity, Btu/ft² /hr/in./°F, 70°F: 121
Electrical resistivity, ohms/cmil/ft,70°F: 370
Curie temperature, °F: annealed, <–210; age hardened, <–150
Permeability (70°F, 200 Oe): annealed, 1.001; age hardened, 1.002

Heat treatments
Solution treat 1,800°F – water quench or fast air cool. Age 1,150°F/2 hr, furnace cool to 1,050°F, hold at 1,050°F/4 hr, furnace cool to 950°F, hold at 950°F/4 hr. Air cool.

Tensile Properties

Annealed and Aged

Temperature, °F	T.S., psi	Y.S., psi, 0.2% offset	Elong., in 2 in., %	Hardness, Brinell
70	141,000	98,000	28	–

Specifications

Federal: QQ-N-286 Class B

Producers and Tradenames

Huntington Alloy Products Division, INCO: Monel alloy 502

Table 1.4.2–6
Ni-Cu ALLOY 410

Chemical composition, percent by weight: C, 0.20; Mn, 0.80; Fe, 1.00; S, 0.008; Si, 1.60; Cu, 30.5; Ni, 66.0

Physical constants and thermal properties
  Density, lb/in.^{3} : 0.312
  Coefficient of thermal expansion, (70–200°F) in./in./°F × 10^{-6} : 7.6
  Modulus of elasticity, psi: tension, 23 × 10^6

Melting range, °F: 2,700–2,900
Specific heat, Btu/lb/°F, 70°F: 0.13
Thermal conductivity, Btu/ft^2/hr/in./°F, 70°F:186
Electrical resistivity, ohms/cmil/ft,70°F: 320

Heat treatments
  Generally used as cast.

Tensile Properties

As Cast

Temperature, °F	T.S., psi	Y.S., psi, 0.2% offset	Elong., in 2 in., %	Hardness, Brinell
70	75,000	35,000	38	150

Fatigue Strength

As Cast

Test temperature, °F	Stress, psi	Cycles to failure
70	17,500	10^8

Impact Strength

Test temperature, °F	Type test	Strength, ft-lb
–210	Charpy–V-notched	28
–110	Charpy–V-notched	31
+ 70	Charpy–V-notched	32

Specifications

Federal: QQ-N-288 Comp. A

Table 1.4.2–7
Ni-Cu ALLOY 505

Chemical composition, percent by weight: C, 0.08; Mn, 0.80; Fe, 2.00; S, 0.008; Si, 4.0; Cu, 29.0; Ni, 64.0

Physical constants and thermal properties
Density, lb/in.^{3} : 0.302
Coefficient of thermal expansion, (70–200°F) in./in./°F × 10^{-6} : 7.5
Modulus of elasticity, psi: tension, 24 × 10^6
Melting range, °F: 2,250–2,350

Specific heat, Btu/lb/°F, 70°F: 0.13
Thermal conductivity, Btu/ft^2/hr/in./°F, 70°F:136
Electrical resistivity, ohms/cmil/ft,70°F: 380

Heat treatments
Annealing: 1,650°F/1 hr per inch thickness plus oil quench. Age hardening: place castings in furnace at 600°F, heat slowly to 1,100°F, hold 4–6 hr, furnace or air cool.

Tensile Properties

Annealed and Aged

Temperature, °F	T.S., psi	Y.S., psi, 0.2% offset	Elong., in 2 in., %	Hardness, Brinell
70	135,000	110,000	2	–

Fatigue Strength

As Cast

Test temperature, °F	Stress, psi	Cycles to failure
70	21,000	10^8

Impact Strength

Test temperature, °F	Type test	Strength, ft-lb
–210	Charpy–V-notched	6
–110	Charpy–V-notched	6
+ 70	Charpy–V-notched	6

Specifications

Federal: QQ-N-288 Comp. D

1.4.3 NICKEL-CHROMIUM ALLOYS

Table 1.4.3−1
NIMONIC® ALLOY 75

Chemical composition, percent by weight: C, 0.10; Mn, 0.45; Fe, 0.50; S, 0.007; Si, 0.45; Cr, 20.5; Ni, 77.6; Cu, 0.05; Ti, 0.35; Al, 0.20

Physical constants and thermal properties
Density, lb/in.³: 0.301
Coefficient of thermal expansion, (70−200°F) in./in./°F × 10⁻⁶: 6.8
Modulus of elasticity, psi: tension, 32 × 10⁶

Heat treatments
Used in annealed condition, 1,850°F/30 min.

Producers and Tradenames

Superseded by Inconel® 600 and related alloys given in Table 1.4.3−2.

Table 1.4.3−2
INCONEL® 600

Chemical composition, percent by weight: C, 0.08; Mn, 0.5; Fe, 8.0; S, 0.008; Si, 0.25; Cr, 15.5; Ni, 76.0 Cu, 0.25; Ti, 0.35; Al, 0.25

Physical constants and thermal properties
Density, lb/in.³: 0.304
Coefficient of thermal expansion, (70−200°F) in./in./°F × 10⁻⁶: 7.4
Modulus of elasticity, psi: tension, 30 × 10⁶; torsion, 11 × 10⁶

Poisson's ratio: 0.29
Melting range, °F: 2,470−2,575
Specific heat, Btu/lb/°F, 70°F: 0.106
Thermal conductivity, Btu/ft²/hr/in./°F, 70°F: 103
Electrical resistivity, ohms/cmil/ft, 70°F: 620
Curie temperature, °F: annealed, −192
Permeability (70°F, 200 Oe): annealed, 1.010

Heat treatments
Used in annealed condition, 1,850°F/30 min.

Tensile Properties

Hot Rolled

Temperature, °F	T.S., psi	Y.S., psi, 0.2% offset	Elong., in 2 in., %	Hardness, Brinell
70	90,500	36,500	47	−
600	90,500	31,000	46	−
800	88,500	29,500	49	−
1,000	84,000	28,500	47	−
1,200	65,000	26,500	39	−
1,400	27,500	17,000	46	−
1,600	15,000	9,000	80	−
1,800	7,500	4,000	118	−

Table 1.4.3–2 (continued)
INCONEL® 600

Rupture Strength, 1,000 hr

Solution Annealed 2,050°F/2 hr

Test temperature, °F	Strength, psi	Elong., in 2 in., %	Reduction of area, %
1,500	5,600	–	–
1,600	3,500	–	–
1,800	1,800	–	–
2,000	920	–	–

Creep Strength (Stress, psi, to Produce 1% Creep)

Solution Annealed 2,050°F/2 hr

Test temperature, °F	10,000 hr	100,000 hr
1,300	5,000	–
1,500	3,200	–
1,600	2,000	–
1,700	1,100	–
1,800	560	–
2,000	270	–

Fatigue Strength

Annealed

Test temperature, °F	Stress, psi	Cycles to failure
70	39,000	10^8

Impact Strength

Annealed

Test temperature, °F	Type test	Strength, ft-lb
+70	Charpy – V-notched	180
800	Charpy – V-notched	187
1,000	Charpy – V-notched	160

Specifications

ASTM: B166, B168, B167, B163
SAE-AMS: 5665, 5540, 5580, 5687, 7232

Producers and Tradenames

Huntington Alloy Products Division, INCO: Inconel alloy 600
Allegheny-Ludlum Steel Co.: Altemp® 600
Armco Steel: Armco® 600
Carpenter Steel Co.: Pyromet® 600
Simonds Steel Co.: Simalloy 600

Table 1.4.3–3
INCONEL® ALLOY 601

Chemical composition, percent by weight: C, 0.05; Mn, 0.50; Fe, 14.1; S, 0.007; Si, 0.25; Cr, 23.0; Ni, 60.5; Cu, 0.50; Al, 1.35

Physical constants and thermal properties
Density, lb/in.³: 0.291
Coefficient of thermal expansion, (70–200°F) in./in./°F × 10⁻⁶ : 7.6
Modulus of elasticity, psi: tension, 29.9 × 10⁶; torsion, 11.8 × 10⁶

Poisson's ratio: 0.267
Melting range, °F: 2,374–2,494
Specific heat, Btu/lb/°F, 70°F: 0.107
Thermal conductivity, Btu/ft² /hr/in./°F, 70°F: 87
Electrical resistivity, ohms/cmil/ft., 70°F: 725
Curie temperature, °F: annealed, <–320
Permeability (70°F, 200 Oe): annealed, 1.003

Heat treatments
Used in annealed condition.

Tensile Properties

Annealed

Temperature, °F	T.S., psi	Y.S., psi, 0.2% offset	Elong., in 2 in., %	Hardness, Brinell
70	107,000	49,000	45	–

Rupture Strength, 1,000 hr

Solution Annealed

Test temperature, °F	Strength, psi	Elong., in 2 in., %	Reduction of area, %
1,000	58,000	–	–
1,100	40,000	–	–
1,200	28,000	–	–
1,300	14,000	–	–
1,400	9,200	–	–
1,500	6,400	–	–
1,600	4,300	–	–
1,800	2,100	–	–
2,000	1,100	–	–

Creep Strength (Stress, psi, to Produce 1% Creep)

Solution Annealed

Test temperature, °F	10,000 hr	100,000 hr
1,000	43,000	–
1,100	28,000	–
1,200	18,000	–
1,300	7,000	–
1,400	4,000	–
1,500	2,700	–
1,600	2,050	–
1,800	750	–

Table 1.4.3–3 (continued)
INCONEL® ALLOY 601

Producers and Tradenames

Huntington Alloy Products Division, INCO: Inconel alloy 601

Table 1.4.3–4
Ni-Cr 50-50

Chemical composition, percent by weight: C, 0.05; Cr, 48.0; Ni, balance; Ti, 0.35

Physical constants and thermal properties
  Density, lb/in.^{3}: 0.284
  Coefficient of thermal expansion, (70–200°F) in./in./°F × 10^{-6}: 6.5
  Melting range, °F: 2,385–2,460
  Specific heat, Btu/lb/°F, 70°F: 0.109
  Thermal conductivity, Btu/ft^2/hr/in./°F, 70°F: 109
  Electrical resistivity, ohms/cmil/ft, 70°F: 523

Heat treatments
  Anneal 2,200°F/1 hr. Air cool.

Tensile Properties

Annealed; 2,200°F/1 hr; Air Cool

Temperature, °F	T. S., psi	Y.S., psi, 0.2% offset	Elong., in 2 in., %	Hardness, Brinell
400	103,000	50,500	19	–
600	102,500	48,300	19	–
800	101,500	47,500	22	–
1,000	96,700	44,000	27	–
1,200	80,300	42,500	37	–
1,400	59,000	43,000	42	–
1,600	27,800	24,000	67	–
1,800	13,100	10,800	61	–
2,000	6,500	5,400	42	–

Rupture Strength, 1,000 hr

Annealed 2,200°F/1 hr; Air Cool

Test temperature, °F	Strength, psi	Elong., in 2 in.,%	Reduction of area, %
1,200	14,000	–	–
1,300	8,200	–	–
1,400	5,000	–	–
1,500	3,000	–	–
1,600	1,800	–	–
1,700	1,300	–	–
1,800	800	–	–

Table 1.4.3—4 (continued)
Ni-Cr 50-50

Impact Strength

Annealed 2,200°F/1 hr; Air Cool

Test temperature, °F	Type test	Strength, ft-lb
+70	Charpy—V-notched	15

Producers and Tradenames

Huntington Alloy Products Division, INCO; Inconel® alloy 671
Universal-Cyclops: Uniloy® 50 Ni-50 Cr

Table 1.4.3—5
Ni-Cr ALLOY 610 (ACI CY-40)

Chemical composition, percent by weight: C, 0.20; Mn, 0.90; Fe, 9.00; S, 0.008; Si, 2.00; Cr, 15.50; Ni, 71.0; Cu, 0.50; Cb, 1.00

Physical constants and thermal properties
Density lb/in.^{3} : 0.300
Coefficient of thermal expansion, (70—200°F) in./in./°F × 10^{-6} : 6.7

Modulus of elasticity, psi: tension, 23 × 10^6
Melting range, °F: 2,540—2,600
Specific heat, Btu/lb/°F, 70°F: 0.11
Thermal conductivity, Btu/ft^2/hr/in./°F, 70°F: 104
Electrical resistivity, ohms/cmil/ft, 70°F: 700

Heat treatments
None

Tensile Properties

As Cast

Temperature, °F	T. S., psi	Y. S., psi, 0.2% offset	Elong., in 2 in., %	Hardness, Brinell
70	70,000	40,000	15	–

Impact Strength

Test temperature, °F	Type test	Strength, ft-lb
–210	Charpy—V-notched	19
–110	Charpy—V-notched	19
+ 70	Charpy—V-notched	24

Specifications

ACI: Cy-40

Table 1.4.3–5 (continued)
Ni-Cr ALLOY 610 (ACI CY-40)

Producers and Tradenames

Alloy Foundries Division: CY-40
Aloyco, Inc.: CY-40
Cannon-Muskegon: Cast Inconel®
Esco Corp.: Esco® 23
Hica Co.: Alloy 22
Lebanon Steel Foundry: (L)INC
Quaker Alloy Casting: Q INC

Table 1.4.3–6
Ni-Cr ALLOY 705

Chemical composition, percent by weight: C, 0.30; Mn, 0.90; Fe, 8.00; S, 0.008; Si, 5.50; Cr, 15.50; Ni, 69.5; Cu, 0.50

Physical constants and thermal properties
Density, lb/in.^{3}: 0.292
Coefficient of thermal expansion, (70–200°F) in./in./°F $\times$ 10^{-6}: 6.5

Modulus of elasticity, psi: tension, 25 $\times$ 10^6
Melting range, °F: 2,540–2,600
Electrical resistivity, ohms/cmil/ft, 70°F: 760

Heat treatments
Anneal 1,650°/1 hr per inch thickness, oil quench.
Age 600°F, heat slowly to 1,100°F, hold 4–6 hr, furnace or air cool.

Tensile Properties

Annealed and Aged

Temperature, °F	T. S., psi	Y. S., psi, 0.2% offset	Elong., in 2 in., %	Hardness, Brinell
70	110,000	95,000	2	–

1.4.4 NICKEL-MOLYBDENUM ALLOYS

Table 1.4.4–1
HASTELLOY® B

Chemical composition, percent by weight: C, 0.05[a]; Mn, 1.00[a]; Fe, 5.00; S, 0.030[a]; Si, 1.00[a]; Cr, 1.00[a]; Ni, balance; Co, 2.50[a]; Mo, 28.00; V, 0.30

[a]Maximum

Physical constants and thermal properties
Density, lb/in.³ : 0.334
Coefficient of thermal expansion, (70–200°F) in./in./°F × 10⁻⁶ : 5.6

Modulus of elasticity, psi: tension, 31.1 × 10⁶
Melting range, °F: 2,375–2,495
Specific heat, Btu/lb/°F, 70°F: 0.091
Thermal conductivity, Btu/ft² /hr/in./°F, 70°F: 85
Electrical resistivity, ohms/cmil/ft, 70°F: 811
Permeability (70°F, 200 Oe): annealed, < 1.001

Heat treatments
Solution treat 2,000–2,150°F, quench or rapid air cool.

Tensile Properties

Solution Annealed 2,000°F, Rapid Air Cool

Temperature, °F	T.S., psi	Y.S., psi, 0.2% offset	Elong., in 2 in., %	Hardness, Brinell
70	134,100	67,000	51	–
1,000	113,500	48,700	55	–
1,200	106,600	50,400	50	–
1,400	85,300	47,800	30	–
1,600	71,600	41,100	22	–
1,800	36,200	14,800	21	–
2,000	25,400	10,100	20	–

Rupture Strength, 1,000 hr

Solution Annealed 2,000°F, Rapid Air Cool

Test temperature, °F	Strength, psi	Elong., in 2 in., %	Reduction of area, %
1,000	74,000	–	–
1,200	36,500	–	–
1,400	15,500	–	–
1,500	9,400	–	–

Fatigue Strength

Solution Annealed, 2,000°F, Water Quench, 1,200°F/4 hr

Test temperature, °F	Stress, psi	Cycles to failure
1,500	66,000	10⁸

Table 1.4.4–1 (continued)
HASTELLOY® B

Impact Strength

Solution Annealed

Test temperature, °F	Type test	Strength, ft-lb
–326	Izod V-notch	53
–148	Izod V-notch	53
–58	Izod V-notch	49
+70	Izod V-notch	60

Specifications

ASTM: B333, B335, A494 (castings); SAE-AMS: 5396 (castings)

Producers and Tradenames

Stellite Division, Cabot Corp.: Hastelloy alloy B

Table 1.4.4–2
HASTELLOY® ALLOY X

Chemical composition, percent by weight: C, 0.10; Mn, 1.00[a]; Fe, 18.50; Si, 1.00[a]; Cr, 22.25; Ni, balance; Co, 1.50; Mo, 9.00; W, 0.60

[a]Maximum

Physical constants and thermal properties
Density, lb/in.^{3} : 0.297
Coefficient of thermal expansion, (70–200° F) in./in./ °F $\times$ 10^{-6} : 7.7

Modulus of elasticity, psi: tension, 28.5 $\times$ 10^6
Poisson's ratio: 0.32
Melting range, °F: 2,300–2,470
Specific heat, Btu/lb/°F, 70°F: 0.116
Thermal conductivity, Btu/ft^2 /hr/in./°F, 70°F: 63
Electrical resistivity, ohms/cmil/ft, 70°F: 712
Permeability (70°F, 200 Oe): annealed, < 1.002

Heat treatments
Solution anneal 2,150°F, rapid cool or quench.

Tensile Properties

Solution Annealed 2,150° F, Rapid Air Cool

Temperature, °F	T.S., psi	Y.S., psi, 0.2% offset	Elong., in 2 in., %	Hardness, Brinell
70	113,900	52,200	43	–
400	103,400	48,700	41	–
800	99,700	43,700	44	–
1,000	94,000	41,500	45	–
1,200	83,000	39,500	37	–
1,400	63,100	37,800	37	–
1,600	36,500	25,700	51	–
1,800	22,500	16,000	45	–
2,000	13,000	8,000	40	–

Table 1.4.4–2 (continued)
HASTELLOY® ALLOY X

Rupture Strength, 1,000 hr

Solution Annealed 2,150°F, Rapid Air Cool

Test temperature, °F	Strength, psi	Elong., in 2 in., %	Reduction of area, %
1,200	32,000	–	–
1,500	9,500	–	–
1,800	2,800	–	–

Creep Strength (Stress, psi, to Produce 1% Creep)

Solution Annealed 2,150°F, Rapid Air Cool

Test temperature, °F	10,000 hr	100,000 hr
1,200	16,500	–
1,500	6,500	–
1,800	1,000	–

Impact Strength

Solution Annealed 2,150°F, Rapid Air Cool

Test temperature, °F	Type test	Strength, ft-lb
–321	Charpy – V-notched	37
–216	Charpy – V-notched	44
+70	Charpy – V-notched	54
1,500	Charpy – V-notched	58

Specifications

SAE-AMS: 5754, 5536, 5390 (castings)

Producers and Tradenames

Stellite Division, Cabot Corp.: Hastelloy alloy X

Table 1.4.4–3
HASTELLOY® ALLOY C-276

Chemical composition, percent by weight: C, 0.02[a]; Mn, 1.00[a]; Fe, 5.50; S, 0.03[a]; Si, 0.05[a]; Cr, 15.50; Ni, balance; Co, 2.50[a]; Mo, 16.00; W, 3.75; V, 0.35[a]; P, 0.03[a]

[a]Maximum

°F × 10^{-6} : 6.2
Modulus of elasticity, psi: tension, 29.8 × 10^6
Melting range, °F: 2,415–2,500
Specific heat, Btu/lb/°F, 70°F: 0.102
Thermal conductivity, Btu/ft² /hr/in./°F, 70°F: 69
Electrical resistivity, ohms/cmil/ft, 70°F: 779

Physical constants and thermal properties
Density, lb/in.³ : 0.321
Coefficient of thermal expansion, (70–200°F) in./in./

Heat treatments
Solution heat treat 2,100°F, rapid quench.

Tensile Properties

Solution Treated 2,100°F, Water Quench

Temperature, °F	T.S., psi	Y.S., psi, 0.2% offset	Elong., in 2 in., %	Hardness, Brinell
70	113,500	52,000	70	–
400	101,700	44,100	71	–
600	95,100	39,100	71	–
800	93,800	33,500	75	–
1,000	89,600	31,700	74	–
1,200	86,900	32,900	73	–
1,400	80,700	30,900	78	–
1,600	63,500	29,900	92	–
1,800	39,000	27,000	127	–

Rupture Strength, 1,000 hr

Solution Treated 2,100°F, Water Quench

Test temperature, °F	Strength, psi	Elong., in 2 in., %	Reduction of area, %
1,200	40,000	–	–
1,400	18,000	–	–
1,600	7,000	–	–
1,800	3,100	–	–

Impact Strength

Solution Treated 2,100°F, Water Quench

Test temperature, °F	Type test	Strength, ft-lb
–320	Charpy – V-notched	181
+70	Charpy – V-notched	238
+392	Charpy – V-notched	239

Producers and Tradenames

Stellite Division, Cabot Corp.: Hastelloy alloy C-276

Table 1.4.4–4
HASTELLOY® ALLOY N

Chemical composition, percent by weight: C, 0.06; Mn, 0.80[a]; Fe, 5.0[a]; Cr, 7.0; Ni, 69.5; Mo, 16.5; other, 1.10[a]

[a]Maximum

Physical constants and thermal properties
Density, lb/in.³ : 0.317
Coefficient of thermal expansion, (70–200°F) in./in./°F × 10⁻⁶ :6.2
Melting range, °F: 2,375–2,550
Specific heat, Btu/lb/°F, 70°F: 0.095
Thermal conductivity, Btu/ft² /hr/in./°F, 70°F: 71
Electrical resistivity, ohms/cmil/ft,70°F:329

Tensile Properties

Solution Annealed 2,150 °F, Rapid Air Cool

Temperature, °F	T. S., psi	Y. S., psi, 0.2% offset	Elong., in 2 in., %	Hardness, Brinell
70	115,000	45	51	–

Rupture Strength, 100 hr

Test temperature, °F	Strength, psi	Elong., in 2 in., %	Reduction of area, %
1,200	40,000	–	–
1,500	9,000	–	–
1,800	2,100	–	–

Producers and Tradenames

Stellite Division, Cabot Corp.: Hastelloy alloy N

Table 1.4.4–5
INCONEL®ALLOY 617

Chemical composition, percent by weight: C, 0.07; Cr, 22.0; Ni, 54.0 Co, 12.5; Mo, 9.0; Al, 1.0

Physical constants and thermal properties
Density, lb/in.^{3} : 0.302
Coefficient of thermal expansion, (70–200°F) in./in./°F × 10^{-6} : 6.4
Melting range, °F: 2,430–2,510
Specific heat, Btu/lb/°F, 70°F: 0.100
Thermal conductivity, Btu/ft^2/hr/in./°F, 70°F: 94
Electrical resistivity, ohms/cmil/ft,70°F: 736

Heat treatments
Solution anneal 2,150°F.

Tensile Properties

Cold Rolled, Solution Treated

Temperature, °F	T. S., psi	Y. S., psi, 0.2% offset	Elong., in 2 in., %	Hardness, Brinell
70	111,000	47,000	54	–

Rupture Strength, 1,000 hr

Solution Treated

Test temperature, °F	Strength, psi	Elong., in 2 in., %	Reduction of area, %
1,200	45,000	–	–
1,400	22,000	–	–
1,600	10,000	–	–
1,800	3,900	–	–
2,000	1,800	–	–

Producers and Tradenames

Huntington Alloy Products Divsion, INCO: Inconel alloy 617

Table 1.4.4–6
INCONEL® ALLOY 625

Chemical composition, percent by weight: C, 0.05; Mn, 0.25; Fe, 2.50; S, 0.008; Si, 0.25; Cr, 21.5; Ni, 61.0; Mo, 9.00; Cb, 3.65; Ti, 0.20; Al, 0.20

Physical constants and thermal properties
　Density, lb/in.^{3}: 0.305
　Coefficient of thermal expansion, (70–200°F) in./in./
　　°F × 10^{-6}: 7.1
　Modulus of elasticity, psi: tension, 30 × 10^6; torsion,
　　11 × 10^6
　Poisson's ratio: 0.31

Melting range, °F: 2,350–2,460
Specific heat, Btu/lb/°F, 70°F: 0.098
Thermal conductivity, Btu/ft^2/hr/in./°F, 70°F: 68
Electrical resistivity, ohms/cmil/ft, 70°F: 776
Curie temperature, °F: annealed, < –320
Permeability (70°F, 200 Oe): annealed, 1.0006

Heat treatments
　Anneal 1,700–1,900°F; solution treat 2,000–2,200°F.

Tensile Properties

Solution Treated 2,100°F

Temperature, °F	T.S., psi	Y.S., psi, 0.2% offset	Elong., in 2 in., %	Hardness, Brinell
70	124,000	52,000	58	–
600	112,000	40,000	60	–
800	110,000	39,000	60	–
1,000	106,000	38,000	60	–
1,200	106,000	38,000	70	–
1,400	72,000	39,000	90	–
1,800	19,000	–	118	–
2,000	10,000	–	125	–

Rupture Strength, 1,000 hr

Solution Treated 2,100°F

Test temperature, °F	Strength, psi	Elong., in 2 in., %	Reduction of area, %
1,200	52,000	–	–
1,300	35,000	–	–
1,400	23,000	–	–
1,500	13,000	–	–
1,600	7,200	–	–
1,800	2,600	–	–

Creep Strength (Stress, psi, to Produce 1% Creep)

Test temperature, °F	10,000 hr	100,000 hr
1,200	45,000	–
1,300	20,000	–
1,400	12,000	–
1,500	7,200	–
1,600	3,400	–

Table 1.4.4–6 (continued)
INCONEL® ALLOY 625

Fatigue Strength

Annealed

Test temperature, °F	Stress, psi	Cycles to failure
70	92,000	10^8
800	92,000	10^8
1,000	80,000	10^8
1,200	68,000	10^8

Impact Strength

Hot Rolled, As Rolled

Test temperature, °F	Type test	Strength, ft-lb
–320	Charpy keyhole notched	35
–110	Charpy keyhole notched	44
85	Charpy keyhole notched	49

Specifications

ASTM: B443, B444, B446; SAE- AMS: 5599, 5666

Producers and Tradenames

Huntington Alloy Products Division, INCO: Inconel alloy 625
Cartech: Pyromet® 625
Special Metals Corp.: Udimet® 625
Stellite Division, Cabot: Haynes® alloy 625

Table 1.4.4–7
ILLIUM® B

Chemical composition, percent by weight: S, 3.50; Cr, 28.00; Ni, 52,00: Cu, 5.50; Mo, 8.50

Physical constants and thermal properties
Density, lb/in.^{3}: 0.310

Heat treatments
Consult manufacturer.

Tensile Properties

Cast Hardened

Temperature, °F	T.S., psi	Y.S., psi, 0.2% offset	Elong., in 2 in., %	Hardness, Brinell
70	61,000	–	1.0	221/420

Table 1.4.4–7 (continued)
ILLIUM® B

Producers and Tradenames

Stainless Foundry and Engineering, Inc.: Illium B

Table 1.4.4–8
ILLIUM® G

Chemical composition, percent by weight: Fe, 6.50; Cr, 22.50; Ni, 56.00; Cu, 6.50; Mo, 6.40

Physical constants and thermal properties
Density, lb/in.^{3}: 0.31
Coefficient of thermal expansion, (70–200°F) in./in./ °F × 10^{-6}: 6.8
Modulus of elasticity, psi: tension, 24.3 × 10^6
Specific heat, Btu/lb/°F, 70°F: 0.105

Heat treatments
Consult manufacturer.

Tensile Properties

As Cast

Temperature, °F	T.S., psi	Y.S., psi, 0.2% offset	Elong., in 2 in., %	Hardness, Brinell
70	68,000	38,900	7.5	159–177

Producers and Tradenames

Stainless Foundry and Engineering, Inc.: Illium G

Table 1.4.4–9
ILLIUM® 98

Chemical composition, percent by weight: Si, 0.70; Cr, 28.00; Ni, 55.00; Cu, 5.50; Mo, 8.50

Physical constants and thermal properties
Density, lb/in.^{3}: 0.302

Heat treatments
Consult manufacturer.

Table 1.4.4–9 (continued)
ILLIUM® 98

Tensile Properties

As Cast

Temperature, °F	T.S., psi	Y.S., psi, 0.2% offset	Elong., in 2 in., %	Hardness, Brinell
70	54,000	–	18	152/167

Producers and Tradenames

Stainless Foundry and Engineering, Inc.: Illium 98

Table 1.4.4–10
CHLORIMET® No. 2

Chemical composition, percent by weight: C, 0.10; Fe, 3.00[a]; Si, 1.00; Ni, 63.00; Mo, 32.00

[a]Maximum

Physical constants and thermal properties
 Density, lb/in.^{3}: 0.333
 Coefficient of thermal expansion, (70–400°F) in./in./°F $\times$ 10^{-6}: 5.0
 Modulus of elasticity, psi: tension, 27 $\times$ 10^6

Heat treatments
 2,050°F/1 hr per inch water quench and 1,290°F/8 hr air cool.

Tensile Properties

Age Hardened

Temperature, °F	T.S., psi	Y.S., psi, 0.2% offset	Elong., in 2 in., %	Hardness, Brinell
70	80,000	55,000	5	–

Rupture Strength, 1,000 hr

2,125°F/½ hr Air Cool and 1,700°F/72 hr Air Cool

Test temperature, °F	Strength, psi	Elong., in 2 in., %	Reduction of area, %
1,500	10,700	–	–

Producers and Tradenames

The Duriron Company: Chlorimet No. 2

Table 1.4.4−11
CHLORIMET® No. 3

Chemical composition, percent by weight: C, 0.07; Fe, 3.00[a]; Si, 1.00; Cr, 18.00; Ni, 60.00; Mo, 18.00

[a]Maximum

Physical constants and thermal properties
  Density, lb/in.³ : 0.320

Coefficient of thermal expansion, (70−600°F) in./in./
  °F × 10^{-6} : 7.1
Modulus of elasticity, psi: tension, 24.5 × 10^6
Thermal conductivity, Btu/ft² /hr/in./°F, 392°F: 78

Heat treatments
  2,050°F/1 hr per inch, water quench.

Tensile Properties

As Cast

Temperature, °F	T.S., psi	Y.S., psi, 0.2% offset	Elong., in 2 in., %	Hardness, Brinell
70	81,000	–	10	–
200	80,000	–	10	–
400	78,000	–	10	–
600	75,000	–	10	–
800	72,000	–	11	–
1,000	68,000	–	13	–
1,200	62,000	–	16	–
1,400	55,000	–	18	–
1,600	46,000	–	16	–
1,800	32,000	–	15	–

Rupture Strength, 1,000 hr

2,225°F/½ hr Air Cool and 2,100°F/8 hr Air Cool

Test temperature, °F	Strength, psi	Elong., in 2 in., %	Reduction of area, %
1,500	10,700	–	–

Producers and Tradenames

The Duriron Company: Chlorimet No. 3

1.4.5 NICKEL-IRON-CHROMIUM ALLOYS

Table 1.4.5–1
INCOLOY® ALLOY 800

Chemical composition, percent by weight: C, 0.05; Mn, 0.75; Fe, 46.00; S, 0.008; Si, 0.50; Cr, 21.00; Ni, 32.5; Ti, 0.38; Al, 0.38

Physical constants and thermal properties
Density, lb/in.^{3} : 0.287
Coefficient of thermal expansion, (70–200°F) in./in./ °F × 10^{-6} : 7.9
Modulus of elasticity, psi: tension, 28.5 × 10^6 ; torsion, 10.6 × 10^6
Poisson's ratio: 0.34

Melting range, °F: 2,475–2,525
Specific heat, Btu/lb/°F, 70°F: 0.12
Thermal conductivity, Btu/ft^2 /hr/in./°F, 70°F: 80
Electrical resistivity, ohms/cmil/ft, 70°F: 595
Curie temperature, °F: annealed, –175
Permeability (70°F, 200 Oe): annealed, 1.009

Heat treatments
Used in annealed (1,800–1,900°) or solution treated (2,000–2,100°) condition.

Tensile Properties

Annealed – 1,800°/15 min

Temperature, °F	T. S., psi	Y. S., psi, 0.2% offset	Elong., in 2 in., %	Hardness, Brinell
70	86,800	42,700	44	–
200	81,700	39,700	42.5	–
400	77,300	36,000	39	–
600	76,200	33,700	40	–
800	74,600	33,300	40	–
1,000	72,000	31,700	38.5	–
1,200	54,000	29,000	55.5	–
1,400	32,100	22,600	85	–

Rupture Strength, 1,000 hr

Solution Treated

Test temperature, °F	Strength, psi	Elong., in 2 in., %	Reduction of area, %
1,000	50,000	–	–
1,100	38,000	–	–
1,200	25,000	–	–
1,300	17,000	–	–
1,400	9,500	–	–
1,500	6,000	–	–
1,600	3,800	–	–
1,700	2,500	–	–
1,800	1,700	–	–
2,000	900	–	–

Table 1.4.5 –1 (continued)
INCOLOY® ALLOY 800

Creep Strength (Stress, psi, to Produce 1% Creep)

Solution Treated

Test temperature, °F	10,000 hr	100,000 hr
1,000	47,000	42,000
1,100	33,000	28,000
1,200	17,000	12,000
1,300	7,800	5,600
1,400	4,800	3,400
1,600	2,200	1,700
1,800	800	–

Fatigue Strength

Annealed

Test temperature, °F	Stress, psi	Cycles to failure
70	42,000	10^8
800	42,000	10^8
1,000	38,000	10^8
1,400	22,000	10^8

Impact Strength

Test temperature, °F	Type test	Strength, ft-lb
+70	Charpy – V-notched	207
1,200	Charpy – V-notched	181

Specifications

ASTM: B163, B407, B408, B409

Producers and Tradenames

Huntington Alloy Products Division, INCO: Incoloy alloy 800
Armco Steel Co.: Armco® 800
Carpenter Technology: Pyromet® 800
Allegheny Ludlum: AL® 332

Table 1.4.5−2
INCOLOY® ALLOY 801

Chemical composition, percent by weight: C, 0.05; Mn, 0.75; Fe, 44.50; S, 0.008; Si, 0.50; Cr, 20.50; Ni, 32.00; Ti, 1.13

Physical constants and thermal properties
Density, lb/in.^{3} : 0.287
Coefficient of thermal expansion, (70−200°F) in./in./°F × 10^{-6} : 7.8
Modulus of elasticity, psi: tension, 30 × 10^8 ; torsion, 10.6 × 10^8

Melting range, °F: 2,475−2,525
Specific heat, Btu/lb/°F, 70°F: 0.108
Thermal conductivity, Btu/ft^2 /hr/in./°F, 70°F: 86
Electrical resistivity, ohms/cmil/ft, 70°F: 609

Heat treatments
Stabilizing anneal: 1,725°F/ 30 min. Age harden: 1,750°F/30 min plus 1,350°F/1 hr. Furnace cool 100°F/hr to 1,200°F. Hold at 1,200°F/4 hr, air cool.

Tensile Properties

Annealed and Aged

Temperature, °F	T. S., psi	Y. S., psi, 0.2% offset	Elong., in 2 in., %	Hardness, Brinell
70	114,500	56,300	30.5	−
800	97,200	51,300	29.0	−
1,000	95,500	44,800	27.5	−
1,200	77,600	44,300	25.5	−

Fatigue Strength

Test temperature, °F	Stress, psi	Cycles to failure
70	37,000	10^8
1,500	11,000	10^8

Specifications

SAE-AMS: 5552, 5742

Producers and Tradenames

Huntington Alloy Products Division, INCO: Incoloy alloy 801

Table 1.4.5–3
INCOLOY® ALLOY 802

Chemical composition, percent by weight: C, 0.35; Mn, 0.75; Fe, 46.0; S, 0.008; Si, 0.38; Cr, 21.0; Ni, 32.5; Ti, 0.75; Al, 0.58

Physical constants and thermal properties
Density, lb/in.³ : 0.283
Modulus of elasticity, psi: tension, 29.7 × 10⁶

Melting range, °F: 2,450–2,500
Specific heat, Btu/lb/°F, 70°F: 0.111
Electrical resistivity, ohms/cmil/ft, 70°F: 669
Permeability (70°F, 200 Oe): annealed, 1.036

Heat treatments
Age harden: solution treat plus 1,400°F/10 hr.

Tensile Properties

Age Hardened

Temperature, °F	T.S., psi	Y.S., psi, 0.2% offset	Elong., in 2 in., %	Hardness, Brinell
70	98,300	56,000	–	215
800	93,300	46,500	–	190
1,000	94,700	52,000	–	190
1,200	77,000	45,500	–	174
1,400	55,400	42,400	–	128

Rupture Strength, 1,000 hr

Solution Treated

Test temperature, °F	Strength, psi	Elong., in 2 in., %	Reduction of area, %
1,200	25,000	–	–
1,400	16,000	–	–
1,600	9,800	–	–
1,800	3,700	–	–
2,000	1,050	–	–

Creep Strength (Stress, psi, to Produce 1% Creep)

Solution Treated

Test temperature, °F	10,000 hr	100,000 hr
1,200	18,000	15,000
1,400	12,000	6,800
1,600	4,400	2,600
1,700	2,200	1,200
1,800	1,100	480
2,000	450	210

Table 1.4.5–3 (continued)
INCOLOY® ALLOY 802

Fatigue Strength

Solution Treated

Test temperature, °F	Stress, psi	Cycles to failure
70	33,000	10^8
1,000	48,000	10^8
1,200	45,000	10^8
1,600	17,000	10^8

Impact Strength

Solution Treated and Aged

Test temperature, °F	Type test	Strength, ft-lb
+70	Charpy – V-notched	28

Producers and Tradenames

Huntington Alloy Products Division, INCO: Incoloy alloy 802
INCO: Incoloy alloy 802

Table 1.4.5–4
INCOLOY® ALLOY 825

Chemical composition, percent by weight: C, 0.03: Mn, 0.50; Fe, 30.0; S, 0.015; Si, 0.25; Cr, 21.5 Ni, 42.0; Cu, 2.25; Mo, 3.0; Ti, 0.90; Al, 0.10

Physical constants and thermal properties
Density, lb/in.³ : 0.294
Coefficient of thermal expansion, (70–200°F) in./in./ °F × 10^{-6} : 7.8
Modulus of elasticity, psi: tension, 28 × 10^6

Melting range, °F: 2,500–2,550
Thermal conductivity, Btu/ft² /hr/in./°F, 70°F: 77
Electrical resistivity, ohms/cmil/ft, 70°F: 678
Curie temperature, °F: annealed, <–320
Permeability (70°F, 200 Oe): annealed, 1.005

Heat treatments
Anneal: 1,725°F.

Tensile Properties

Annealed

Temperature, °F	T.S., psi	Y.S., psi, 0.2% offset	Elong., in 2 in., %	Hardness, Brinell
70	100,500	43,700	43.0	–
200	95,000	40,400	44.0	–
400	92,400	35,600	43.0	–
600	91,700	33,600	46.0	–
800	88,500	33,000	43.5	–
1,000	85,900	33,200	43.0	–
1,200	67,500	30,900	62.0	–
1,400	39,700	26,500	87.0	–
1,600	19,600	17,000	102.0	–
1,800	10,900	6,800	173.0	–
2,000	6,100	3,300	105.5	–

Table 1.4.5—4 (continued)
INCOLOY® ALLOY 825

Impact Strength

Test temperature, °F	Type test	Strength, ft-lb
−423	Charpy keyhole notched	68
−110	Charpy keyhole notched	78
+70	Charpy keyhole notched	81

Specifications

ASTM: B163, B423, B424, B425

Producers and Tradenames

Huntington Alloy Products Division,
INCO: Incoloy alloy 825

Table 1.4.5—5
20 Cb-3

Chemical composition, percent by weight: C, 0.03; Mn, 1.00; S, 0.015; Si, 0.50; Cr, 20.00; Ni, 33.75; Cu, 3.50; Mo, 2.50; Cb, 0.50

Physical constants and thermal properties
Density, lb/in.³ : 0.291
Coefficient of thermal expansion, (70−200°F) in./in./ °F × 10⁻⁶ : 8.3

Modulus of elasticity, psi: tension, 28×10^6; torsion, 11×10^6
Specific heat, Btu/lb/°F, 70°F: 0.12
Electrical resistivity, ohms/cmil/ft, 70°F: 625

Heat treatments
Used in annealed (1,700−1,750°F) condition.

Tensile Properties

Annealed

Temperature, °F	T.S., psi	Y.S., psi, 0.2% offset	Elong., in 2 in., %	Hardness, Brinell
70	91,000	45,000	45	−
200	86,000	40,000	46	−
400	83,000	35,000	44	−
600	80,000	33,000	42	−
800	79,000	30,000	40	−
1,000	77,000	28,000	38	−

Table 1.4.5–5 (continued)
20 Cb-3

Rupture Strength, 1,000 hr

Annealed

Test temperature, °F	Strength, psi	Elong., in 2 in., %	Reduction of area, %
1,000	57,000	–	–
1,100	43,000	–	–
1,200	31,000	–	–
1,300	20,000	–	–
1,400	13,000	–	–
1,500	7,500	–	–
1,600	3,500	–	–
1,700	2,000	–	–

Specifications

ASTM: B462, B463, B464, B468, B471, B472,
B473, B474, B475

Producers and Tradenames

Carpenter Technology: 20 Cb-3

Table 1.4.5–6
JS®700

Chemical composition, percent by weight: C, 0.03; Mn, 1.70; Fe, 47.0; Si, 0.50; Cr, 21.00; Ni, 25.00; Mo, 4.50; Cb, 0.30

Physical constants and thermal properties
Density, lb/in.³ : 0.29
Coefficient of thermal expansion, (70–200°F) in./in./°F × 10⁻⁶ : 9.15
Thermal conductivity, Btu/ft²/hr/in./°F, 70°F: 102

Tensile Properties

Temperature, °F	T. S., psi	Y. S., psi, 0.2% offset	Elong., in 2 in., %	Hardness, Brinell
70	85,000	39,000	45	170

Producers and Tradenames

Jessop Steel Company (Washington, Pa.): JS 700

Table 1.4.5–7
RA®330

Chemical composition, percent by weight: C, 0.05; Mn, 1.50; Fe, 43.00; S, 0.015; Si, 1.25; Cr, 19.00; Ni, 35.00

Physical constants and thermal properties
  Density, lb/in.³ : 0.289

Modulus of elasticity, psi: tension, 28.5×10^6
Poisson's ratio: 0.303
Melting range, °F: 2,550–2,600
Specific heat, Btu/lb/°F, 70°F: 0.11
Thermal conductivity, Btu/ft² /hr/in./°F, 70°F: 86
Electrical resistivity, ohms/cmil/ft, 70°F: 607
Permeability (70°F, 200 Oe): annealed, 1.02

Tensile Properties

Mill Anneal

Temperature, °F	T. S., psi	Y. S., psi, 0.2% offset	Elong., in 2 in., %	Hardness, Brinell
70	85,000	42,000	45	–
1,200	55,700	21,500	51	–
1,400	34,000	18,800	65	–
1,600	18,700	15,900	69	–
1,800	10,700	9,000	74	–

Rupture Strength, 1,000 hr

Solution Treated

Test temperature, °F	Strength, psi	Elong., in 2 in., %	Reduction of area, %
1,400	8,000	–	–
1,600	3,700	–	–
1,800	1,500	–	–

Creep Strength (Stress, psi, to Produce 1% Creep)

Solution Treated

Test temperature, °F	10,000 hr	100,000 hr
1,400	5,700	–
1,500	4,000	–
1,600	2,700	–
1,700	1,700	–
1,800	900	–

Specifications

ASTM: 511, 512, 535; SAE-AMS: 5592, 5716

Producers and Tradenames

Rolled Alloys, Inc.: RA 330

Table 1.4.5–8
RA®333

Chemical composition, percent by weight: C, 0.05; Mn, 1.50; Fe, 18.00; S, 0.015; Si, 1.25; Cr, 25.00; Ni, 45.00; Co, 3.00; Mo, 3.00; W, 3.00

Physical constants and thermal properties
Density, lb/in.³ : 0.298
Modulus of elasticity, psi: tension, 28.0
Poisson's ratio: 0.315

Melting range, °F: 2,420–2,465
Specific heat, Btu/lb/°F, 70°F: 0.11
Thermal conductivity, Btu/ft²/hr/in./°F, 70°F: 72
Electrical resistivity, ohms/cmil/ft, 70°F: 687
Permeability (70°F, 200 Oe): annealed, 1.004

Heat treatments
Used in annealed condition (1,900–1,950°F).

Tensile Properties

Annealed

Temperature, °F	T. S., psi	Y.S., psi 0.2 % offset	Elong., in 2 in., %	Hardness, Brinell
70	100,000	50,000	50	–
1,400	55,800	34,900	43.5	–
1,600	33,300	29,700	70.9	–
1,800	18,400	15,400	53.4	–

Rupture Strength, 1,000 hr

Solution Treated

Test temperature, °F	Strength, psi	Elong., in 2 in., %	Reduction of area, %
1,400	15,000	–	–
1,500	10,500	–	–
1,600	7,000	–	–
1,700	4,400	–	–
1,800	2,600	–	–

Creep Strength (Stress, psi, to Produce 1% Creep)

Solution Treated

Test temperature, °F	10,000 hr	100,000 hr
1,400	11,000	–
1,500	8,200	–
1,600	4,300	–
1,700	2,800	–
1,800	1,300	–

Producers and Tradenames

Rolled Alloys, Inc: RA 333

Specifications

SAE-AMS: 5593, 5717

Table 1.4.5–9
ACI TYPE HT

Chemical composition, percent by weight: C, 0.55; Mn, 1.00; Fe, 47.00; S, 0.020; Si, 1.25; Cr, 15.00; Ni, 35.00

Modulus of elasticity, psi: tension, 27×10^6
Thermal conductivity, $Btu/ft^2/hr/in./°F$, 212°F: 92
Electrical resistivity, ohms/cmil/ft, 70°F: 602

Physical constants and thermal properties
  Density, $lb/in.^3$: 0.286
  Coefficient of thermal expansion, (70–1,000°F) in./in./°F $\times 10^{-6}$: 8.5

Heat treatments
  Age: 1,400°F/24 hr. Air cool.

Tensile Properties

Cast and Aged

Temperature, °F	T. S., psi	Y. S., psi, 0.2% offset	Elong., in 2 in., %	Hardness, Brinell
70	75,000	45,000	5	200

Specifications

ASTM: A297, A448

Producers and Tradenames

The Hica Corp.: HT
Oklahoma Steel Castings Corp.: HT
Sandusky Foundry and Machine Co.: HT
Stainless Foundry and Eng. Co.: HT
U.S. Pipe and Foundry Co.: HT
Lebanon Steel Foundry:(L)HT
Waukesha Foundry Co.: 35Ni-15 Cr

Table 1.4.5–10
ACI TYPE HV

Chemical composition, percent by weight: C, 0.55; Mn, 1.00; Fe, 39.00; S, 0.02; Si, 1.25; Cr, 19.00; Ni, 39.00

Physical constants and thermal properties
Density, $lb/in.^3$: 0.290
Coefficient of thermal expansion, (70–1,000°F) in./in./°F $\times 10^{-6}$: 8.8
Modulus of elasticity, psi: tension, 27×10^6
Electrical resistivity, ohms/cmil/ft, 70°F: 632

Heat treatments
Cast plus aged 1,800°F/48 hr. Air cool.

Tensile Properties

Cast and Aged

Temperature, °F	T. S., psi	Y. S., psi, 0.2% offset	Elong., in 2 in., %	Hardness, Brinell
70	73,000	43,000	5	190

Specifications

ASTM: A297

Producers and Tradenames

ESCO Corp.: ESCO® 55
The Hica Corp.: HV
Sandusky Foundry and Machine Co.: HV
Stainless Foundry and Eng. Co.: HV
U.S. Pipe and Foundry Co.: HV

Table 1.4.5–11
ACI TYPE HW

Chemical composition, percent by weight: C, 0.55; Mn, 1.00; Fe, 25.00; S, 0.02; Si, 1.25; Cr, 12.00; Ni, 60.00

Physical constants and thermal properties
Density, lb/in.^{3}: 0.294
Modulus of elasticity, psi: tension, 25×10^6
Thermal conductivity, Btu/ft^2/hr/in./°F, 212°F: 92
Electrical resistivity, ohms/cmil/ft, 70°F: 674

Heat treatments
Cast plus aged 1,800°F/ 48 hr. Air cool.

Tensile Properties

Cast and Aged

Temperature, °F	T. S., psi	Y. S., psi, 0.2% offset	Elong., in 2 in., %	Hardness, Brinell
70	84,000	52,000	4	205

Specifications

ASTM: A297

Producers and Tradenames

ESCO Corp.: ESCO® 56
The Hica Corp.: HW
Sandusky Foundry and Machine Co.: HW
Stainless Foundry and Eng. Co.: HW

Table 1.4.5–12
ACI TYPE HX

Chemical composition, percent by weight: C, 0.55; Mn, 1.00; Fe, 14.00; S, 0.02; Si, 1.25; Cr, 17.00; Ni, 66.00

Physical constants and thermal properties
Density, lb/in.^{3}: 0.294
Coefficient of thermal expansion, (70–1,000°F) in./in./°F × 10^{-6}: 7.8
Modulus of elasticity, psi: tension, 25 × 10^6

Heat treatments
Cast plus aged 1,800°F/48 hr. Air cool.

Tensile Properties

Cast and Aged

Temperature, °F	T. S., psi	Y. S., psi, 0.2% offset	Elong., in 2 in., %	Hardness, Brinell
70	73,000	44,000	9	185

Specifications
ASTM: A297

Producers and Tradenames

ESCO Corp.: ESCO® 57
The Hica Corp.: HX
Sandusky Foundry and Machine Co.: HX
Stainless Foundry and Eng. Co.: HX
Lebanon Steel Foundry:(L)HX

1.4.6 HIGH TEMPERATURE-HIGH STRENGTH ALLOYS

Table 1.4.6—1
ASTROLOY

Chemical composition, percent by weight: C, 0.06; Cr,
15.00; Ni, 55.00; Co, 17.00; Mo, 5.30; Ti, 3.50; Al,
4.00; B, 0.030; Zr, 0.06

Physical constants and thermal properties
Density, lb/in.^{3} : 0.286
Coefficient of thermal expansion, (70–200°F) in./in./
°F × 10^{-6} : 7.5

Heat treatments
Age: 2,150°F/4 hr/air cool plus 1,975°F/4 hr/ air cool
plus 1,550°F/24 hr/air cool plus 1,400°F/16 hr/
air cool.

Tensile Properties

Aged

Temperature, °F	T.S., psi	Y.S., psi, 0.2% offset	Elong., in 2 in., %	Hardness, Brinell
70	205,000	152,000	16	–
1,000	180,000	140,000	16	–
1,200	190,000	140,000	18	–
1,400	168,000	132,000	21	–
1,600	112,000	100,000	25	–
1,800	60,000	40,000	30	–

Rupture Strength, 1,000 hr

Test temperature, °F	Strength, psi	Elong., in 2 in., %	Reduction of area, %
1,200	112,000	–	–
1,300	84,000	–	–
1,400	62,000	–	–
1,500	42,000	–	–
1,600	25,000	–	–
1,700	16,000	–	–
1,800	8,000	–	–

Producers and Tradenames

Teledyne Allvac: Astroloy

Table 1.4.6–2
D-979, AISI 664

Chemical composition, percent by weight: C, 0.05; Mn, 0.25; Fe, 27.00; Si, 0.70; Cr, 15.00; Ni, 45.00; Mo, 4.00; W, 4.00; Ti, 3.00; Al, 1.00; B, 0.010

Physical constants and thermal properties
Density, lb/in.^{3} : 0.296
Coefficient of thermal expansion, (70–200°F) in./in./
  °F × 10^{-6} : 7.6

Modulus of elasticity, psi: tension, 30.0 × 10^6; torsion, 11.6 × 10^6
Thermal conductivity, Btu/ft^2/hr/in./°F, 70°F: 87

Heat treatments
1,900°F/oil quench plus 1,550°F/6 hr/air cool plus 1,300°F/16 hr/air cool.

Tensile Properties

Aged

Temperature, °F	T.S., psi	Y.S., psi, 0.2% offset	Elong., in 2 in., %	Hardness, Brinell
70	204,000	146,000	15	–
1,000	188,000	134,000	15	–
1,200	160,000	129,000	21	–
1,400	104,000	95,000	17	–
1,600	50,000	44,000	18	–

Rupture Strength, 1,000 hr

Test temperature, °F	Strength, psi	Elong., in 2 in., %	Reduction of area, %
1,200	75,000	–	–
1,300	55,000	–	–
1,400	36,000	–	–
1,500	21,000	–	–
1,600	10,000	–	–

Specifications

SAE-AMS: 5509, 5746

Producers and Tradenames

Allegheny Ludlum Steel Co.: Altemp® D-979
Latrobe Steel Co.: Lescalloy® D-979
Special Metals Corp.: Udimet® D-979

Table 1.4.6–3
INCONEL® 706

Chemical composition, percent by weight: C, 0.03; Mn, 0.18; Fe, 40.0; S, 0.008; Si, 0.18; Cr, 16.0; Ni, 41.5; Cb, 2.9; Ti, 1.75; Al, 0.20

Physical constants and thermal properties
Density, $lb/in.^3$: 0.291
Coefficient of thermal expansion, (70–200°F) in./in./°F × 10^{-6} : 7.8
Modulus of elasticity, psi: tension, 30.4 × 10^6; torsion, 10.9 × 10^6
Poisson's ratio: 0.28

Melting range, °F: 2,434–2,499
Specific heat, Btu/lb/°F, 70°F: 0.106
Thermal conductivity, $Btu/ft^2/hr/in./°F$, 70°F: 87
Electrical resistivity, ohms/cmil/ft, 70°F: 592

Heat treatments
A. 1,800°F/1 hr/air cool plus 1,550°F/3 hr air cool plus 1,325°F/8 hr fast cool to 1,150°F/18 hr air cool.
B. 1,800°F/1 hr/air cool plus 1,350°F/8 hr/fast cool to 1,150°F/18 hr/ fast cool.

Tensile Properties

A

Temperature, °F	T.S., psi	Y.S., psi, 0.2% offset	Elong., in 2 in., %	Hardness, Brinell
70	188,000	142,000	19	–
1,000	163,000	130,000	19	–
1,200	147,000	120,000	21	–
1,400	100,000	98,000	32	–

Rupture Strength, 1,000 hr

A

Test temperature, °F	Strength, psi	Elong., in 2 in., %	Reduction of area, %
1,200	84,000	–	–
1,300	54,000	–	–
1,400	24,500	–	–

Fatigue Strength

Test temperature, °F	Stress, psi	Cycles to failure
70	61,500	10^8
1,000	61,500	10^8
1,200	61,500	10^8

Specifications

SAE-AMS: 5605, 5606, 5701, 5702, 5703

Producers and Tradenames

Huntington Alloy Products Division, INCO: Inconel alloy 706

Table 1.4.6–4
INCONEL® 718

Chemical composition, percent by weight: C, 0.04; Mn, 0.18; Fe, 18.5; S, 0.008; Si, 0.18; Cr, 19.0; Ni, 52.5; Mo, 3.05; Cb, 5.13; Ti, 0.90; Al, 0.50

Physical constants and thermal properties
Density, lb/in.^{3}: 0.296
Coefficient of thermal expansion, (70–200°F) in./in./°F × 10^{-6}: 7.2
Modulus of elasticity, psi: tension, 29.8 × 10^6; torsion, 11.6 × 10^6
Poisson's ratio: 0.28
Melting range, °F: 2,300–2,437
Specific heat, Btu/lb/°F, 70°F: 0.104

Thermal conductivity, Btu/ft^2/hr/in./°F, 70°F: 78
Electrical resistivity, ohms/cmil/ft, 70°F: 751
Curie temperature, °F: annealed, <–320; age hardened, <–170
Permeability (70°F, 200 Oe): annealed, 1.001; age hardened, 1.001

Heat treatments
A. 1,800°F/1 hr/air cool plus 1,325°F/8 hr/fast cool to 1,150°F/18 hr/air cool.
B. 1,900°F/1 hr/air cool plus 1,400°F/10 hr/fast cool to 1,200°F/20 hr/air cool.

Tensile Properties

A

Temperature, °F	T.S., psi	Y.S., psi, 0.2% offset	Elong., in 2 in., %	Hardness, Brinell
70	208,000	172,000	21	–
1,000	185,000	154,000	18	–
1,200	178,000	148,000	19	–
1,400	138,000	107,000	25	–
1,600	49,000	48,000	88	–
1,800	15,000	15,000	170	–
2,000	8,000	8,000	125	–

Rupture Strength, 1,000 hr

A

Test temperature, °F	Strength, psi	Elong., in 2 in., %	Reduction of area, %
1,200	86,000	–	–
1,300	53,000	–	–
1,400	25,000	–	–

Creep Strength (Stress, psi, to Produce 1% Creep)

Aged – A

Test temperature, °F	10,000 hr	100,000 hr
1,100	98,000	–
1,200	75,000	–
1,300	42,000	–

Table 1.4.6–4 (continued)
INCONEL® 718

Fatigue Strength

Aged – A

Test temperature, °F	Stress, psi	Cycles to failure
70	90,000	10^8
600	110,000	10^8
1,000	90,000	10^8
1,200	72,000	10^8

Impact Strength

Aged – B

Test temperature, °F	Type test	Strength, ft-lb
+70	Charpy – V-notched	35

Specifications

SAE-AMS: 5589, 5590, 5596, 5597, 5662, 5663, 5664, 5832

Producers and Tradenames

Huntington Alloy Products Division, INCO: Inconel® alloy 718
Allvac Teledyne: Allvac® 718
Special Metals Corp.: Udimet® 718
Stellite Division: Cabot Corp.: Haynes® 718
Universal-Cyclops: Unitemp® 718

Table 1.4.6–5
INCONEL® X-750

Chemical composition, percent by weight: C, 0.04; Mn, 0.50; Fe, 7.00; S, 0.005; Si, 0.25; Cr, 15.5; Ni, 73.0; Cb, 0.95; Ti, 2.50; Al, 0.70

Physical constants and thermal properties
Density, lb/in.³: 0.298
Coefficient of thermal expansion, (70–200°F) in./in./°F × 10^{-6}: 5.7
Modulus of elasticity, psi: tension, 31 × 10^6; torsion, 11 × 10^6
Poisson's ratio: 0.29
Melting range, °F: 2,540–2,600

Specific heat, Btu/lb/°F, 70°F: 0.103
Thermal conductivity, Btu/ft² /hr/in./°F, 70°F: 83
Electrical resistivity, ohms/cmil/ft, 70°F: 731
Curie temperature, °F: annealed, –225; age hardened, –193
Permeability (70°F, 200 Oe): annealed, 1.002; age hardened, 1.0035

Heat treatments
A. 1,800°F/1 hr/air cool plus 1,350°F/8 hr/fast cool to 1,150°F/18 hr/air cool.
B. 2,100°F/2–4 hr/air cool plus 1,550°F/24 hr/air cool plus 1,300°F/20 hr/air cool.

Table 1.4.6–5 (continued)
INCONEL® X-750

Tensile Properties

A

Temperature, °F	T.S., psi	Y.S., psi, 0.2% offset	Elong., in 2 in., %	Hardness, Brinell
70	195,500	140,000	24	–
600	178,000	131,500	21	–
800	173,000	131,500	21	–
1,000	168,500	128,000	13	–
1,200	143,000	122,500	6	–

Rupture Strength, 1,000 hr

B

Test temperature, °F	Strength, psi	Elong., in 2 in., %	Reduction of area, %
1,200	68,000	–	–
1,300	47,000	–	–
1,400	30,000	–	–
1,500	16,000	–	–
1,600	6,800	–	–
1,700	3,100	–	–

Creep Strength (Stress, psi, to Produce 1% Creep)

B

Test temperature, °F	10,000 hr	100,000 hr
1,200	75,000	–
1,500	21,000	–

Fatigue Strength

A

Test temperature, °F	Stress, psi	Cycles to failure
70	80,000	10^8

Impact Strength

B

Test temperature, °F	Type test	Strength, ft-lb
–320	Charpy–V-notched	33
+70	Charpy–V-notched	37
1,000	Charpy–V-notched	49

Table 1.4.6–5 (continued)
INCONEL® X-750

Specifications

SAE-AMS: 5542, 5598, 5668, 5670, 5671, 5582, 5698, 5699

Producers and Tradenames

Huntington Alloy Products Division, INCO: Inconel alloy X-750
Allvac Teledyne: Nickelvac® X-750
Simonds Steel: Simalloy 750
Special Metals: Udimet® X-750
Carpenter Technology: Pyromet® X-750

Table 1.4.6–6
NIMONIC® ALLOY 80A

Chemical composition, percent by weight: C, 0.06; Mn, 0.30; Si, 0.30; Cr, 19.5; Ni, 76.0; Ti, 2.4; Al, 1.4; B, 0.003; Zr, 0.06

Physical constants and thermal properties
Density, lb/in.³ : 0.295
Coefficient of thermal expansion, (70–200°F) in./in./
 °F × 10⁻⁶ : 7.0
Modulus of elasticity, psi: tension, 31.8×10^6

Melting range, °F: 2,480–2,535
Specific heat, Btu/lb/°F, 70°F: 0.11
Thermal conductivity, Btu/ft²/hr/in./°F, 70°F: 60
Electrical resistivity, ohms/cmil/ft, 70°F: 735

Heat treatments
Age harden: 1,975°F/8 hr/air cool plus 1,300°F/16
 hr/air cool.

Tensile Properties

Temperature, °F	T.S., psi	Y.S., psi, 0.2% offset	Elong., in 2 in., %	Hardness, Brinell
70	145,000	90,000	39	–
1,000	127,000	77,000	37	–
1,200	115,000	80,000	21	–
1,400	87,000	73,000	17	–
1,600	45,000	38,000	30	–
1,800	11,000	9,000	–	–

Rupture Strength, 1,000 hr

Test temperature, °F	Strength, psi	Elong., in 2 in., %	Reduction of area, %
1,200	61,000	–	–
1,300	39,000	–	–
1,400	23,000	–	–
1,500	12,000	–	–

Producers and Tradenames

Huntington Alloy Products Division, INCO: Nimonic alloy 80A
Carpenter Technology: Pyromet® 80A
Special Metals Corp.: Udimet® 80A

Table 1.4.6–7
NIMONIC® ALLOY 90

Chemical composition, percent by weight: C, 0.07; Mn, 0.30; Si, 0.30; Cr, 19.5; Ni, 59.0; Co, 16.5; Ti, 2.45; Al, 1.45; B, 0.003; Zr, 0.06

Physical constants and thermal properties
Density, lb/in.^{3}: 0.296
Coefficient of thermal expansion, (70–200°F) in./in./°F × 10^{-6}: 6.4
Modulus of elasticity, psi: tension, 32.7 × 10^6

Melting range, °F: 2,480–2,535
Specific heat, Btu/lb/°F, 70°F: 0.11
Thermal conductivity, Btu/ft^2/hr/in./°F, 70°F: 68
Electrical resistivity, ohms/cmil/ft, 70°F: 690

Heat treatments
Age harden: 1,975°F/8 hr/air cool plus 1,300°F/16 hr/air cool.

Tensile Properties

Aged

Temperature, °F	T.S., psi	Y.S., psi, 0.2% offset	Elong., in 2 in., %	Hardness, Brinell
70	179,000	117,000	33	–
1,000	156,000	105,000	28	–
1,200	136,000	99,000	14	–
1,400	95,000	78,000	12	–
1,600	48,000	38,000	23	–
1,800	11,000	9,000	70	–

Rupture Strength, 1,000 hr

Aged

Test temperature, °F	Strength, psi	Elong., in 2 in., %	Reduction of area, %
1,200	66,000	–	–
1,300	47,000	–	–
1,400	30,000	–	–
1,500	16,000	–	–
1,600	8,800	–	–

Producers and Tradenames

Huntington Alloy Products Division, INCO: Nimonic alloy 90
Carpenter Technology: Pyromet® 90

Table 1.4.6–8
PYROMET® 860

Chemical composition, percent by weight: C, 0.05; Mn, 0.05; Fe, 30.0; Si, 0.05; Cr, 12.6; Ni, 43.0; Co, 4.0; Mo, 6.0; Ti, 3.0; Al, 1.25; B, 0.010

Modulus of elasticity, psi: tension, 29.0 × 10^6

Physical constants and thermal properties
Density, lb/in.^{3}: 0.297
Coefficient of thermal expansion, (70–200°F) in./in./°F × 10^{-6}: 7.8

Heat treatments
Age harden: 2,000°F/2 hr/water quench plus 1,525°F/2 hr/air cool plus 1,400°F/24 hr/air cool.

Table 1.4.6–8 (continued)
PYROMET® 860

Tensile Properties

Aged

Temperature, °F	T.S., psi	Y.S., psi, 0.2% offset	Elong., in 2 in., %	Hardness, Brinell
70	188,000	121,000	22	–
1,000	182,000	122,000	15	–
1,200	161,000	123,000	17	–
1,400	132,000	121,000	18	–

Rupture Strength, 1,000 hr

Aged

Test temperature, °F	Strength, psi	Elong., in 2 in., %	Reduction of area, %
1,200	79,000	–	–
1,300	53,000	–	–
1,400	36,000	–	–
1,500	20,000	–	–

Producers and Tradenames

Carpenter Technology Corp.: Pyromet 860

Table 1.4.6–9
RENE 41 (AISI No. 683)

Chemical composition, percent by weight: C, 0.09; Cr, 19.0; Ni, 55.0; Co, 11.0; Mo, 10.0; Ti, 3.1; Al, 1.5; B, 0.005

Physical constants and thermal properties
Density, lb/in.^{3}: 0.298
Coefficient of thermal expansion, (70–200°F) in./in./°F × 10^{-6}: 6.6

Modulus of elasticity, psi: tension 31.9 × 10^6
Melting range, °F: 2,385–2,450
Specific heat, Btu/lb/°F, 70°F: 0.108
Thermal conductivity, Btu/ft^2/hr/in./°F, 70°F: 62

Heat treatments
Age harden: 1,950°F/4 hr/air cool plus 1,400°F/16 hr/air cool.

Tensile Properties

Age Hardened

Temperature, °F	T.S., psi	Y.S., psi, 0.2% offset	Elong., in 2 in., %	Hardness, Brinell
70	206,000	154,000	14	–
1,000	203,000	147,000	14	–
1,200	194,000	145,000	14	–
1,400	160,000	136,000	11	–
1,600	90,000	80,000	19	–
1,800	42,000	38,000	36	–

Table 1.4.6−9 (continued)
RENE 41 (AISI No. 683)

Rupture Strength, 1,000 hr

Age Hardened

Test temperature, °F	Strength, psi	Elong., in 2 in., %	Reduction of area, %
1,200	100,000	−	−
1,300	74,000	−	−
1,400	40,000	−	−
1,500	24,000	−	−
1,600	14,000	−	−

Specifications

SAE-AMS: 5399, 5545, 5712, 5713, 5800, 7469

Producers and Tradenames

Teledyne Allvac: Allvac® René 41
Simonds Steel: Simalloy René 41
Carpenter Technology: Vacumeltrol® 41
Stellite Division, Cabot: Haynes® alloy R-41
Special Metals, Inc.: Udimet® 41

Table 1.4.6−10
M-252 (AISI 689)

Chemical composition, percent by weight: C, 0.15; Mn, 0.50; Si, 0.50; Cr, 20.0; Ni, 55.0; Co, 10.0; Mo, 10.0; Ti, 2.6; Al, 1.0; B, 0.005

in./in./°F × 10^{-6} : 5.9
Modulus of elasticity, psi: tension, 29.8 × 10^6
Thermal conductivity, Btu/ft²/hr/in./°F, 70°F: 82

Physical constants and thermal properties
Density, lb/in.³: 0.298
Coefficient of thermal expansion, (70−200°F)

Heat treatments
Age harden: 1,900°F/4 hr/air cool plus 1,400°F/16 hr/air cool.

Tensile Properties

Temperature, °F	T.S., psi	Y.S., psi, 0.2% offset	Elong., in 2 in., %	Hardness, Brinell
70	180,000	122,000	16	−
1,000	178,000	111,000	15	−
1,200	168,000	108,000	11	−
1,400	137,000	104,000	10	−
1,600	74,000	70,000	18	−

Rupture Strength, 1,000 hr

Test temperature, °F	Strength, psi	Elong., in 2 in., %	Reduction of area, %
1,200	82,000	−	−
1,300	60,000	−	−
1,400	39,000	−	−
1,500	23,000	−	−
1,600	14,000	−	−

Table 1.4.6–10 (continued)
M-252 (AISI 689)

Specifications

ASTM: A461; SAE-AMS: 5756, 5757, 5551

Producers and Tradenames

Teledyne Allvac: Allvac® M-252
Special Metals Inc.: Udimet® M-252
Carpenter Technology: Carpenter® M-252
Allegheny-Ludlum: Altemp® M-252
Universal-Cyclops: Unitemp® M-252

Table 1.4.6–11
UDIMET® 500 (AISI 684)

Chemical composition, percent by weight: C, 0.08; Cr, 18.0; Ni, 54.0; Co, 18.5; Mo, 4.0; Ti, 2.9; Al, 2.9; B, 0.006; Zr, 0.05

$^\circ$F $\times$ 10^{-6}: 6.8
Modulus of elasticity, psi: tension, 32.1 $\times$ 10^6
Thermal conductivity, Btu/ft^2/hr/in./$^\circ$F, 70°F: 77

Physical constants and thermal properties
Density, lb/in.^{3}: 0.290
Coefficient of thermal expansion, (70–200°F) in./in./

Heat treatments
Age harden: 1,975°F/4 hr/air cool plus 1,550°F/24 hr/air cool plus 1,400°F/16 hr/air cool.

Tensile Properties

Temperature, $^\circ$F	T.S., psi	Y.S., psi, 0.2% offset	Elong., in 2 in., %	Hardness, Brinell
70	190,000	122,000	32	–
1,000	180,000	115,000	28	–
1,200	176,000	110,000	28	–
1,400	151,000	106,000	39	–
1,600	93,000	72,000	20	–
1,800	42,000	33,000	22	–

Rupture Strength, 1,000 hr

Test temperature, $^\circ$F	Strength, psi	Elong., in 2 in., %	Reduction of area, %
1,200	110,000	–	–
1,300	74,000	–	–
1,400	47,000	–	–
1,500	30,000	–	–
1,600	18,000	–	–
1,700	12,000	–	–

Specifications

ASTM: A461; SAE-AMS: 5384, 5751, 5753

Producers and Tradenames

Teledyne Allvac: Allvac® 500
Special Metals Inc.: Udimet 500

Table 1.4.6–12
UDIMET® 700 (AISI 687)

Chemical composition, percent by weight: C, 0.08; Cr, 15.0; Ni, 53.0; Co, 18.5; Mo, 5.2; Ti, 3.5; Al, 4.3; B, 0.030

Modulus of elasticity, psi: tension, 32.4×10^6
Thermal conductivity, Btu/ft²/hr/in./°F, 70°F: 136

Physical constants and thermal properties
Density, lb/in.³: 0.286
Coefficient of thermal expansion, (70–200°F) in./in./°F $\times 10^{-6}$: 7.5

Heat treatments
Age harden: 2,150°F/4 hr/air cool plus 1,975°F/4 hr/air cool plus 1,550°F/24 hr/air cool plus 1,400°F/16 hr/air cool.

Tensile Properties

Temperature, °F	T.S., psi	Y.S., psi, 0.2% offset	Elong, in. 2 in., %	Hardness, Brinell
70	204,000	140,000	17	–
1,000	185,000	130,000	16	–
1,200	180,000	124,000	16	–
1,400	150,000	120,000	20	–
1,600	100,000	92,000	27	–
1,800	52,000	44,000	32	–
2,000	15,000	12,000	35	–

Rupture Strength, 1,000 hr

Test temperature, °F	Strength, psi	Elong., in 2 in., %	Reduction of area, %
1,200	102,000	–	–
1,300	83,000	–	–
1,400	62,000	–	–
1,500	43,000	–	–
1,600	29,000	–	–
1,700	16,000	–	–
1,800	7,500	–	–

Producers and Tradenames

Special Metals Corp.: Udimet 700
Teledyne Allvac: Allvac® 700

Table 1.4.6–13
WASPALLOY (AISI 685)

Chemical composition, percent by weight: C, 0.08; Cr, 19.5; Ni, 58.0; Co, 13.5; Mo, 4.3; Ti, 3.0; Al, 1.3; B, 0.006; Zr, 0.06

Modulus of elasticity, psi: tension, 30.9×10^6
Thermal conductivity, Btu/ft²/hr/in./°F, 70°F: 74

Physical constants and thermal properties
Density, lb/in.³: 0.296
Coefficient of thermal expansion, (70–200°F) in./in./°F $\times 10^{-6}$: 6.8

Heat treatments
Age harden: 1,975°F/4 hr/air cool plus 1,550°F/24 hr/air cool plus 1,400°F/16 hr/air cool.

Table 1.4.6–13 (continued)
WASPALLOY (AISI 685)

Tensile Properties

Temperature, °F	T.S., psi	Y.S., psi, 0.2% offset	Elong., in 2 in., %	Hardness, Brinell
70	185,000	115,000	25	–
1,000	170,000	105,000	23	–
1,200	162,000	100,000	34	–
1,400	115,000	98,000	28	–
1,600	76,000	75,000	35	–
1,800	28,000	20,000	–	–

Rupture Strength, 1,000 hr

Test temperature, °F	Strength, psi	Elong., in 2 in., %	Reduction of area, %
1,200	89,000	–	–
1,300	65,000	–	–
1,400	42,000	–	–
1,500	26,000	–	–
1,600	16,000	–	–

Fatigue Strength

Age Hardened

Test temperature, °F	Stress, psi	Cycles to failure
70	63,000	10^7
1,200	52,000	10^7
1,500	40,000	10^7

Impact Strength

Age Hardened

Test temperature, °F	Type test	Strength, ft-lb
+70	Charpy – V-notched	14

Specifications

ASTM: A461; SAE-AMS: 5544, 5586, 5704, 5706, 5707, 5708, 5709, 5828, 7471

Producers and Tradenames

Allegheny Ludlum: Altemp® Waspalloy
Teledyne Allvac: Allvac® Waspalloy
Carpenter Technology: Carpenter® Waspalloy
Crucible Steel: Crucible® Waspalloy
Special Metals Corp.: Udimet® Waspalloy
Universal-Cyclops: Unitemp® Waspalloy

Table 1.4.6–14
INCOLOY® 901 (AISI 681 and 682)

Chemical composition, percent by weight: C, 0.05; Mn, 0.10; Fe, 36.0; Si, 0.10; Cr, 12.5; Ni, 42.5; Mo, 5.7; Ti, 2.8; Al, 0.2; B, 0.015

Physical constants and thermal properties
Density, lb/in.^{3}: 0.297
Coefficient of thermal expansion, (70–200°) in./in./°F × 10^{-6}: 7.8

Modulus of elasticity, psi: tension, 29.9 × 10^6
Poisson's ratio: 0.29
Thermal conductivity, Btu/ft^2/hr/in./°F, 70°F: 92

Heat treatments
Age harden: 2,000°F/2 hr/water quench plus 1,450°F/2 hr/air cool plus 1,325°F/24 hr/air cool.

Tensile Properties

Age Hardened

Temperature, °F	T.S., psi	Y.S., psi, 0.2% offset	Elong., in 2 in., %	Hardness, Brinell
70	175,000	130,000	14	–
1,000	149,000	113,000	14	–
1,200	139,000	110,000	13	–
1,400	105,000	92,000	19	–

Rupture Strength, 1,000 hr

Age Hardened

Test temperature, °F	Strength, psi	Elong., in 2 in., %	Reduction of area, %
1,200	76,000	–	–
1,300	53,000	–	–
1,400	30,000	–	–
1,500	12,000	–	–

Specifications

SAE-AMS: 5660, 5661

Producers and Tradenames

Allegheny Ludlum: Altemp® 901
Carpenter Technology: Pyromet® 901
Huntington Alloy Products Division, INCO: Incoloy® alloy 901
Special Metals Corp.: Udimet® 200
Simonds Steel: Simonds® 901
Cameron Iron Works: Camvac® 901
Universal-Cyclops: Unitemp® 901

Table 1.4.6–15
ALLOY 713C

Chemical composition, percent by weight: C, 0.12; Cr, 12.5; Ni, 74.0; Mo, 4.2; Cb, 2.0; Ti, 0.8; Al, 6.1; B, 0.012; Zr, 0.10

Coefficient of thermal expansion, (70–200°F) in./in./°F × 10⁻⁶ : 5.9

Coefficient of thermal expansion, $(70-200°F)$ in./in./ $°F \times 10^{-6}$: 5.9
Modulus of elasticity, psi: tension, 29.9×10^6
Specific heat, Btu/lb/°F, 70°F: 0.10

Physical constants and thermal properties
Density, lb/in.³ : 0.286

Density, lb/in.^{3} : 0.286

Heat treatments
As cast.

Tensile Properties

As Cast

Temperature, °F	T.S., psi	Y.S., psi, 0.2% offset	Elong., in 2 in., %	Hardness, Brinell
70	123,000	107,000	8	–
1,000	125,000	102,000	10	–
1,200	126,000	104,000	7	–
1,400	136,000	108,000	6	–
1,600	105,000	72,000	14	–
1,800	68,000	44,000	20	–

Rupture Strength, 1,000 hr

As Cast

Test temperature, °F	Strength, psi	Elong., in 2 in., %	Reduction of area, %
1,300	88,000	–	–
1,400	65,000	–	–
1,500	44,000	–	–
1,600	28,000	–	–
1,700	18,000	–	–
1,800	13,000	–	–
1,900	7,700	–	–

Specifications

SAE-AMS: 5391

Producers and Tradenames

Austenal Special Products: Austenal 655
Cannon-Muskegon: AMS 5391
ESCO Corp.: Esco® 713 C-E
Special Metals: Udimet® 713C
Stellite Division, Cabot: Haynes® alloy 713C

Table 1.4.6–16
ALLOY 713LC

Chemical composition, percent by weight: C, 0.05; Cr, 12.0; Ni, 75.0; Mo, 4.5; Cb, 2.0; Ti, 0.6; Al, 5.9; B, 0.010; Zr, 0.10

Coefficient of thermal expansion, (70–200°F) in./in./°F × 10^{-6}: 5.6
Modulus of elasticity, psi: tension, 28.6 × 10^6
Specific heat, Btu/lb/°F, 70°F: 0.105

Physical constants and thermal properties
Density, lb/in.^{3}: 0.289

Heat treatments
As cast.

Tensile Properties

As Cast

Temperature, °F	T.S., psi	Y.S., psi, 0.2% offset	Elong., in 2 in., %	Hardness, Brinell
70	130,000	109,000	15	–
1,000	130,000	110,000	11	–
1,200	131,000	114,000	11	–
1,400	135,000	110,000	11	–
1,600	109,000	84,000	12	–
1,800	68,000	41,000	22	–

Rupture Strength, 1,000 hr

As Cast

Test temperature, °F	Strength, psi	Elong., in 2 in., %	Reduction of area, %
1,400	60,000	–	–
1,500	45,000	–	–
1,600	30,000	–	–
1,700	19,000	–	–
1,800	13,000	–	–

Producers and Tradenames

Austenal Special Products: Austenal 579M
Cannon Muskegon: LC 713C
Stellite Division, Cabot; Haynes® 713C (low carbon)

Table 1.4.6–17
B-1900

Chemical composition, percent by weight: C, 0.10; Cr, 8.0; Ni, 64.0; Co, 10.0; Mo, 6.0; Ti, 1.0; Al, 6.0; B, 0.015; Zr, 0.10; Ta, 4.0

Coefficient of thermal expansion, (70–200°F) in./in./°F × 10^{-6}: 6.5
Modulus of elasticity, psi: tension, 31.0 × 10^6

Physical constants and thermal properties
Density, lb/in.^{3}: 0.297

Heat treatments
As cast.

Table 1.4.6–17 (continued)
B-1900

Tensile Properties

As Cast

Temperature, °F	T.S., psi	Y.S., psi, 0.2% offset	Elong., in 2 in., %	Hardness, Brinell
70	141,000	120,000	8	–
1,000	146,000	126,000	7	–
1,200	147,000	134,000	6	–
1,400	136,000	117,000	4	–
1,600	115,000	101,000	4	–
1,800	80,000	60,000	7	–
2,000	38,000	28,000	11	–

Rupture Strength, 1,000 hr

As Cast

Test temperature, °F	Strength, psi	Elong., in 2 in., %	Reduction of area, %
1,500	55,000	–	–
1,600	37,000	–	–
1,700	25,000	–	–
1,800	15,000	–	–
1,900	8,800	–	–
2,000	4,900	–	–

Producers and Tradenames

Austenal Special Products: Austenal 579M
Cannon Muskegon: LC 713C
Stellite Division, Cabot: Haynes® 713C (low carbon)

Table 1.4.6–18
IN 100

Chemical composition, percent by weight: C, 0.18; Cr, 10.0; Ni, 60.0; Co, 15.0; Mo, 3.0; Ti, 4.7; Al, 5.5; B, 0.014; Zr, 0.06; V, 1.0

Coefficient of thermal expansion, (70–200°F) in./in./°F × 10^{-6}: 7.2
Modulus of elasticity, psi: tension, 31.2 × 10^6
Melting range, °F: 2,305–2,435

Physical constants and thermal properties
Density, lb/in.^{3}: 0.280

Heat treatments
As cast.

Tensile Properties

As Cast

Temperature, °F	T.S., psi	Y.S., psi, 0.2% offset	Elong., in 2 in., %	Hardness, Brinell
70	147,000	123,000	9	–
1,000	158,000	128,000	9	–
1,200	161,000	129,000	6	–
1,400	155,000	125,000	6	–
1,600	128,000	101,000	6	–
1,800	82,000	54,000	6	–

Table 1.4.6–18 (continued)
IN 100

Rupture Strength, 1,000 hr

As Cast

Test temperature, °F	Strength, psi	Elong., in 2 in., %	Reduction of area, %
1,400	75,000	–	–
1,500	55,000	–	–
1,600	37,000	–	–
1,700	25,000	–	–
1,800	15,000	–	–
1,900	8,500	–	–

Specifications

SAE-AMS: 5397

Producers and Tradenames

Austenal Special Products: Austenal 100
Cannon Muskegon: AMS 5397 IN-100
Special Metals Corp.: Udimet® IN-100
TRW Metals Division: IN-100
Stellite Division, Cabot: Haynes® alloy IN-100

Table 1.4.6–19
IN 738

Chemical composition, percent by weight: C, 0.17; Cr, 16.0; Ni, 61.0; Co, 8.5; Mo, 1.7; W, 2.6; Cb, 0.9; Ta, 1.7; Ti, 3.4; Al, 3.4; B, 0.010; Zr, 0.10

Modulus of elasticity, psi: tension, 29.2×10^6
Specific heat, Btu/lb/°F, 70°F: 0.10

Physical constants and thermal properties
Density, lb/in.³: 0.293

Heat treatments
2,050°F/2 hr/air cool plus 1,550°F/24 hr/air cool.

Tensile Properties

Age Hardened

Temperature, °F	T.S., psi	Y.S., psi, 0.2% offset	Elong., in 2 in., %	Hardness, Brinell
70	159,000	138,000	5	–
1,200	151,000	118,000	7	–
1,400	140,000	115,000	6.5	–
1,600	112,000	80,000	11	–
1,800	66,000	50,000	13	–

Table 1.4.6–19 (continued)
IN 738

Rupture Strength, 1,000 hr

Age Hardened

Test temperature, °F	Strength, psi	Elong., in 2 in., %	Reduction of area, %
1,400	69,000	–	–
1,500	48,000	–	–
1,600	31,500	–	–
1,700	20,000	–	–
1,800	12,000	–	–

Producers and Tradenames

Austenal Special Products: Austenal 738X

Table 1.4.6–20
GMR 235-D

Chemical composition, percent by weight: C, 0.15; Fe, 4.5; Cr, 15.5; Ni, 68.0; Mo, 5.0; Ti, 2.5; Al, 3.5; B, 0.050

Modulus of elasticity, psi: tension, 28.7×10^6

Physical constants and thermal properties
Density, lb/in.^{3}: 0.291

Heat treatments
As cast.

Tensile Properties

As Cast

Temperature, °F	T.S., psi	Y.S., psi, 0.2% offset	Elong., in 2 in., %	Hardness, Brinell
70	112,000	103,000	3.5	–
1,000	104,000	82,000	5	–
1,200	110,000	85,000	6	–
1,400	115,000	86,000	3	–
1,600	105,000	65,000	6	–

Rupture Strength, 1,000 hr

As Cast

Test temperature, °F	Strength, psi	Elong., in 2 in., %	Reduction of area, %
1,300	85,000	–	–
1,400	60,000	–	–
1,500	33,000	–	–
1,600	23,000	–	–
1,700	17,000	–	–
1,800	12,000	–	–

Producers and Tradenames
Cannon-Muskegon: GMR 235-D

Table 1.4.6–21
MAR-M® 200

Chemical composition, percent by weight: C, 0.15; Cr, 9.0; Ni, 60.0; Co, 10.0; W, 12.0; Cb, 1.0; Ti, 2.0; Al, 5.0; B, 0.015; Zr, 0.05

Physical constants and thermal properties
Density, lb/in.^{3}: 0.308

Modulus of elasticity, psi: tension, 31.6×10^6
Melting range, °F: 2,400–2,500
Thermal conductivity, Btu/ft^2/hr/in./°F, 70°F: 88

Heat treatments
Age harden: 1,600°F/50 hr/air cool.

Tensile Properties

Age Hardened

Temperature, °F	T.S., psi	Y.S., psi, 0.2% offset	Elong., in 2 in., %	Hardness, Brinell
70	135,000	120,000	7	–
1,000	137,000	123,000	5	–
1,200	138,000	124,000	4	–
1,400	135,000	122,000	3	–
1,600	122,000	110,000	4	–
1,800	80,000	68,000	5	–

Rupture Strength, 1,000 hr

Age Hardened

Test temperature, °F	Strength, psi	Elong., in 2 in., %	Reduction of area, %
1,400	84,000	–	–
1,500	60,000	–	–
1,600	43,000	–	–
1,700	29,000	–	–
1,800	18,000	–	–

Producers and Tradenames

TRW Metals Division: MAR-M 200
Stellite Division, Cabot: Haynes® alloy 200

Table 1.4.6–22
MAR-M® 246

Chemical composition, percent by weight: C, 0.15; Cr, 9.0; Ni, 60.0; Co, 10.0; Mo, 2.5; W, 10.0; Ta, 1.5; Ti, 1.5; Al, 5.5; B, 0.015; Zr, 0.05

Physical constants and thermal properties
Density, lb/in.^{3}: 0.305

Coefficient of thermal expansion, (70–200°F) in./in./°F $\times 10^{-6}$: 6.3
Modulus of elasticity, psi: tension, 29.8×10^6

Heat treatments
Age harden: 1,550°F/50 hr/air cool.

Table 1.4.6–22 (continued)
MAR-M® 246

Tensile Properties

Temperature, °F	T.S., psi	Y.S., psi, 0.2% offset	Elong., in 2 in., %	Hardness, Brinell,
70	140,000	125,000	5	–
1,000	145,000	125,000	5	–
1,200	150,000	125,000	5	–
1,400	150,000	125,000	5	–
1,600	125,000	100,000	5	–
1,800	80,000	55,000	8	–

Rupture Strength, 1,000 hr

Test temperature, °F	Strength, psi	Elong., in 2 in., %	Reduction of area, %
1,400	86,000	–	–
1,500	62,000	–	–
1,600	44,000	–	–
1,700	30,000	–	–
1,800	19,000	–	–

Table 1.4.6–23
MAR-M® 421

Chemical composition, percent by weight: C, 0.15; Cr, 15.8; Ni, 61.0; Co, 9.5; Mo, 2.0; W, 3.8; Cb, 2.0; Ti, 1.8; Al, 4.3; B, 0.015; Zr, 0.05

Modulus of elasticity, psi: tension, 29.4×10^6

Physical constants and thermal properties
Density, lb/in.³: 0.292

Heat treatments
Age harden: 2,100°F/2 hr/air cool plus 1,950°F/4 hr/air cool plus 1,400°F/16 hr/air cool.

Tensile Properties

Age Hardened

Temperature, °F	T.S., psi	Y.S., psi, 0.2% offset	Elong., in 2 in., %	Hardness, Brinell
70	157,000	135,000	4.5	–
1,000	147,000	118,000	3	–
1,200	140,000	119,000	4	–
1,400	138,000	125,000	2.5	–
1,600	109,000	94,000	6	–
1,800	55,000	39,000	22	–

Table 1.4.6–23 (continued)
MAR-M® 421

Rupture Strength, 1,000 hr

Age Hardened

Test temperature, °F	Strength, psi	Elong., in 2 in., %	Reduction of area, %
1,400	63,000	–	–
1,500	44,000	–	–
1,600	31,000	–	–
1,700	20,000	–	–
1,800	12,000	–	–
1,900	7,000	–	–
2,000	4,000	–	–

Table 1.4.6–24
MAR-M® 432

Chemical composition, percent by weight: C, 0.15; Cr, 15.5; Ni, 50.0; Co, 20.0; W, 3.0; Cb, 2.0; Ta, 2.0; Ti, 4.3; Al, 2.8; B, 0.015; Zr, 0.05

Heat treatments
Age harden: 1,975°F/4 hr/air cool plus 1,400°F/16 hr/air cool.

Physical constants and thermal properties
Density, lb/in.³: 0.295

Tensile Properties

Age Hardened

Temperature, °F	T.S., psi	Y.S., psi, 0.2% offset	Elong., in 2 in., %	Hardness, Brinell
70	180,000	155,000	6	–
1,000	160,000	132,000	–	–
1,200	158,000	132,000	–	–
1,400	156,000	132,000	3.5	–
1,600	106,000	88,000	8	–
1,800	54,000	41,000	21	–

Rupture Strength, 1,000 hr

Age Hardened

Test temperature, °F	Strength, psi	Elong., in 2 in., %	Reduction of area, %
1,400	70,000	–	–
1,500	48,000	–	–
1,600	31,000	–	–
1,700	22,000	–	–
1,800	14,000	–	–
1,900	7,500	–	–

Table 1.4.6—25
INCONEL® ALLOY MA 753

Chemical composition, percent by weight: Cr, 20.00; Ni, 75.00; Ti, 2.50; Al, 1.50; Y_2O_3, 1.3

Coefficient of thermal expansion, (70–400°F) in./in./°F $\times$ 10^{-6} : 7.05
Modulus of elasticity, psi: tension, 32.4 $\times$ 10^6
Melting range, °F: 2,425–2,450

Physical constants and thermal properties
Density, lb/in.^{3}: 0.292

Heat treatments
Age harden: 1,300°F/24 hr.

Tensile Properties

Aged

Temperature, °F	T.S., psi	Y.S., psi, 0.2% offset	Elong., in 2 in., %	Hardness, Brinell
70	175,000	129,000	9	–
1,000	141,000	107,000	10	–
1,200	120,000	101,000	18	–
1,400	84,000	81,000	25	–
1,600	33,000	31,000	25	–
1,800	23,000	21,000	14	–

Rupture Strength, 1,000 hr

Aged

Test temperature, °F	Strength, psi	Elong., in 2 in., %	Reduction of area, %
1,200	60,000	–	–
1,300	47,000	–	–
1,400	34,000	–	–
1,500	25,000	–	–
1,600	22,000	–	–
1,700	19,500	–	–
1,800	17,500	–	–
1,900	14,000	–	–
2,000	11,500	–	–

Producers and Tradenames

Huntington Alloy Products Division, INCO: Inconel alloy MA 753

1.5 OTHER METALS AND MISCELLANEOUS PROPERTIES

Table 1.5–1
MECHANICAL PROPERTIES OF COPPER ALLOYS

Because of their corrosion resistance and the fact that copper alloys have been used for many thousands of years, the number of copper alloys available is second only to the ferrous alloys. In general, copper alloys do not have the high-strength qualities of the ferrous alloys, while their density is comparable. The cost per strength-weight ratio is high; however, they have the advantage of ease of joining by soldering, which is not shared by other metals that have reasonable corrosion resistance.

Material	Nominal composition	Form and condition	Typical mechanical properties				Comments
			Yield strength (0.2% offset), 1000 lb/sq in.	Tensile strength, 1000 lb/sq in.	Elongation, in 2 in., %	Hardness, Brinell	
Copper ASTM B152 ASTM B124, B133 ASTM B1, B2, B3	Cu 99.9 plus	Annealed Cold-drawn Cold-rolled	10 40 40	32 45 46	45 15 5	42 90 100	Bus-bars, switches, architectural, roofing, screens
Gilding metal ASTM B36	Cu 95.0 Zn 5.0	Cold-rolled	50	56	5	114	Coinage, ammunition
Cartridge 70–30 brass ASTM B14 ASTM B19 ASTM B36 ASTM B134 ASTM B135	Cu 70.0 Zn 30.0	Cold-rolled	63	76	8	155	Good cold-working properties; radiator covers, hardware, electrical
Phosphor bronze 10% ASTM B103 ASTM B139 ASTM B159	Cu 90.0 Sn 10.0 P 0.25	Spring temper	—	122	4	241	Good spring qualities, high-fatigue strength
Yellow brass (high brass) ASTM B36 ASTM B134 ASTM B135	Cu 65.0 Zn 35.0	Annealed Cold-drawn Cold-rolled (HT)	18 55 60	48 70 74	60 15 10	55 115 180	Good corrosion resistance; plumbing, architectural
Manganese bronze ASTM B138	Cu 58.5 Zn 39.2 Fe 1.0 Sn 1.0 Mn 0.3	Annealed Cold-drawn	30 50	60 80	30 20	95 180	Forgings

Table 1.5—1 (continued)
MECHANICAL PROPERTIES OF COPPER ALLOYS

Material	Nominal composition			Form and condition	Typical mechanical properties				Comments
					Yield strength (0.2% offset), 1000 lb/sq in.	Tensile strength, 1000 lb/sq in.	Elongation, in 2 in., %	Hardness, Brinell	
Naval brass ASTM B21	Cu 60.0	Zn 39.25		Annealed Cold-drawn	22 40	56 65	40 35	90 150	Condensor tubing; high resistance to salt-water corrosion
	Sn 0.75								
Muntz metal ASTM B111	Cu 60.0	Zn 40.0		Annealed	20	54	45	80	Condensor tubes; valve stress
Aluminum bronze ASTM B169, alloy A ASTM B124 ASTM B150	Cu 92.0	Al 8.0		Annealed Hard	25 65	70 105	60 7	80 210	
Beryllium copper 25 ASTM B194 ASTM B197 ASTM B196	Be 1.9 Co or Ni 0.25	Cu bal.		Annealed, solution-treated Cold-rolled Cold-rolled	32 104 70	70 110 190	45 5 3	B60 (Rockwell) B81 C40	Bellows, fuse clips, electrical relay parts, valves, pumps
Free-cutting brass	Cu 62.0 Pb 2.5	Zn 35.5		Cold-drawn	44	70	18	B80 (Rockwell)	Screws, nuts, gears, keys
Nickel silver 18% Alloy A (wrought) ASTM B122, No. 2	Cu 65.0 Ni 18.0	Zn 17.0		Annealed Cold-rolled Cold-drawn wire	25 70 —	58 85 105	40 4 —	70 170	Hardware, optical goods, camera parts
Nickel silver 13% (cast) 10A ASTM B149, No. 10A	Ni 12.5 Sn 2.0 Zn 20.0	Pb 9.0 Cu bal.		Cast	18	35	15	55	Ornamental castings, plumbing; good machining qualities
Cupronickel 55-45 (Constantan)	Cu 55.0	Ni 45.0		Annealed Cold-drawn Cold-rolled	30 50 65	60 65 85	45 30 20	— —	Electrical-resistance wire; low temperature coefficient, high resistivity
Cupronickel	Cu 70.0	Ni 30.0		Wrought					Heat-exchanger process equipment, valves
Cupronickel 10% ASTM B111 ASTM B171	Cu 88.35 Fe 1.25	Ni 10.0 Mn 0.4		Annealed Cold-drawn tube	22 57	44 60	45 15	— —	Condensor, salt-water piping

Table 1.5-1 (continued)
MECHANICAL PROPERTIES OF COPPER ALLOYS

Material	Nominal composition		Form and condition	Typical mechanical properties				Comments
				Yield strength (0.2% offset), 1000 lb/sq in.	Tensile strength, 1000 lb/sq in.	Elongation, in 2 in., %	Hardness, Brinell	
Red brass (cast) ASTM B30, No. 4A	Cu 85.0 Pb 5.0	Zn 5.0 Sn 5.0	As-cast	17	35	25	60	
Silicon bronze ASTM B30, alloy 12A	Si 4.0 Zn 4.0 Mn 1.0	Fe 2.0 Al 1.0	Castings					Cheaper substitute for tin bronze
Tin bronze ASTM B30, alloy 1B	Sn 8%	Zn 4.0	Castings					Bearings, high-pressure bushings, pump impellers
Navy bronze			Cast					

From Bolz, R. E. and Tuve, G. L., Eds., *Handbook of Tables for Applied Engineering Science*, 2nd ed., CRC Press, Cleveland, 1973, 109.

Table 1.5–2
MECHANICAL PROPERTIES OF TIN AND LEAD-BASE ALLOYS

Major uses for these alloys are as "white"-metal bearing alloys, extruded cable sheathing, and solders. Tin forms the basis of pewter used for culinary applications.

Material	Nominal composition	Form and condition	Typical mechanical properties				Comments
			Yield strength (0.2% offset), 1000 lb/sq in.	Tensile strength, 1000 lb/sq in.	Elongation, in 2 in., %	Hardness, Brinell	
Lead-base Babbitt ASTM B23, alloy 19	Pb 85.0 Sn 5.0 Sb 10.0 As 0.6 Cu 0.5	Chill cast	—	10	5	19	Bearings, light loads and low speeds
Arsenical-lead Babbitt ASTM B23, alloy 15	Pb 83.0 Sn 1.0 Sb 16.0 As 1.1 Cu 0.6	Chill cast	—	10.3	2	20	Bearings, high loads and speeds, diesel engines, steel mills
Chemical lead	Pb 99.9 Cu 0.06 Bi 0.005 max	Rolled 95%	1.9	2.5	50	5	
Antimonial lead (hard lead)	Pb 94.0 Sb 6.0	Chill cast Rolled 95%	— —	6.8 4.1	22 47	(500 kg) 9	Good corrosion resistance and strength
Calcium lead	Pb 99.9 Ca 0.025 Cu 0.10	Extruded and aged	—	4.5	25	—	Cable sheathing, creep-resistant pipe
Tin Babbitt alloy ASTM B23-61, grade 1	Sb 4.5 Sn bal. Cu 4.5	Chill cast	—	9.3	2	17	General bearings and die casting
Tin die-casting alloy ASTM B102-52	Sb 13.0 Sn bal. Cu 5.0	Die-cast	—	10	1	29	Die-casting alloy
Pewter	Sn 91.0 Sb 7.0 Cu 2.0	Rolled sheet, annealed	—	8.6	40	9.5	Ornamental and household items
Solder 50-50	Sn 50.0 Pb 50.0	Cast	4.8	6.1	60	14	General-purpose solder
Solder	Sn 20.0 Pb 80.0	Cast	3.6	5.8	16	11	Coating and joining, filling seams on automobile bodies

From Bolz, R. E. and Tuve, G. L., Eds., *Handbook of Tables for Applied Engineering Science*, 2nd ed., CRC Press, Cleveland, 1973, 111.

Table 1.5–3
MECHANICAL PROPERTIES OF TITANIUM ALLOYS

The main application for these alloys is in the aerospace industry. Because of the low density and high strength of titanium alloys, they present excellent strength-to-weight ratios.

Material	Nominal composition	Form and condition	Typical mechanical properties				Comments
			Yield strength (0.2% offset), 1000 lb/sq in.	Tensile strength, 1000 lb/sq in.	Elongation, in 2 in., %	Hardness, Brinell	
Commercial titanium ASTM B265–58T	Ti 99.4	Annealed at 1100 to 1350°F (593 to 732°C)	70	80	20	—	Moderate strength, excellent fabricability; chemical industry pipes
Titanium alloy ASTM B265–58T–5 Ti–6 Al–4V		Water-quenched from 1750°F (954°C); aged at 1000°F (538°C) for 2 hr	160	170	13	—	High-temperature strength needed in gas-turbine compressor blades
Titanium alloy Ti–4 Al–4Mn		Water-quenched from 1450°F (788°C); aged at 900°F (482°C) for 8 hr	170	185	13	—	Aircraft forgings and compressor parts
Ti–Mn alloy ASTM B265–58T–7	Fe 0.5 Ti bal. Mn 7.0–8.0	Sheet	140	150	18	—	Good formability, moderate high-temperature strength; aircraft skin

From Bolz, R. E. and Tuve, G. L., Eds., *Handbook of Tables for Applied Engineering Science*, 2nd ed., CRC Press, Cleveland, 1973, 114.

Table 1.5—4
MECHANICAL PROPERTIES OF ZINC ALLOYS

A major use for these alloys is for low-cost die-cast products such as household fixtures, automotive parts, and trim.

Material	Nominal composition	Form and condition	Typical mechanical properties				Comments
			Yield strength (0.2% offset), 1000 lb/sq in.	Tensile strength, 1000 lb/sq in.	Elongation, in 2 in., %	Hardness, Brinell	
Zinc ASTM B69	Cd 0.35 Zn bal. Pb 0.08	Hot-rolled	—	19.5	65	38	Battery cans, grommets, lithographer's sheet
Zilloy-15	Cu 1.00 Zn bal. Mg 0.010	Hot-rolled Cold-rolled	— —	29 36	20 25	61 80	Corrugated roofs, articles with maximum stiffness
Zilloy-40	Cu 1.00 Zn bal.	Hot-rolled Cold-rolled	— —	24 31	50 40	52 60	Weatherstrip, spun articles
Zamac-5 ASTM 25	Zn (99.99% pure remainder) Al 3.5– 4.3 Cu 0.75– 1.25 Mg 0.03– 0.08	Die-cast	—	47.6	7	91	Die casting for automobile parts, padlocks; used also for die material

From Bolz, R. E. and Tuve, G. L., Eds., *Handbook of Tables for Applied Engineering Science*, 2nd ed., CRC Press, Cleveland, 1973, 114.

Table 1.5–5
MECHANICAL PROPERTIES OF ZIRCONIUM ALLOYS

These alloys have good corrosion resistance but are easily oxidized at elevated temperatures in air. The major application is for use in nuclear reactors.

Material	Nominal composition		Form and condition	Typical mechanical properties				Comments
				Yield strength (0.2% offset), 1000 lb/sq in.	Tensile strength, 1000 lb/sq in.	Elongation, in 2 in., %	Hardness, Brinell	
Zirconium, commercial	O₂ 0.07	C 0.15	Annealed	40	65	27	B80 (Rockwell)	
	Hf 1.90	Zr bal.						
Zircaloy 2	Hf 0.02	Ni 0.05	Annealed	50	75	22	B90 (Rockwell)	Nuclear power-reactor cores at elevated temperatures
	Fe 0.15	Other 0.25						
	Sn 1.46	Zr bal.						

From Bolz, R. E. and Tuve, G. L., Eds., *Handbook of Tables for Applied Engineering Science*, 2nd ed., CRC Press, Cleveland, 1973, 115.

Table 1.5—6
TRADE NAMES OF ALLOYS

Trademark	Owner
ALUMEL	Hoskins Manufacturing Company
CARPENTER STAINLESS NO. 20	The Carpenter Steel Company
CARPENTER 426	The Carpenter Steel Company
CHROMEL	Hoskins Manufacturing Company
COBENIUM	Wilbur B. Driver Company
CONPERNIK	Westinghouse Electric Corporation
COR-TEN	United States Steel Corporation
CUFENLOY	Phelps Dodge Corporation
DISCALOY	Westinghouse Electric Corporation
DURANICKEL	The International Nickel Company, Inc.
DYNALLOY	Alan Wood Steel Corporation
DYNAVAR	Precision Metals Division
ELGILOY	Elgin National Watch Company
ELINVAR	Hamilton Watch Company
GEMINOL	Driver-Harris Company
HASTELLOY	Union Carbide Corporation
HI-STEEL	Inland Steel Corporation
HIPERNIK	Westinghouse Electric Corporation
HP	Republic Steel Company
HY-TUF	Crucible Steel Company
ILLIUM	Stainless Foundry & Engineering Inc.
INCOLOY	The International Nickel Company, Inc.
INCONEL	The International Nickel Company, Inc.
INVAR	Soc. Anon. de Commentry-Fourchambault et Decaziville (Acieries d'Imphy)
KANTHAL	The Kanthal Corporation
KOVAR	Westinghouse Electric Corporation
MAGARI-R	Bethlehem Steel Corporation
MANGANIN	Driver-Harris Company
MINOVAR	The International Nickel Company, Inc.
MONEL	The International Nickel Company, Inc.
MONIMAX	Allegheny Ludlum Steel Corporation
NICROTUNG	Westinghouse Electric Corporation
NIMOCAST	The International Nickel Company, Inc.
NIMONIC	The International Nickel Company, Inc.
NISILOY	The International Nickel Company, Inc.
NI-SPAN-C	The International Nickel Company, Inc.
PERMALLOY	Allegheny Ludlum Steel Corporation
PERMANICKEL	The International Nickel Company, Inc.
REFRACTALOY	Westinghouse Electric Corporation
RENE 41	Allvac Metals Corporation (Division of Teledyne)
RODAR	Wilbur B. Harris Company
SD	The International Nickel Company, Inc.
SIMINEX	Allegheny Ludlum Steel Corporation
SEALMET	Allegheny Ludlum Steel Corporation
STAINLESS STEEL W	United States Steel Corporation
STAINLESS STEEL 17-4PH	Armco Steel Corporation
SUPERMALLOY	Allegheny Ludlum Steel Corporation
T-1	United States Steel Corporation
TRI-TEN	United States Steel Corporation
TRW	TRW, Inc.
UDIMET	Special Metals Corporation
UNITEMP	Universal Cyclops Speciality Steel Division, Cyclops Corporation
USS STRUX	United States Steel Corporation
WASPALLOY	Pratt and Whitney Aircraft
WELCON	Japanese Steel Works, Ltd.
WEL-TEN	Yawata Iron & Steel Company, Ltd.
YOLOY	Youngstown Sheet & Tube Company

From Bolz, R. E. and Tuve, G. L., Eds., *Handbook of Tables for Applied Engineering Science,* 2nd ed., CRC Press, Cleveland, 1973, 116.

Table 1.5−7
PROPERTIES OF HIGH-TEMPERATURE METALS[a]

Approximate or Typical Values

Metals are listed in order of their melting points. Actual properties depend on purity, condition, form, and treatment.

Melting point		Metal	Properties at room temperatures						At 1000°C (1830°F)		
°C	°F		Specific gravity	Specific heat	Thermal conductivity, $\frac{Btu}{hr\,ft\,°F}$	Electrical resistivity, microhm-cm	Coef of linear expansion, $(\times 10^6)$ $(°C)^{-1}$	Tensile strength, kpsi	Specific heat, $\frac{Btu}{lb\,°F}$	Thermal conductivity, $\frac{Btu}{hr\,ft\,°F}$	Tensile strength, kpsi
3400	6150	Tungsten (Wolfram)	19.3	0.032	103	5.65	4.5	430	0.037	66	100
3180	5760	Rhenium	21.0	0.033	41.1	19	7	337	0.039		120
3025	5477	Osmium	22.6	0.031	35	9	5		0.036		
2980	5400	Tantalum	16.6	0.034	31.2	12.4	6.5	55	0.037	35	30
2620	4750	Molybdenum	10.2	0.060	81	5.2	5	85	0.075	60	
2470	4470	Niobium (Columbium)	8.57	0.064	30.1	13	7	40	0.075	40	8
2450	4440	Iridium	22.4	0.031	85	5.3	6	90	0.038		48
2400	4350	Ruthenium	12.4	0.057	—	7.5	9	78	0.074		
2220	4030	Hafnium	13.3	0.035	12.7	35	6	65	0.044		
1965	3569	Rhodium	12.4	0.058	86.7	4.6	8	138	0.080		
1900	3450	Vanadium	6.1	0.116	34.7	25	8	78	0.158		
1860	3380	Chromium	7.20	0.11	52	13	6	12	0.169	36	
1850	3370	Zirconium	6.53	0.067	12.2	41	5.5	49	0.082	15	
1770	3220	Platinum	21.5	0.032	42.2	10.5	9	22	0.038	45.5	
1750	3180	Thorium	11.7	0.03	23.7	18	12	32	0.041		
1670	3040	Titanium	4.54	0.125	11.6	43	8.5	35	0.152	13	
1550	2820	Palladium	12.0	0.058	41.1	10.8	12	25	0.071		
1536	2797	Iron	7.87	0.108	46.4	9.7	12	40	0.148	17.0	
1495	2723	Cobalt	8.9	0.10	39.9	9	12	137	0.192		
1453	2647	Nickel	8.9	0.106	52	6.85	13	46	0.145	45	

[a]See Tables 1.5−3 and 1.5−5 for other properties.

From Bolz, R. E. and Tuve, G. L., Eds., *Handbook of Tables for Applied Engineering Science*, 2nd ed., CRC Press, Cleveland, 1973, 120.

Table 1.5–8
PROPERTIES OF HIGH-TEMPERATURE METALS – SI UNITS

Approximate or Typical Values

Metals are listed in order of their melting points. Actual properties depend on purity, condition, form, and treatment.

Melting point (deg C)	Melting point (K)	Melting point (deg F)	Metal	Specific gravity	Specific heat, J/kg·K	Thermal conductivity, W/m·K	Electrical resistivity, μΩ·m	Coefficient of linear expansion, μm/m·K	Tensile strength, MN/m²	Specific heat, J/kg·K	Thermal conductivity, W/m·K	Tensile strength, MN/m²
						Properties at room temperatures				At 1 000 deg C (1 830 deg F, 1 275 K)		
3 400	3 675	6 150	Tungsten (Wolfram)	19.3	134	178	0.056 5	4.5	3 000	155	114	690
3 180	3 455	5 760	Rhenium	21.0	138	71.1	0.19	7	2 300	163		830
3 025	3 300	5 477	Osmium	22.6	130	60.5	0.09	5		151		
2 980	3 255	5 400	Tantalum	16.6	142	54.0	0.124	6.5	380	155	60.5	210
2 620	2 895	4 750	Molybdenum	10.2	251	140	0.052	5	590	314	104	
2 470	2 745	4 470	Niobium (Columbium)	8.57	268	52.1	0.13	7	280	314	69.2	55
2 450	2 725	4 440	Iridium	22.4	130	147	0.053	6	620	159		330
2 400	2 675	4 350	Ruthenium	12.4	238		0.075	9	540	310		
2 220	2 493	4 030	Hafnium	13.3	146	22.0	0.35	6	450	184		
1 965	2 240	3 569	Rhodium	12.4	243	150	0.046	8	950	335		
1 900	2 175	3 450	Vanadium	6.1	485	60.0	0.25	8	540	661	62.3	
1 860	2 135	3 380	Chromium	7.20	460	89.9	0.13	6	83	707	25.9	
1 850	2 125	3 370	Zirconium	6.53	280	21.1	0.41	5.5	340	343	78.7	
1 770	2 045	3 220	Platinum	21.5	134	73.0	0.105	9	150	159		
1 750	2 025	3 180	Thorium	11.7	126	41.0	0.18	12	220	172		
1 670	1 945	3 040	Titanium	4.54	523	20.1	0.43	8.5	240	636	22.5	
1 550	1 825	2 820	Palladium	12.0	243	71.1	0.108	12	170	297		
1 536	1 810	2 797	Iron	7.87	452	80.2	0.097	12	280	619	29.4	
1 495	1 770	2 723	Cobalt	8.9	418	69.0	0.09	12	940	803		
1 453	1 725	2 647	Nickel	8.9	444	89.9	0.068 5	13	320	607	77.8	

From Bolz, R. E. and Tuve, G. L., Eds., *Handbook of Tables for Applied Engineering Science*, 2nd ed., CRC Press, Cleveland, 1973, 121.

Table 1.5–9

HEAT CAPACITIES OF COPPER, SILVER, AND GOLD AT LOW TEMPERATURES

Specific heat values are in cal/mol·K; for J/mol·K multiply by 4.184. Enthalpy values are in cal/mol: for J/mol multiply by 4.184. Gram atomic weights are Cu = 63.540; Ag = 107.87; Au = 196.97. One Btu = 252 calories = 1 054.4 J.

Temperature		Copper		Silver		Gold	
°K	°R	c_pa	H	c_pa	H	c_pa	H
1.0	1.8	0.000177	0.000086	0.000196	0.000088	0.000282	0.000114
2.0	3.6	0.000423	0.000378	0.000633	0.000472	0.00121	0.000779
3.0	5.4	0.000805	0.000978	0.00155	0.00151	0.00337	0.00294
4.0	7.2	0.00139	0.00206	0.00320	0.00382	0.00732	0.00811
5.0	9.0	0.00225	0.00385	0.00581	0.00823	0.0136	0.0184
6.0	10.8	0.00346	0.00668	0.00964	0.0158	0.0228	0.0363
7.0	12.6	0.00508	0.0109	0.0150	0.0280	0.0356	0.0652
8.0	14.4	0.00720	0.0170	0.0222	0.0464	0.0526	0.109
9.0	16.2	0.00990	0.0255	0.0316	0.0731	0.0748	0.172
10.0	18.0	0.0133	0.0370	0.0437	0.110	0.103	0.261
11.0	19.8	0.0174	0.0523	0.0590	0.162	0.138	0.380
12.0	21.6	0.0224	0.0721	0.0778	0.230	0.180	0.539
13.0	23.4	0.0284	0.0974	0.101	0.318	0.230	0.744
14.0	25.2	0.0355	0.129	0.128	0.432	0.288	1.002
15.0	27.0	0.0439	0.169	0.160	0.576	0.352	1.321
16.0	28.8	0.0538	0.218	0.197	0.754	0.424	1.709
17.0	30.6	0.0651	0.277	0.240	0.972	0.501	2.170
18.0	32.4	0.0783	0.348	0.287	1.235	0.584	2.712
19.0	34.2	0.0933	0.434	0.338	1.547	0.671	3.339
20.0	36.0	0.110	0.536	0.394	1.912	0.762	4.055
25.0	45.0	0.230	1.363	0.733	4.689	1.254	9.074
30.0	54.0	0.405	2.928	1.141	9.354	1.763	16.62
35.0	63.0	0.631	5.496	1.580	16.15	2.246	26.66
40.0	72.0	0.894	9.296	2.012	25.14	2.682	38.99
45.0	81.0	1.178	14.47	2.417	36.23	3.073	53.40
50.0	90.0	1.471	21.09	2.786	49.25	3.415	69.64
55.0	99.0	1.765	29.18	3.117	64.03	3.709	87.47
60.0	108.0	2.054	38.73	3.411	80.36	3.964	106.7
65.0	117.0	2.332	49.70	3.669	98.08	4.185	127.1
70.0	126.0	2.596	62.03	3.896	117.0	4.377	148.5
75.0	135.0	2.843	75.63	4.096	137.0	4.544	170.8
80.0	144.0	3.072	90.43	4.272	157.9	4.691	193.9
85.0	153.0	3.285	106.3	4.429	179.7	4.820	217.7
90.0	162.0	3.480	123.2	4.568	202.2	4.934	242.1
95.0	171.0	3.660	141.1	4.692	225.3	5.034	267.0
100.0	180.0	3.825	159.8	4.804	249.1	5.123	292.4
120.0	216.0	4.362	242.0	5.148	348.8	5.393	397.7
140.0	252.0	4.749	333.3	5.382	454.3	5.571	507.5
160.0	288.0	5.032	431.3	5.550	563.7	5.691	620.2
180.0	324.0	5.244	534.1	5.676	676.0	5.772	734.9

Table 1.5–9 (continued)

HEAT CAPACITIES OF COPPER, SILVER, AND GOLD AT LOW TEMPERATURES

Temperature		Copper		Silver		Gold	
°K	°R	c_p[a]	H	c_p[a]	H	c_p[a]	H
200.0	360.0	5.408	640.7	5.773	790.5	5.835	850.9
220.0	396.0	5.537	750.2	5.852	906.8	5.892	968.2
240.0	432.0	5.640	862.0	5.917	1025.	5.944	1087.
260.0	468.0	5.722	975.7	5.973	1143.	5.992	1206.
280.0	504.0	5.788	1091.	6.020	1263.	6.037	1326.
273.15	491.7	5.767	1051.	6.005	1222.	6.022	1285.
298.15	536.7	5.840	1196.	6.059	1373.	6.075	1436.

[a]This is the instantaneous specific heat at the given temperature. Mean specific heat over any range may be computed from the enthalpies.

From Furukawa, G. T., Saba, W. G., and Reilly, M. L., Critical Analysis of the Heat-capacity Data of the Literature and Evaluation of Thermodynamic Properties of Copper, Silver, and Gold from 0 to 300 K, NSRDS-NBS-18, National Bureau of Standards, 1968.

Table 1.5–10

SILVER AND SILVER ALLOYS

Typical Silver Alloys

Typical composition, %	Names and uses	Important property[a]
Ag 99.9 +	Fine silver; contacts, chemical uses	Conductivity; ductility
Ag 99 +, Mg 0.25, Ni 0.2 +	High temperature contacts and spring contacts	Hardness; high conductivity
Ag 99 + (layer, 0.025 in.)	High-service machine bearings	Fatigue strength; thermal conductivity
Ag 92.5, Cu 7.5	Sterling silver; tableware	Appearance; value
Ag 90, Cu 10	Coin silver (to 1966)	Value; corrosion resistance
Ag 90 (outside layer)	Pressure-bonded laminates; coins	Appearance; conductivity
Ag 90, Pd 10	Contacts; brazing alloys	High conductivity
Ag 90, CdO 10 (sintered)	Non-sticking contacts	Heat resistance; hardness
Ag 85, Cd 15	Arc-quenching contacts	Wear resistance; ductility
Ag 80, Cu 20	Laminated coins; tableware	Appearance; value
Ag 80, In 15, Cd 5	Nuclear reactor control	Neutron absorption
Ag 77, Cd 22.6, Ni 0.4	Spring contacts	Wear resistance; ductility
Ag 72, Cu 28	Brazing alloy	Eutectic alloy; hardness
Ag 60, Ni 40 (sintered)	Circuit-breaker contacts	Hardness; heat resistance
Ag 49, Au 41.7, Cu 9, Zn 0.3	10-karat green gold; jewelry	Appearance
Ag 40, Mo or W 60 (sintered)	Switching contacts	Hardness; wear resistance
Ag 40, Au 30, Pd 30	Corrosive-chemical apparatus (e.g., for halogens and nitric acid)	Corrosion resistance; strength
Ag 35, Cu 26, Zn 21, Cd 18	Thin-joint brazing	Wetting ability; ductility
Ag 33, Hg 52, Sn 12.5, Cu 2, Zn 0.5	Dental amalgam fillings	Amalgam hardening
Ag 30.5, Au 50, Cu 17.5, Zn 2	Gold solder; jewelry	Appearance; strength

Table 1.5–10 (continued)
SILVER AND SILVER ALLOYS

Typical composition, %	Names and uses	Important property[a]
Ag 15, Au 60, Cu 10, Pd 10, Pt 4, Zn 1	White gold denture metal	Strength; wear resistance
Ag 15, Au 60, Cu 15	Precious metal solder	Appearance; strength
Ag 10, Au 58.3, Cu 29.7, Zn 2	14-karat yellow gold	Appearance; wear
Ag 2.5, Pb 97.5	Soft solder	Eutectic alloy

[a]Properties of pure silver and the high-silver alloys vary with temperature; they are subject to work hardening, and the room-temperature properties depend on the annealing temperature.

Chemical Uses of Silver

Type of use	Form of silver	Processes and remarks
Batteries	Powder; oxide; chloride	Silver oxide with zinc or cadmium; silver chloride with magnesium
Catalysts	Shapes, screens, powder, salts	In dehydrogenation, oxidation, desulfurization
Coatings	Powder; vapor	Air-drying paints and fired-glass conductive coatings
Electroplating	Cyanide; anode shapes	Usual plate less than 0.001 5 in. thickness
Medicine	Nitrate; metal, colloidal	Sterilization by metal, nitrate, or colloidal suspension
Photography	Nitrate; halide colloidal solutions	Metallic silver from salt by radiation and "developer"
Reflectors (not ultraviolet)	Nitrate; metal vapor	Silver nitrate reacts with reducing solution; vacuum evaporation

REFERENCES

Metals Handbook, Vol. I, 8th ed., American Society for Metals, Metals Park, Oh., 1961, 1174.
Butts, A. and Coxe, C. D., Eds., *Silver, Economics, Metallurgy and Use,* Van Nostrand, Princeton, N.J., 1967.
Addicks, L., *Silver in Industry,* Reinhold, New York, 1940.
ASTM Standards B253, B413, and E56.

From Bolz, R. E. and Tuve, G. L., Eds., *Handbook of Tables for Applied Engineering Science,* 2nd ed., CRC Press, Cleveland, 1973, 123.

Table 1.5–11
PROPERTIES OF RARE-EARTH METALS

To convert density from g/cm³ to kg/m³, multiply by 1000. To convert Young's modulus from kg/cm² to N/m², multiply by 98,067. Values in parentheses are estimates.

Element	Melting point, °C	Boiling point, °C	Heat of sublimation, kcal/mole ΔH 298°K	Density, g/cm³ 298°K	Atomic volume, cm³/mole	Metallic radius, Å	Electrical resistivity at 298°K, microhm-cm	Residual resistivity at 4.2°K, microhm-cm	Compressibility, cm²/kg·a	Young's modulus kg/cm², millions	Poisson's ratio
Scandium	1539	2832	91.0	2.989	15.04	1.641	50.9	3.7	2.26	0.809	(0.269)
Yttrium	1523	3337	99.6	4.457	19.95	1.803	59.6	3.2	2.68	0.663	0.265
Lanthanum	920	3454	103.0	6.166	22.53	1.877	79.8	S.C.[b]	4.04	0.384	0.288
Cerium	798	3257	111.60	6.771	20.69	1.824	75.3	0.7	4.10	0.306	0.248
Praseodymium	931	3212	89.09	6.772	20.81	1.828	68.0		3.21	0.332	0.305
Neodymium	1010	3127	77.3	7.003	20.60	1.822	64.3	6.8	3.0	0.387	0.306
Promethium	1080	(2460)	(64)						(2.8)	(0.430)	(0.278)
Samarium	1072	1778	49.3	7.537	19.95	1.802	105.0	6.2	3.34	0.348	0.352
Europium	822	1597	42.5	5.253	28.93	1.983	91.0	0.6	8.29	0.150	(0.286)
Gadolinium	1311	3233	95.75	7.898	19.91	1.801	131.0	4.4	2.56	0.573	0.259
Terbium	1360	3041	93.96	8.234	19.30	1.783	114.5	3.5	2.45	0.586	0.261
Dysprosium	1409	2335	71.2	8.540	19.03	1.775	92.6	2.4	2.55	0.644	0.243
Holmium	1470	2720	71.7	8.781	18.78	1.767	81.4	7.0	2.47	0.684	0.255
Erbium	1522	2510	74.5	9.045	18.49	1.758	86.0	4.7	2.39	0.748	0.238
Thulium	1545	1727	58.3	9.314	18.14	1.747	67.6	5.6	2.47	(0.770)	(0.235)
Ytterbium	824	1193	38.2	6.972	24.82	1.939	25.1	0.29	7.39	0.182	0.284
Lutetium	1656	3315	102.16	9.835	17.79	1.735	58.2	4.5	2.38	(0.860)	(0.233)

[a] All values in this column should be divided by 10⁶.
[b] S.C.: superconductor.

From Bolz, R. E. and Tuve, G. L., Eds., *Handbook of Tables for Applied Engineering Science*, 2nd ed., CRC Press, Cleveland, 1973, 129.

Table 1.5–12
SPECIFIC STIFFNESS OF METALS, ALLOYS, AND CERTAIN NONMETALLICS

Specific stiffness is usually expressed as the modulus of elasticity (in tension) per unit weight-density, i.e., E/ρ, in units of pounds and inches. While the stiffness of similar alloys varies considerably, there are definite ranges and groups to be recognized. Since the specific stiffness of steel is about 100 million, the values in the following table are also approximately the percentage stiffness, referred to steel.

Material	Specific stiffness, millions
Beryllium	650
Silicon carbide	600
Alumina ceramics	400
Mica	350
Titanium carbide cermet	250
Alumina cermet	200
Molybdenum and alloys; silica glass	130
Titanium and alloys; cobalt superalloys; soda-lime glass	110
Carbon and low-alloy steel; wrought iron	105
Stainless steel; nodular cast iron; magnesium and alloys; aluminum and alloys	100
Nickel and alloys; malleable iron	95
Iron silicon alloys (cast); iridium; vanadium	90
Monel alloys; tungsten	80
Gray cast iron; columbium alloys	70
Aluminum bronze; beryllium copper	65
Nickel silver; cupronickel; zirconium	55
Yellow brass; nickel cast iron; bronze; Muntz metal; antimony	50
Copper; red brass; tantalum	45
Silver and alloys; pewter; platinum and alloys; white gold	30
Tin; thorium	25
Gold	20
Tin-lead alloy	10
Lead	5

From Bolz, R. E. and Tuve, G. L., Eds., *Handbook of Tables for Applied Engineering Science,* 2nd ed., CRC Press, Cleveland, 1973, 130.

Table 1.5–13
APPROXIMATE EFFECT OF TEMPERATURES ON STRENGTH-WEIGHT RATIO OF METALS

Material	Temperature, °F					
	200°	400°	600°	800°	1000°	1200°
Titanium MST6A1–4V	773	696	627	550	454	
Titanium RC–130	727	636	554	455	236	
Stainless 17–7PH	654	632	582	496	250	
Magnesium AZ80X	491	196	77.3			
Inconel X	435	426	418	400	368	322
Aluminum SAP	350	306	225	155		
Molybdenum	468	400	350	309	273	245
Aluminum 7075–T6	650	127	61.8			
L–605	196	159	127	109	109	109
Stainless Type 321	112	100	91.0	81.7	70.0	59.0

Table 1.5–13 (continued)
APPROXIMATE EFFECT OF TEMPERATURES ON STRENGTH-WEIGHT RATIO OF METALS

REFERENCE

For tensile strengths of 40 superalloys at 1200–2000°F, see Simmons, W. F., Guide to selection of superalloys, *Metal Progress*, 91:86-86C, June 1967.

From Bolz, R. E. and Tuve, G. L., Eds., *Handbook of Tables for Applied Engineering Science*, 2nd ed., CRC Press, Cleveland, 1973, 130.

Table 1.5–14
METAL POWDERS

Typical Metal Powders and Granules

Varieties and uses — The largest use of metal powders is for production of parts, shapes, and electrodes by powder metallurgy. Other major uses are in metal coatings, applied by flame spraying, or as paints, lacquers or inks, chemical catalysts, reducing agents, fuels, and explosives. Many composite materials contain metal powder fillers. Annual U.S. consumption of iron powder and alloys is well over 200 million pounds. Some 200 kinds are available. Copper and its alloys represent almost as great a variety, but the total consumption is less than one third that of iron. Aluminum, nickel, chromium, tungsten, and silver powders are used in quantity, but there are very important uses for some of the other metals also. The following table gives only a few typical examples.

Particle size and surface — Actual dimensions and size-distribution are not yet independent of the methods of measurement. For instance, fine particles that pass a 100-mesh sieve can be dimensionally analyzed by three or four methods with divergent results. The apparent density of a powder is not independent of the method of handling, and the constant-volume "tap density" is affected by the shape of particles. Even a single method such as a sieve analysis depends on the condition and quality of the sieves. Particle sizes

above 40 μm in the table are based on standard ASTM sieve analyses and their metric equivalents in micrometers (25,400 μm = 1 in.). Particle size distribution and surface area of particles are not given in this table, but they are important for applications such as those involving either surface coverage or gas adsorption by the particles. Surface areas greater than 1 m^2/g are attainable for metals. Integration of the area under the size distribution curve gives a low value of surface area, since it is only practical to assume spherical particles. Other methods depend on air permeability, liquid surface spread, liquid mixture turbidity, or gas adsorption. Comparable results from the same sample are very difficult to obtain if more than one test method is used.

Purity and analysis — Metal purity to 99.9999% is often quoted, but the purity should be specified according to the use. Metals of purity 99% or less are much cheaper. A metal powder can be produced from an alloy of almost any composition. Although most of the common and proprietary alloys are commercially available in powder form, a special order may be required if the particle-size specification is unusual.

For suppliers of metal powders, consult the Metal Powder Industries Federation, 201 E. 42nd Street, New York, N.Y. 10017.

Metal or alloy	Purity, %	Largest impurity	Particle size		Specific gravity		Designation
			Micrometers (av. or range)	% passing-sieve No.	Powder	Solid	
Aluminum	99.9+	Fe	75–150	100–100	2.5	2.7	Pure
Aluminum	99.5	Fe or Si	<60	95–325	2.5	2.7	Atomized, fine
Aluminum	99.3	Oxygen	30–75	90–325	1.7	2.7	Atomized, spherical
Aluminum		Oxygen	6–600	95–40	1.2	2.7	Granular
Aluminum	95	Oxygen	0.03	—	0.2	2.7	Fine, spheroidal

Table 1.5–14 (continued)
METAL POWDERS

Metal or alloy	Purity, %	Largest impurity	Particle size		Specific gravity		Designation
			Micrometers (av. or range)	% passing-sieve No.	Powder	Solid	
Antimony	99.5		<100	95–325	2.42	6.7	Fine
Beryllium	99.0	Oxygen	<80	98–200		1.85	Structural
Bismuth	99.6	Sb	40–150	80–325	4.29	9.75	Commercial
Cadmium	99.9	Pb	40–150	95–325		8.65	Fine
Chromium	99.5	Fe	<45	100–325		7.2	Fine
Cobalt	99.8	Ni	1.2	99.9–400	0.7	8.9	Very fine
Copper	99.9	Sn	>1 000	—	5.3	8.95	Granular
Copper	99.8	Sn	15	95–325	4.55	8.95	Atomized shot
Copper	99.5	Sn	40–850	60–325	2.7	8.95	Electrolytic, coarse
Brass	90 Cu	Pb	40–250	55–325	3.0	8.5	10% Zn, atomized
Bronze	90 Cu		40–150	55–325	2.7	8.5	10% Sn
Gold	99.9+		<35	100–400	—	19.3	Pure, reduced
Gold	99.+		0.5 × 10	100–325	2.0	19.3	Flake 0.5 × 10 μm
Iron	99.8	Oxygen	7–10	—	3.2	7.9	Very fine
Iron	98.5	SiO$_2$	40–150	30–325	2.9	7.9	Reduced, annealed
Iron	91.	C	100–900	95–100	2.3	7.9	Ground cast iron
Stainless steel	71.	Si	20–150	40–325	3.0	8.0	Stainless steel 18-8
Lead	99.8	Bi	10–45	100–325	5.0	11.35	Very fine
Magnesium	99.8	Mn	<75	100–200	0.6	1.74	Commercial
Manganese	99.8	Oxygen	<100	80–325	2.8	7.3	Commercial
Ferromanganese	80 Mn	C	40–300	30–325			Medium carbon
Molybdenum	99.9	Oxygen	3–6	—	1.4	10.2	Hydrogen reduced
Nickel	99.9+	Co	10–100	80–250	2.0	8.9	High purity
Nickel	99.5	Co	<150	50–325	3.4	8.9	Electrolytic
Niobium	99.6	Oxygen	<75	100–200		8.57	High purity
Palladium	99.9+	Pt	<45	100–325		12.0	Pure
Platinum	99.0	Ag	<420	100–40		21.4	Sponge
Silver	99.9+	Si	<150	90–325	1.5	10.5	Precipitated, spongy
Silver braze	70.	Optional	<175	100–80			28% Cu brazing alloy
Tin	99.5	Oxygen	<150	30–250	3.4	7.3	Commercial
Titanium	99.		8	98–325		4.54	Commercial
Tungsten	99.9	Mo	<175	100–80	9.8	19.3	Reduced
Vanadium	99.5	Fe	<800	100–20		6.1	Commercial
Zinc	99.	Pb	<50	98–325	7.0	7.0	Reduced
Zirconium	99.+	Oxygen	<175	100–80	3.5	6.53	Reactor

REFERENCES

Barth, V. D. and McIntire, H. O., *Tungsten Powder Metallurgy*, NASA SP-5035, National Aeronautics and Space Administration, 1965.

Goetzel, C. G., *Treatise on Powder Metallurgy*, Interscience Books, New York, 1949.

Hausner, H. H., Ed., *Modern Developments in Powder Metallurgy*, Plenum Publishing, New York, 1966.

Leszynski, W., Ed., *Powder Metallurgy*, Interscience Books, New York, 1961.

Poster, A. R., *Handbook of Metal Powders*, Reinhold, New York, 1966.

Sands, R. L. and Shakespeare, C. R., *Powder Metallurgy*, Newnes, London, 1966 (distributed in the U.S.A. by The Chemical Rubber Co., Cleveland).

From Bolz, R. E. and Tuve, G. L., Eds., *Handbook of Tables for Applied Engineering Science*, 2nd ed., CRC Press, Cleveland, 1973, 131.

Table 1.5–15
PRODUCTS OF POWDER METALLURGY[a]

Powder metallurgy refers to the production of parts by a process of molding metal powders and agglomerating the form by heat. The powder mixture is often hot molded under pressure (10,000 to 100,000 psi) and is sintered in an inert or a reducing atmosphere, at a temperature between 400 to 2,000°F, depending on the metal mixture. For the refractory metals higher temperatures are necessary. The methods of powder metallurgy provide a close control of the composition and allow use of mixtures that could not be fabricated by any other process. As dimensions are determined by the mold, finish machining or grinding is often eliminated, thereby reducing cost and handling, especially for large lots. Special properties of the finished product such as porosity, friction coefficient, and electrical conductivity can be varied somewhat by changing the proportions of the powder components.

Class	Composition or constituents	Applications and uses	Desirable properties and advantages
Small, finished parts	Various ferrous, copper, and nickel alloys	Complex shapes; small parts not requiring high strength or ductility; plain bearings	Control of dimensions and finish; two-phase bearing metals; low cost in large production lots
Refractory metals	Pure W, Mo, Ta, Nb, Re, Ti alloys	Production of high-purity tungsten, molybdenum, tantalum, niobium, etc.; beryllium; cobalt alloys	Metals used in high-temperature service; electrical, electronic, and nuclear applications
Porous metals	Copper; copper-lead; bronze; stainless steel	Porous bearings, oil-impregnated, or with graphite or plastic; friction materials; metal filters; porous electrodes; catalysts; throttle plates	Interconnected pores in the size range $5-50$ μm; porosity about $20-30\%$
Composite metals	Al, Cu, etc. with W, Mo, Co, or stainless steel reinforcing; reactor fuel elements	Services requiring high strength with lightness, high electrical and thermal conductivities; nuclear reactor components	High-strength materials from common metals; durability of nuclear materials
Metal-nonmetal composites	Filament-reinforced ceramics; dispersion strengthening by oxides	Ceramics with good structural properties; lightweight materials for high temperature (*e.g.*, SAP)	Strengthened ceramics; heat-resistant aluminum
Magnetic materials	Nickel-iron; cobalt mixtures; ferrites	High-permeability materials; permanent magnets; ferrite cores; magnetic storage	Very high magnetic properties and close control of magnetic properties
Cermets, oxide	Al_2O_3-Cr; Al_2O_3-Cr-W; Al_2O_3-Cr-Mo; ThO_2-W	Combustion and rocket nozzles; furnace muffler, tubes, seals, extrusion dies; power-tube cathodes	High-temperature strength (2 000 deg F and above); resistance to thermal shock; high thermal conductivity; corrosion resistance

Table 1.5–15 (continued)
PRODUCTS OF POWDER METALLURGY

Class	Composition or constituents	Applications and uses	Desirable properties and advantages
Cermets, carbide	TiC-Ni; TiC-Fe-Cr; TiC-Co-Cr-W; Cr_3C_2-Ni-W	High-temperature bearings, seals, and dies; gage blocks	Strength, toughness, and corrosion resistance at high temperatures (to 1 700 deg F); hardness
Cemented carbides	WC-Co; WC-TaC-Co; TiC-Ni; Cr_3C_2-WC-Ni	Tips for cutting tools, lathe centers, gages; wire-drawing dies; rock drills; crushers; blast nozzles	Very high hardness, compressive strength, and elastic modulus; wear and corrosion resistance; high conductivity; high-temperature strength

[a]For other data and references on powder metallurgy, see Table 1.5–14.

From Bolz, R. E. and Tuve, G. L., Eds., *Handbook of Tables for Applied Engineering Science,* 2nd ed., CRC Press, Cleveland, 1973, 133.

Table 1.5–16
CORROSION OF METALS

Metal	Subject to corrosion by	Resistant to
Aluminum and alloys	Acid solutions (except concentrated nitric and acetic); caustic and mild alkalies; sea water; saturated halogen vapors; mercury and its compounds; carbon tetrachloride; cobalt; copper and nickel compounds in solution	Air; water; ammonia; combustion products; halide refrigerants; dry steam; sulfur and its compounds; concentrated ammonium hydroxide; organic acids; most organic compounds
Cast iron	All water solutions, moist gases; dilute acids; acid-salt solutions	Dry gases except halogens; dry air; neutral water; dry soil; concentrated acids (nitric, sulfuric, phosphoric); weak or strong alkalies; organic acids
Chromium and high-chrome steels	Most strong acids (limited use with acetic, nitric, sulfuric, and phosphoric); most chlorides	Air; water; steam; weak acids; most inorganic salts; most alkalies; ammonia
Copper, red brass, and bronze	Mercury and its salts; aqueous ammonia; saturated halogen vapors; sulfur and sulfides; oxidizing acids (nitric, concentrated sulfuric, sulfurous); oxidizing salts (Hg, Ag, Cr, Fe, Cu); cyanides	Air; water; sea water; steam; sulfate and carbonate solutions; dry halogens; moist soils; alkaline solutions; refrigerants; petrochemicals; non-oxidizing acids (acetic, hydrochloric, sulfuric)
Lead	Caustic solutions; halogens; acetic acid; calcium hydroxide; magnesium chloride; ferric chloride; sodium hypochlorite	Air; water; moist soil; ammonia; alcohols; sulfuric acid; ferrous chloride
Magnesium and alloys	Heavy metal; salts; all mineral acids (except hydrofluoric and chromic); sea water; fruit juices; milk	Most alkalies and organic compounds; air; water; soil; dry refrigerants; dry halogens
Monel metal and Ni-Cu alloys containing in excess of 50% nickel	Inorganic acids; sulfur; chlorine; acid solutions of ferric, stannic, or mercuric salts	Air; sea water; steam; food acids; neutral and alkaline salt solutions; dry gases; most alkalies; ammonia
Nickel and high-nickel steels	Inorganic acids; ammonia; mercury; oxidizing salts (Fe, Cu, Hg)	Air; water; steam; caustic and mild alkalies; organic acids; neutral and alkaline organic compounds; dry gases
Silver	Halogens and halogen acids; sulfur compounds; ammonia	Alkalies, including high-temperature caustic alkalies; hot concentrated organic acids; phosphoric and hydrofluoric acids

Table 1.5–16 (continued)
CORROSION OF METALS

Metal	Subject to corrosion by	Resistant to
Steel, mild, low-alloy steels	Most acids; strong alkalies; salt water; sulfur and its compounds	Air; steam; ammonia; most alkalies; concentrated nitric acid; halide refrigerants
Tantalum	Hydrofluoric acid; concentrated sulfuric acid; strong alkalies	Nearly all salts; most acids; water; sea water; air; alcohols; hydrocarbons; sulfur
Tin	Inorganic acids; caustic solutions; halides	Most food acids; ammonia; neutral solutions
Titanium	Hydrochloric acid; sulfuric acid; hydrofluoric, oxalic, and formic acids; dangerously explosive in presence of nitric acid or liquid oxygen	Oxidizing media; air; water; sea water; aqueous chloride solutions; moist chlorine gas; sodium hypochlorite
Zinc	Acid or strong alkali solutions; sulfur dioxide; chlorides	Air; water; ammonia; dry common gases; refrigerants; gasoline
Zirconium	Concentrated sulfuric acid (hot); hydrofluoric acid; cupric and ferric chlorides	Solutions of alkalies and acids; aqua regia

Notes: Polished surfaces resist corrosion.
Nonuniformity within a metal tends to increase corrosion.
Stress (especially alternating stress) tends to increase corrosion.
Dissolved gases in water (especially oxygen) accelerate corrosion.

REFERENCES

LaQue, F. L. and Copson, H. R., Eds., *Corrosion Resistance of Metals and Alloys,* Reinhold, New York, 1963.
Romanoff, M., *Underground Corrosion,* National Bureau of Standards Circular 579, 1957.
Schweitzer, P. A., *Handbook of Corrosion Resistant Piping,* Industrial Press, New York, 1969.
Smithells, C. J., *Metals Reference Book,* Vol. 2, 4th ed., Butterworths, London, 1967.
Uhlig, H. H., *The Corrosion Handbook,* John Wiley & Sons, New York, 1948.

From Bolz, R. E. and Tuve, G. L., Eds., *Handbook of Tables for Applied Engineering Science,* 2nd ed., CRC Press, Cleveland, 1973, 350.

Table 1.5–17
CORROSION-RATE RANGES EXPRESSED IN MILS PENETRATION PER YEAR

1 mil = 0.001 in.

Metal	Acid solutions — Non-oxidizing — Sulfuric, 5%	Acid solutions — Non-oxidizing — Acetic, 5%	Acid solutions — Oxidizing — Nitric, 5%	Alkaline solutions — Sodium hydroxide, 5%	Neutral solutions — Fresh water	Neutral solutions — Sea water	Air — Normal outdoor urban exposure
Aluminum	8–100	0.5–5	15–80	13,000	0.1	1–50	0–0.5
Zinc	High	600–800	High	15–200	0.5–10	0.5–10[a]	0–0.5
Tin	2–500[a]	2–500[a]	100–400	5–20	0–0.5	0.1	0–0.2
Lead	0–2	10–150[a]	100–6,000	5–500[a]	0.1–2	0.2–15	0–0.2
Iron	15–400[a]	10–400	1,000–10,000	0–0.2	0.1–10[a]	0.1–10[a]	1–8
Silicon iron	0–5	0–0.2	0–20	0–10	0–0.2	0–3	0–0.2
Stainless steel	0–100[b]	0–0.5	0–2	0–0.2	0–0.2	0–200[b]	0–0.2
Copper alloys	2–50[a]	2–15[a]	150–1500	2–5	0–1	0.2–15[a]	0–0.2
Nickel alloys	2–35[a]	2–10[a]	0.1–1500	0–0.2	0–0.2	0–1	0–0.2
Titanium	10–100	<0.1	0.1–1	<0.2	<0.1	<0.1	<0.1
Molybdenum	0–0.2	<0.1	High	<0.1	<0.1	<0.1	<0.1

Table 1.5–17 (continued)
CORROSION-RATE RANGES EXPRESSED IN MILS PENETRATION PER YEAR

Metal	Acid solutions			Alkaline solutions	Neutral solutions		Air
	Non-oxidizing		Oxidizing	Sodium hydroxide, 5%	Fresh water	Sea water	Normal outdoor urban exposure
	Sulfuric, 5%	Acetic, 5%	Nitric, 5%				
Zirconium	<0.5	<0.1	<0.1	<0.1	<0.1	<0.1	<0.1
Tantalum	<0.1	<0.1	<0.1	<1	<0.1	<0.1	<0.1
Silver	0–1	<0.1	High	<0.1	<0.1	<0.1	<0.1
Platinum	<0.1	<0.1	<0.1	<0.1	<0.1	<0.1	<0.1
Gold	<0.1	<0.1	<0.1	<0.1	<0.1	<0.1	<0.1

[a] Aeration leads to the higher rates in the range.
[b] Aeration leads to passivity, scarcity of dissolved air to activity.

Note: The corrosion-rate ranges for the solutions are based on temperatures up to 212°F.

From LaQue, F. and Copson, H., Eds., *Corrosion Resistance of Metals and Alloys,* © 1963 by Litton Educational Publishing, Inc., New York. Reprinted by permission of Van Nostrand Reinhold Company.

Table 1.5–18
CORROSION RESISTANCE OF COPPER ALLOYS

Table A lists specific materials for which copper and its alloys are suitable; Table B lists materials for which these metals are not suitable in any use where direct contact is involved. These lists apply to pure copper and to the alloys of copper with zinc, tin, aluminum, and nickel. Included are such classes as the bronzes, red and yellow brass, Muntz metal, aluminum brass, copper nickel, and nickel silver.

Most common chemicals and fluids that are not included in the following lists should not be used with copper alloys without a careful check of suitability. A helpful guide is the list of 22 alloys and almost 200 materials in the bulletin "Copper and Copper Alloys for the Process Industries," published by the Copper Development Association, New York, N.Y.

Table A. Materials Suitable for Use with Copper Alloys

Acetone	Coffee	Methyl alcohol
Alcohols	Ethers	Methyl chloride, dry
Alumina	Ethyl alcohol	Nitrogen
Aluminum hydroxide	Fluorine refrigerants	Oxygen
Amyl alcohol	Gasoline	Paraffin
Asphalt	Gelatine	Potassium chromate
Barium carbonate	Glucose	Potassium sulfate
Barium sulfate	Glycerine	Propane
Benzine	Hydrocarbons, pure	Rosin
Benzol	Hydrogen	Sodium chromate
Borax	Hydrogen sulfide, dry	Sulfur chloride, dry
Butane	Kerosene	Sulfur dioxide, dry
Butyl alcohol	Lacquers	Sulfur trioxide, dry
Carbon dioxide, dry	Lacquer solvents	Toluene
Castor oil	Lime	Trichloroethylene, dry
Chloroform, dry	Magnesium hydroxide	Varnish

Table 1.5–18 (continued)
CORROSION RESISTANCE OF COPPER ALLOYS

Table B. Materials Unsuitable for Use with Copper Alloys

Ammonia, moist	Ferric sulfate	Potassium dichromate, acid
Ammonium chloride	Hydrocyanic acid	Silver salts
Ammonium hydroxide	Mercury	Sodium cyanide
Ammonium nitrate	Mercury salts	Sodium dichromate, acid
Chromic acid	Nitric acid	Sulfur, molten
Ferric chloride	Potassium cyanide	

From Bolz, R. E. and Tuve, G. L., Eds., *Handbook of Tables for Applied Engineering Science*, 2nd ed., CRC Press, Cleveland, 1973, 351.

Table 1.5–19
METHODS FOR BONDING DISSIMILAR METALS[a]

Union of	Procedure
Aluminum–beryllium	Direct dip braze
Aluminum–stainless steel	Dip braze after tin-coating stainless steel
Aluminum–stainless steel	Soldering after nickel-plating aluminum
Aluminum–stainless steel	Diffusion bonding after interface-coating application
Aluminum–stainless steel	Dip braze after Ag-plating stainless steel
Aluminum–titanium	Tungsten-arc inert-gas shielded brazing with aluminum brazing alloy
Aluminum–titanium	Soldering after plating Ni on Ti and Al
Aluminum–tungsten	Tungsten-arc inert-gas brazing with aluminum brazing alloy
Beryllium–aluminum	Direct dip braze
Beryllium–stainless steel	Vacuum braze
Columbium–molybdenum	Electron-beam welding
Columbium–stainless steel	Inert-gas braze
Copper–nickel	Diffusion bonding after tin soldering
Copper wire–nickel wire	Capacitor discharge resistance micro-welder
Copper–tungsten	Tungsten-arc inert-gas brazing with aluminum brazing alloy
Molybdenum–columbium	Electron-beam welding
Molybdenum–stainless steel	Electron-beam welding
Molybdenum–stainless steel	Vacuum braze
Molybdenum–tungsten	Electron-beam welding
Nickel–copper	Diffusion bonding after tin soldering
Nickel wire–copper wire	Capacitor discharge resistance micro-welder
Stainless steel–aluminum	Dip braze after tin-coating the stainless steel
Stainless steel–aluminum	Soldering after nickel-plating the aluminum
Stainless steel–aluminum	Diffusion bonding after interface-coating application
Stainless steel–aluminum	Dip braze after Ag-plating stainless steel
Stainless steel–beryllium	Vacuum braze
Stainless steel–columbium	Inert-gas braze
Stainless steel–molybdenum	Electron-beam welding
Stainless steel–molybdenum	Vacuum braze
Stainless steel–steel (low alloy)	Percussion-stud welding
Stainless steel–titanium	Resistance-welding-machine braze
Stainless steel–tungsten	Tungsten-arc inert-gas brazing with aluminum brazing alloy
Titanium–aluminum	Tungsten-arc inert-gas shielded brazing with aluminum brazing alloy
Titanium–aluminum	Soldering after plating Ni on Ti and Al
Titanium–stainless steel	Resistance-welding-machine braze

Table 1.5–19 (continued)
METHODS FOR BONDING DISSIMILAR METALS

Union of	Procedure
Titanium–tungsten	Tungsten-arc inert-gas brazing with aluminum brazing alloy
Tungsten–aluminum	Tungsten-arc inert-gas brazing with aluminum brazing alloy
Tungsten–copper	Tungsten-arc inert-gas brazing with aluminum brazing alloy
Tungsten–molybdenum	Electron-beam welding
Tungsten–stainless steel	Tungsten-arc inert-gas brazing with aluminum brazing alloy
Tungsten–titanium	Tungsten-arc inert-gas brazing with aluminum brazing alloy

[a]Recommendations based on "NASA Contributions to Metal Joining," NASA SP-5064, 1967.

Note: Ultrasonic welding, where adhesion is accomplished entirely by vibratory energy, may be used for joining thin sections of the following metals:

Gold to copper, germanium, Kovar®, nickel, platinum, or silicon.
Kovar to copper, gold, nickel, or platinum.
Nickel to aluminum, copper, gold, Kovar, or molybdenum.
Platinum to aluminum, copper, gold, Kovar, nickel, or steel.
Silicon to aluminum and gold.
Steel to aluminum, copper, molybdenum, nickel, platinum, or zirconium.
Zirconium to aluminum, copper, and steel.

With the exception of germanium and silicon, these metals have also been successfully bonded to themselves by ultrasonic welding.

From Bolz, R. E. and Tuve, G. L., Eds., *Handbook of Tables for Applied Engineering Science,* 2nd ed., CRC Press, Cleveland, 1973, 1043.

Table 1.5–20
WELDING PROCESS CLASSIFICATIONS

Identifying Names for Welding, Brazing, and Related Processes;
Terminology of the American Welding Society

Arc welding	Gas welding	Other welding processes
Carbon arc	Oxyacetylene	Thermit
Shielded arc	Oxyhydrogen	Induction
Flux-cored arc	Pressure gas	Electron beam
Gas metal arc		Electroslag
Gas tungsten arc	**Solid state welding**	Laser beam
Submerged arc		
Plasma arc	Forge	**Brazing**
Stud	Friction	
	Explosion	Furnace, torch, induction,
Resistance welding	Diffusion	resistance, dip, infrared
	Ultrasonic	
Spot		**Soldering**
Seam		
Projection		Oven, torch, induction,
Flash		resistance, dip
Upset		
Percussion		**Cutting**
		Oxygen, arc, laser beam

From Bolz, R. E. and Tuve, G. L., Eds., *Handbook of Tables for Applied Engineering Science,* 2nd ed., CRC Press, 1973, 1044.

Table 1.5–21
CLAD OR WELDED METAL LAMINATES

Base metal	Surface layer Name	Typical composition, percent	Bonding method	Typical usage
CLAD SHEETS OR SHAPES				
Steel	Stainless steel	18 Cr, 8 Ni	Heat and pressure[a]	Chemical pressure vessels
Aluminum	Stainless steel	18 Cr, 8 Ni	Heat and pressure[a]	Cookware, heat exchange
Aluminum	Alclad[b] (1100)	99 Al	Heat and pressure[a]	High-finish parts
Aluminum	Alclad (3003)	1.2 Mn	Heat and pressure[a]	Building siding
Aluminum	Alclad (2024)	4.4 Cu, 1.5 Mg	Heat and pressure[a]	Airplane frames
Steel	Copper	99.9 Cu	Heat and pressure[a]	Heat exchangers
Steel	Copper-nickel	70 Cu, 30 Ni	Heat and pressure[a]	Marine condensers
Steel	Monel	65 Ni, 33 Cu	Heat and pressure[a]	Marine and saline uses
Steel	Nickel-chromium	72 Ni, 16 Cr	Heat and pressure[a]	Tanks for oxidizers
Steel	Nickel	99 Ni	Heat and pressure[a]	Caustic tanks
HARD-FACING OF PARTS				
Steel, iron, alloys	Stainless steel	19 Cr, 9 Ni	Welding deposition	High-temperature wear
Steel, iron, alloys	Martensitic steel	Cr 4, Mn 1, Ni 0.5	Welding deposition	Dies
Steel, iron, alloys	Nickel-chromium steel	19 Cr, 9 Ni	Welding deposition	Exhaust valves
Steel, iron, alloys	High-speed steel	18 W, 4 Cr, 1 V	Welding deposition	High-temperature tools
Steel, iron, alloys	Austenitic manganese steel	12 Mn, 3.5 Ni	Welding deposition	Rail frogs; dipper teeth
Steel, iron, alloys	High-chromium iron	25 Cr, 6 Mn, 4 C	Welding deposition	Corrosive wear

[a]Includes hot rolling, explosive bonding, and friction bonding; see Table 1.5–22. Similar metal bonding methods, by static or dynamic pressure, are used for other purposes, such as thermostat metal elements (see ASTM B-388).
[b]The Aluminum Association has adopted *alclad* as a generic name applied to any aluminum alloy product with a coating that will electrochemically protect the core.

Note: The above table lists a few typical metal coatings only. Many other combinations are available, including surface layers of various bronzes, cobalt-base alloys, and proprietary special alloys. Very hard materials such as the carbides (e.g., cemented tungsten carbide) can also be provided as surface materials or welded inserts; see Table 4–36 of *CRC Handbook of Materials Science,* Vol. I.

From Bolz, R. E. and Tuve, G. L., Eds., *Handbook of Tables for Applied Engineering Science,* 2nd ed., CRC Press, 1973, 1044.

Table 1.5–22
DIFFUSION WELDING AND ADHESION OF METALS

Engineering problems involving the welding of metals at temperatures well below the melting point are of two opposite kinds: (1) promotion of cold welding and its use as a fabrication process, and (2) prevention of cold welding, or adhesion, or high friction between metal surfaces in use. The following tables summarize the physical and chemical conditions for each case.

A. Diffusion Bonding Processes for Metal Fabrication

Applications	Fabrication methods for laminar metal composites by diffusion bonding. Hot-platen press or rolls are used for flat or near-flat surfaces. Explosive bonding and friction bonding are other related diffusion-welding processes.
Metals	Most common is a noncorrosive metal, a refractory metal, or a hard metal bonded to steel. Aluminum and copper cladding are common.

Table 1.5–22 (continued)
DIFFUSION WELDING AND ADHESION OF METALS

Surface character   Mating surfaces must conform accurately with very smooth surfaces. Metal plating or intermediate foil is sometimes used to promote diffusion. Surfaces may be mated by pressure, heat, and friction, with simultaneous size reduction, as in lap joints or stud attachment.

Surface chemistry   Prior surface cleaning is very important; among the processes used here are scraping, heating, chemical reaction and contaminant decomposition, crushing, cleaving, and abrading. High vacuum for degassing of surfaces is favored. Friction bonding involves cleaning action.

Atmosphere   Nonoxidizing conditions are important; inert gases (helium, nitrogen, argon) may be used at $10^{-3}-10^{-6}$ torr. Vacuum of $10^{-3}-10^{-6}$ torr is desirable, in vacuum furnace or with vacuum envelope welded to workpiece. For high-pressure bonding the atmosphere is less important.

Pressure, duration, and dynamics   Direct and uniform pressure, timed microseconds to several hours. Mating forces are produced by clamping, by inserted ram or deadweight, or in pressure container. Pressure depends on materials and sections, but 10,000–20,000 psi is often reached. Fluid pressure contact or sand packing is used with pressures to 50,000 psi or higher. Duration is momentary in roll or impact bonding. Explosive bonding uses high detonation-velocity charge, proportioned to metal mass and yield strength and producing ballistic velocities.

Temperature   Elevated temperature, but somewhat less than halfway to melting point. Most common processing range is 900–2,500° F, depending on metals and process. Hot-wall or radiation elements, friction or impact are used for heating.

B. Prevention of Cold Welding of Metals in High Vacuum

Applications   Most acute cold-welding and high-friction problems are in high-vacuum equipment or space applications; included here are other cases of mating metal surfaces in which adhesion or high friction must be avoided.

Metals   Problems occur with all metals, but tendency to cold weld is high with copper and aluminum, lower with cobalt and titanium. Coefficients of sliding friction of dry, clean metals in vacuum are very high.

Surface character   Slight surface roughness and contamination reduce cold welding under high-vacuum conditions. Grain size, orientation, and hardness affect dry friction and cold welding; softer metals weld more readily.

Surface chemistry   Even normal chemical contamination on dry metal surfaces reduces the tendency to cold welding in vacuum and reduces the coefficient of sliding friction. Usual liquid or solid lubricants prevent cold welding. Slight surface oxidation on metals almost cancels dry-adhesion tendencies.

Atmosphere   Gas adsorption and atmospheric particulates function as contaminants to reduce cold welding and adhesion. Very high vacuum and degassing are required to definitely initiate cold welding. Prolonged surface exposure to normal oxygen atmosphere reduces adhesion tendencies.

Pressure, duration, and dynamics   To avoid cold welding the surface contact pressure must be kept well below that producing any deformation of metal-surface irregularities. Danger of cold welding is increased by long-period contact, even at low-contact pressure.

Temperature   Temperature rise above ambient should be avoided. Cold welding occurs more readily as temperature is increased.

Table 1.5—22 (continued)
DIFFUSION WELDING AND ADHESION OF METALS

REFERENCES

American Society for Testing and Materials, Adhesion of Cold Welding of Materials in Space Environments (Symposium), ASTM-STP 431, ASTM, Philadelphia, 1967.

Defense Metals Information Center, Explosive Bonding, DMIC Memorandum 225, Battelle Memorial Institute, Columbus, Oh., 1967.

Dietz, A. G. H., Ed., *Composite Engineering Laminates,* M.I.T. Press, Cambridge, Mass., 1969.

Kammer, P. A., Monroe, R. E., and Martin, D. C., *Welding J.,* 48(3), 114, 1969.

From Bolz, R. E. and Tuve, G. L., Eds., *Handbook of Tables for Applied Engineering Science,* 2nd ed., CRC Press Cleveland, 1973, 1045.

Table 1.5—23
SOLDERING ALLOYS[a]

Tin-lead Solders

These most common solders are available in at least 15 compositions, from 2—70% tin, balance lead. High-tin solders are expensive.

Percent tin	Approximate melting range[b]		Tensile strength, kpsi	Electrical resistivity,[c] $Cu = 1$	Corroded by	Typical uses
	Liquid, °F	Solid, °F				
2.5	580	578			Sodium hypochlorite	Seams in cans; cable sheath
5	594	518	3.4	11.7	Chlorides	Filler metal; wiping solder
15	550	440			Potassium permanganate	Plumbing
20	531	370	5.8	10.5		Radiators; tubing joints
25[d]	510	362				Torch soldering
30	491	361				Machine soldering
40	460	360			Air, will tarnish	General-purpose joining
50	420	360	6.1	9.3	Nitric acid	Sheet metal
60	374	361				Electrical
63[e]	361	361	7.5			Electronic parts
70	378	361	6.8	8.7		Coating metals

Table 1.5—23 (continued)
SOLDERING ALLOYS[a]

Antimony Solders

High strength at high temperature; harder than tin-lead solders; should not be used on metals containing zinc.

Percent tin	Percent lead	Percent antimony	Approximate melting range[b]		Tensile strength, kpsi	Electrical resistiv-ity,[c] $Cu = 1$	Typical uses
			Liquid, °F	Solid, °F			
20	79.	1.0	517	363	3.5		Cable sheathing; radiators
30	68.4	1.6	482	364		13.	Machine soldering
40	58.	2.0	448	365			General-purpose joining
95	0.	5.0	467	458	11.	15.	Electrical
0	95.	5.0	554	486			Metal coating and filler; batteries

[a]For data on hard solders, see Table 1.5—25.
[b]The number of degrees between melting and freezing is especially important for wiping and filling solders.
[c]Electrical resistivity is expressed in terms of the resistivity of copper as unity. Thermal conductivity is also roughly one tenth that of copper. Specific heat is less than that of copper.
[d]A typical low-melting solder is 25% tin, 25% lead, and 50% bismuth (liquid at 266°F).
[e]Eutectic composition, lowest melting point for tin-lead alloys.

REFERENCES

American Society for Testing and Materials, ASTM Special Publication 189—1956 (Symposium on Soldering), ASTM, Philadelphia.
American Society for Testing and Materials, ASTM Specification for Solder Metal, B-32-66T, ASTM, Philadelphia, 1966.
American Society for Testing and Materials, ASTM Standards, ASTM, Philadelphia.
American Welding Society, Soldering Manual, Miami, Fl., 1959.
Metals Handbook Committee, Metals Handbook, Vol. I, 8th ed., American Society for Metals, Metals Park, Oh., 1961.
National Bureau of Standards, Solders and Soldering, NBS Circular 492, National Bureau of Standards, U.S. Government Printing Office, Washington, D.C., 1950.
Weast, R. C., Ed., CRC Handbook of Chemistry and Physics, 55th ed., CRC Press, Cleveland, 1974.

From Bolz, R. E. and Tuve, G. L., Eds., Handbook of Tables for Applied Engineering Science, 2nd ed., CRC Press, Cleveland, 1973, 1046.

Table 1.5—24
CONTROLLED ATMOSPHERES FOR BRAZING

Controlled atmospheres are used to prevent oxides and scale, especially for high-quality furnace brazing.

AWS brazing atmosphere type number	Source	Maximum dew point, °F, incoming gas	Approximate composition, percent				Application		Remarks
			H_2	N_2	CO	CO_2	Filler metals	Base metals	
1	Combusted fuel gas (low hydrogen)	Room temp	.5–1	87	.5–1	11–12	BAg[a], BCuP, RBCuZn[a]	Copper, brass[a]	
2	Combusted fuel gas (decarburizing)	Room temp	14–15	70–71	9–10	5–6	BCu, BAg[a], RBCuZn[a], BCuP	Copper[b], brass[a], low-carbon steel, nickel, monel, medium carbon steel[c]	Decarburizes
3	Combusted fuel gas, dried	−40	15–16	73–75	10–11		Same as 2	Same as 2 plus medium and high-carbon steels, monel, nickel alloys	
4	Combusted fuel gas, dried (carburizing)	−40	38–40	41–45	17–19		Same as 2	Same as 2 plus medium and high-carbon steels	Carburizes
5	Dissociated ammonia	−65	75	25			BAg[a], BCuP, RBCuZn[a], BCu, BNi	Same as for 1, 2, 3, 4 plus alloys containing chromium[d]	
6	Cylinder hydrogen	Room temp	97–100				Same as 2	Same as 2	Decarburizes
7	Deoxygenated and dried hydrogen	−75	100				Same as 5	Same as 5 plus cobalt, chromium, tungsten alloys and carbides[‡]	Decarburizes
8	Heated volatile materials	Inorganic vapors (i.e., zinc, cadmium, lithium, volatile fluorides)					BAg	Brasses	Special purpose. May be used in conjunction with 1 thru 7 to avoid use of flux

Table 1.5–24 (continued)
CONTROLLED ATMOSPHERES FOR BRAZING

AWS brazing atmosphere type number	Source	Maximum dew point, °F, incoming gas	Approximate composition, percent				Application		Remarks
			H_2	N_2	CO	CO_2	Filler metals	Base metals	
9	Purified inert gas	Inert gas (e.g., helium, argon, etc.)					Same as 5	Same as 5 plus titanium, zirconium, hafnium	Special purpose. Parts must be *very* clean and atmosphere must be pure
10	Vacuum pumping	Vacuum					Any metal that does not vaporize	Any metal that does not vaporize	Special purpose. Elaborate equipment and procedure

[a] Flux required in addition to atmosphere when alloys that contain volatile components are used.

[b] Copper should be fully deoxidized or oxygen-free.

[c] Heating time should be kept minimal to avoid objectionable decarburization.

[d] Flux must be used in addition to the atmosphere if appreciable quantities of aluminum, titanium, silicon, or beryllium are present.

From *Brazing Manual*, American Welding Society, Miami, Fl, 1963. With permission.

Table 1.5–25
BRAZING METALS AND APPLICATIONS

AWS–ASTM class B–	Composition, percent							Brazing range, °F
	Ag	Cu	Au	Al	Ni	Zn	Other	
Ag–1	45	15	—	—	—	16	Cd 24	1145–1400
Ag–2	35	26	—	—	—	21	Cd 18	1295–1550
Ag–4	40	30	—	—	2	28	—	1435–1650
Ag–7	56	22	—	—	—	17	Sn 5	1205–1400
Ag–18	54	40	—	—	—	5	Ni 1	1575–1775
Au–1	—	62.5	37.5	—	—	—	—	1860–2000
Au–4	—	—	82	—	18	—	—	1740–1840
Cu–1	—	99.9	—	—	—	—	—	2000–2100
Cu P–1	—	95	—	—	—	—	P 5	1450–1700
Cu P–3	5	89	—	—	—	—	P 6	1300–1500
CuZn–D	—	48	—	—	10	42	—	1720–1800
Ni–6	—	—	—	—	89	—	P 11	1700–1875
Ni–2	—	—	—	—	82.4	—	Cr 7, B 3.1, Si 4.5, Fe 3	1850–2150
AlSi 3	—	4	—	86	—	—	Si 10	1060–1120
Mg 2	—	—	—	12	—	5	Mg 83	1080–1130

Notes: Flux for brazing — boric acid, borax, and borates are used on most metals except aluminum and magnesium. For these two metals chlorides and fluorides are used. High-temperature service — Au-filler metals may be used for continuous service temperatures to 800°F, Ni-filler metals to 1,200°F. Other brazing metals are limited to 300–400°F, except Mg, which should not be used over 250°F.

From *Brazing Manual,* American Welding Society, Miami, Fl., 1963. With permission.

Table 1.5–26
TYPICAL METAL FINISHES FOR APPEARANCE AND PROTECTION

Material	Finish	Remarks
Aluminum alloy	Anodizing	An electrochemical-oxidation surface treatment, for improving corrosion resistance; not an electroplating process. For riveted or welded assemblies specify chromic acid anodizing. Do not anodize parts with nonaluminum inserts. Colors vary: yellow-green, gray, or black.
	Alrok®	Chemical-dip oxide treatment. Cheap. Inferior in abrasion and corrosion resistance to the anodizing process, but applicable to assemblies of aluminum and nonaluminum materials.
Copper and zinc alloys	Bright acid dip	Immersion of parts in acid solution. Clear lacquer applied to prevent tarnish.
Brass, bronze, zinc die-casting alloys	Brass, chrome, nickel, tin	As discussed under steel.
Magnesium alloy	Dichromate treatment	Corrosion-preventive dichromate dip. Yellow color.

Table 1.5–26 (continued)
TYPICAL METAL FINISHES FOR APPEARANCE AND PROTECTION

Material	Finish	Remarks
Stainless steel	Passivating treatment	Nitric-acid immunizing dip.
Steel	Cadmium	Electroplate, dull-white color, good corrosion resistance, easily scratched, good thread antiseize. Poor wear and galling resistance.
	Chromium	Electroplate, excellent corrosion resistance, and lustrous appearance. Relatively expensive. Specify hard chrome plate for exceptionally hard abrasion-resistive surface. Has low coefficient of friction. Used to some extent on nonferrous metals particularly when die-cast. Chrome-plated objects usually receive a base electroplate of copper, then nickel, followed by chromium. Used for buildup of parts that are undersized. Do not use on parts with deep recesses.
	Blueing	Immersion of cleaned and polished steel into heated saltpeter or carbonaceous material. Part then rubbed with linseed oil. Cheap. Poor corrosion resistance.
	Silver plate	Electroplate, frosted appearance; buff to brighten. Tarnishes readily. Good bearing lining. For electrical contacts, reflectors.
	Zinc plate	Dip in molten zinc (galvanizing) or electroplate of low-carbon or low-alloy steels. Low cost. Generally inferior to cadmium plate. Poor appearance. Poor wear resistance; electroplate has better adherence to base metal than hot-dip coating. For improving corrosion resistance, zinc-plated parts are given special inhibiting treatments.
	Nickel plate	Electroplate, dull white. Does not protect steel from galvanic corrosion. If plating is broken, corrosion of base metal will be hastened. Finishes in dull white, polished, or black. Do not use on parts with deep recesses.
	Black-oxide dip	Nonmetallic chemical black oxidizing treatment for steel, cast iron, and wrought iron. Inferior to electroplate. No buildup. Suitable for parts with close dimensional requirements as gears, worms, and guides. Poor abrasion resistance.
	Phosphate treatment	Nonmetallic chemical treatment for steel and iron products. Suitable for protection of internal surfaces of hollow parts. Small amount of surface buildup. Inferior to metallic electroplate. Poor abrasion resistance. Good paint base.
	Tin plate	Hot dip or electroplate. Excellent corrosion resistance, but if broken, will not protect steel from galvanic corrosion. Also used for copper, brass, and bronze parts that must be soldered after plating. Tin-plated parts can be severely worked and deformed without rupture of plating.
	Brass plate	Electroplate of copper and zinc. Applied to brass and steel parts where uniform appearance is desired. Applied to steel parts when bonding to rubber is desired.
	Copper plate	Electroplate applied before nickel or chrome plates. Also for parts to be brazed or protected against carburization. Tarnishes readily.

REFERENCE

For a discussion and data on surface protection methods including metal spraying, hard facing, and surface hardening, see Lipson, C. and Colwell, L. V., Eds., *Handbook of Mechanical Wear,* University of Michigan Press, Ann Arbor, 1961.

Table 1.5–27
COATINGS FOR ALUMINUM

Chemically deposited coatings on aluminum and aluminum alloys; for corrosion resistance, paint base, or decoration.

A. Nine Processes

Treatment	Purpose	For use on	Operation	Finish and thickness
Zinc phosphate coating	Paint base	Wrought alloys	Power spray or dip. For light to medium coats, 1–3 min at 130–135°F.	Crystalline, 100–200 mg/ft².
Chromium phosphate coating	Paint base or corrosion protection	Wrought or cast alloys	Power spray, dip, brush, or spray. For light to medium coats, 20 sec–2 min at 110–120°F.	Crystalline, 100–250 mg/ft².
Sulfuric acid anodizing	Corrosion and abrasion resistance, paint base	All alloys; uses limited on assemblies with other metals	15–60 min, 12–14 amp/ft², 18–20 V, 68–74°F. Tank lining of plastic, rubber, lead, or brick.	Very hard, dense, clear. 0.0002–0.0008 in. thick. Withstands 250–1,000-hr salt spray.
Chromic acid anodizing	Corrosion resistance, paint base; also as inspection technique with dyed coatings	All alloys except those with more than 5% Cu	30–40 min, 1–3 A/ft², ft², 40 V DC, 95°F, steel tanks and cathode, aluminum racks.	0.00002 to 0.00006 in. thick, 250-hr minimum salt spray.
Chromate conversion coating	Corrosion resistance, paint adhesion, and decorative effect	All alloys	10 sec–6 min depending on thickness, by immersion, spray, or brush, 70°F, in tanks of stainless, plastic, acid-resistant brick or chemical stoneware.	Electrically conductive, clear to yellow and brown in color, 0.00002 in. or less thick, 150–2,000-hr salt spray depending on alloy composition and coating thickness.
Chemical oxidizing	Corrosion resistance, paint base	All alloys, less satisfactory on copper-bearing alloys	Basket or barrel immersion 15–20 min, 150–212°F.	May be dyed, 250-hr minimum salt spray.
Electropolishing	Increases smoothness and brilliance of paint of plating base	Most wrought alloys, some sand-cast and die-cast alloys	15 min, 30–50 A/ft², 50–100 V, less than 120°F, aluminum cathode.	35–85 RMS depending on treatment.
Zinc immersion	Preplate for subsequent deposition of most plating metals, improves solderability	Many alloys, modifications for others particularly regarding silicon, copper and magnesium content	30–60 sec, 60–80°F, agitated steel or rubber-lined tank.	Thin film.
Electroplating[a]	Decorative appeal and/or function	Most alloys after proper preplating		Same as on steel.

[a]See Table 4–35 of *CRC Handbook of Materials Science,* Vol. I.

Table 1.5–27 (continued)
COATINGS FOR ALUMINUM

B. Electroplating

Metals	Operation
Chromium	Directly over zinc immersion coat, $65-70°F$, $6-8$ V, $200-225$ A/ft². Transfer to bath at $120-125°F$ if copper, or copper and nickel have been applied.
Copper	Directly over zinc, or follow with copper strike, then plate in conventional copper bath.
Brass	Directly over zinc, $80-90°F$, $2-3$ V, $3-5$ A/ft².
Nickel	Directly over zinc, or follow with copper strike, then plate in conventional nickel bath.
Cadmium	Directly over zinc, or follow with copper or nickel strike, or preferably cadmium strike, then plate in conventional cadmium bath.
Silver	Copper strike over zinc using copper cyanide bath, low pH, low temperature, 24 A/ft² for 2 min, drop to 12 A/ft² for $3-5$ min; plate in silver cyanide bath, $75-80°F$, 1 V, $5-15$ A/ft².
Zinc	Directly over zinc immersion coating.
Tin	Directly over zinc immersion coating.
Gold	Copper strike over zinc as for silver, then plate in conventional bath.

From *Iron Age,* June 28, 1956, copyright 1956, Chilton Book Co. With permission.

Glasses and Glass-ceramics

GLASSES AND GLASS-CERAMICS

D. E. Campbell and H. E. Hagy
Corning Glass Works

Oxide glasses and glass-ceramics comprise a host of materials of widely diverse compositions and properties. It is important to remember that although the properties of a glass or glass-ceramic are intrinsically a function of composition, other factors can often have important, sometimes overriding, effects on observed properties. Such factors include atmospheric weathering and thermal history. Exposure to weathering conditions obviously can drastically modify the surface chemistry behavior of these materials. Thermal history, on the other hand, can alter molecular structure and phase constitution. Therefore, chemical durability, density, refractive index, electrical resistivity, etc. can be measurably affected by the time-temperature relationship a glass has experienced on cooling from high temperatures. An important instance of the effect of thermal history is found on prolonged heating of certain glasses, e.g., borosilicates, above the annealing temperature, when degradation of the durability is observed, owing to phase separation into two or more glasses, one of which is highly soluble. Glass-ceramics represent an extreme case because the material, originally a glass, is deliberately heat treated to transform it into a new material whose polycrystalline structure gives rise to a totally different set of properties. Because of these dependencies, glass properties are listed for glasses in the annealed state, and the properties of glass-ceramics are given on the basis of the manufacturer's standard production process. In addition, it should be recognized that the compositions shown are nominal, and, owing to a variety of circumstances such as limitations on raw materials, environmental restrictions, and normal manufacturing fluctuations, changes from the stated compositions are quite possible.

To assist the materials scientist and technologist, the following tables have been organized as follows. Compositions have been listed in Table 2–1 primarily according to application with sublisting according to manufacturer, and manufacturer's code (if it exists). In Table 2–2, properties have been listed primarily according to manufacturer. The materials selected for inclusion were chosen largely on the basis of their commercial importance.

PHYSICAL PROPERTIES

Selected physical properties of major interest are presented in Table 2–2. Some explanation is required for those properties unique to glass. In addition, general glass behavior is discussed with regard to the more universally applicable properties.

1. Viscosity Reference Tests

Softening point — the temperature at which a glass fiber viscously extends under its own weight at a rate of 1 mm/min in a specific test apparatus as described in ASTM Designation C-338. The approximate viscosity level for the temperature is $10^{7.6}$ P.

Annealing point — the temperature at which a fiber subjected to a tensile stress viscously elongates, or, alternately, a beam of glass subjected to simple three-point loading viscously bends at specified rates, according to ASTM Designations C-336 and C-598, respectively. The viscosity at the annealing point is approximately 10^{13} P. Stresses in glass are relieved by viscous flow at the annealing point in a matter of minutes.

Strain point — a temperature derived by extrapolation of the data obtained in the annealing point tests, where the elongation or bending rates are a factor of 31.6 smaller than those obtained at the annealing point. The corresponding temperature is indicative of a viscosity of approximately $10^{14.5}$ P for the test glass. Stresses are substantially relieved by viscous flow at the strain point in a matter of hours.

2. Thermal Expansion

The value tabulated represents the average coefficient for the glass over the temperature range of 0 to 300°C. Values quoted are for glass in the well-annealed state, since thermal expansion, like many other physical properties, is thermal history sensitive.

Table 2–1
COMPOSITIONS OF GLASSES AND GLASS-CERAMICS

Electrical

Type	Manufacturer	Manufacturing code	Application	Description	SiO$_2$	Al$_2$O$_3$	B$_2$O$_3$
Sealing	Corning	0120	Dumet; Ni-Fe	Potash-soda-lead	56.2	1.6	
	Corning	1720	Series; W; Mo	Lime-magnesia-alumino silicate	60.7	17.3	5.0
	Corning	1990	Fe	Lead-potash	41.0		
	Corning	3320	W	Borosilicate	76.5	2.6	14.0
	Corning	7040	Mo	Soda-potash-boro-silicate	66.9	2.6	22.4
	Corning	7050	Mo; Ni-Fe-Co	Soda-potash-boro-silicate	68.0	1.9	23.6
	Corning	7052	Ni-Fe-Co	Alkali-barium-boro-silicate	65.3	7.4	17.8
	Corning	7056	Ni-Fe-Co	Alkali borosilicate	69.2	3.6	17.6
	Corning	7070	Series; W	Borosilicate	71.0	1.2	25.7
	Corning	7570	Solder glass	Lead borosilicate	3.7	10.7	11.2
	Corning	7581	Color TV seal	Lead-barium-boro-silicate	2.0		8.3
	Corning	7720	W	Soda-lead-borosilicate	73.5	1.6	14.4
	Owens-Illinois	EN-1	Ni-Fe-Co	Borosilicate			
	Owens-Illinois	KG-12	Dumet; Ni-Fe	Lead			
	Owens-Illinois	EZ-1	Series; W; Mo	Alkali boroalumino-silicate			
	Owens-Illinois	K-704	Mo	Borosilicate			
	Owens-Illinois	K-705	Mo; Ni-Fe-Co	Borosilicate			
	Owens-Illinois	ES-1	Series; W	Borosilicate			
	Owens-Illinois	K-772	W	Lead borosilicate			
	Owens-Illinois	KG-1	Dumet; Ni-Fe, Pt	Potash-soda-lead			
TV	Corning	0137	Color, neck	Potash lead	51.2	0.7	
	Corning	0138	Color, funnel	Alkali-alkaline earth-lead	50.3	4.7	
	Corning	9008	Black and white	Alkali-lead-barium	65.5	3.5	
	Corning	9068	Color panel	Strontium	63.2	2.0	
	Owens-Illinois	EG-19	Color neck	Lead potash			
	Owens-Illinois	TH-6	Color	Lead			
	Owens-Illinois	TL-10	Color	Strontium			
	Owens-Illinois	TM-5K	Black and white	Barium-lead			
Components	Corning	1723	Resistor	Alkaline earth aluminosilicate	57.3	15.7	3.9
	Corning	7059	Substrates	Barium borosilicate	49.5	11.0	14.2
	Corning	8871	Capacitor	Soda-potash-lead	41.6	0.2	
	Corning	9362	Reed switch	Alkali lead	50.2	1.7	

Table 2–1 (continued)
COMPOSITIONS OF GLASSES AND GLASS-CERAMICS

Electrical

Type	Na_2O	K_2O	Li_2O	MgO/CaO	BaO	PbO	$As_2O_3/$ Sb_2O_3	Other
Sealing	4.1	8.8				28.1	0.3/0.5	
	1.0			7.4/8.6				
	4.8	11.5	2.0			40.0	/1.0	
	4.2	1.5					/0.5	$0.7\ U_3O_8$
	4.8	3.0					0.2/	
	6.1	0.1					0.3/	
	2.2	3.1	0.6		2.8			0.3 F; 0.1 Cl
	0.7	8.0	0.7				0.2/	
	0.5	0.8	0.7					0.1 Cl
						74.4		
					2.0	75.2		12.4 ZnO
	3.8					5.7	0.9/	
TV	0.5	12.7				28.6	0.4/0.4	5.2 SrO
	6.1	8.4		2.9/4.3	0.2	22.5	0.2/0.1	0.1 SrO; 0.1 F
	6.2	6.6	0.5		12.1	3.5	0.2/0.4	$0.3\ Rb_2O$; $0.2\ TiO_2$, $0.05\ MnO_2$; 0.9 F
	7.1	8.8		0.8/1.8	2.4	2.2	0.2/0.4	10.2 SrO; $0.5\ TiO_2$, $0.5\ CeO_2$; 0.2 F
Components	0.1			6.6/9.4	5.9		0.1/	
					25.0		0.4/	
	2.3	6.1	1.0			48.8		
	4.2	8.6				27.2	0.1/	$6.95\ Fe_3O_4$; 1.1 FeO

Table 2–1 (continued)
COMPOSITIONS OF GLASSES AND GLASS-CERAMICS

Optical

Type	Manufacturer	Manufacturing code	Major use	Description	SiO₂	Al₂O₃	B₂O₃	P₂O₅
Transmitting	Corning	2403	Sharp cut red filter	Soda-zinc	67.5		11.5	
	Corning	4602	Heat absorbing	Aluminophosphate	18.5	14.1		57.3
	Corning	5543	Blue filter	Soda-potash-lead	65.5	1.3		
	Corning	7910	UV transmitting	96% silica	96.6	0.3	3.0	
	Corning	7913	IR transmitting	96% silica	96.6	0.3	3.0	
	Corning	7940	Window-heat shock resistant	100% silica	99.9			
	Corning	8463	Radiation absorbing	Lead borosilicate	5.0	3.0	9.7	
	Corning	9741	UV transmitting	Soda-alumina-borosilicate	66.5	5.6	23.7	
	Corning	9863	UV transmitting visible-absorb.	Alkaline earth phosphate		5.4		66.4
Refracting	Corning	8039	Ophthalmic flint	Alkali-lead	34.8	2.0		
	Corning	8097	Photochromic ophthalmic	Lithium-lead-barium-borosilicate	55.9	8.6	16.5	
	Corning	8260	Camera lens	Alkali borosilicate	68.8		11.4	
	Corning	8316	Ophthalmic Ba flint	Baria-lead-soda zirconia	39.0		3.5	
	Corning	8361	Ophthalmic flint	Potash-soda-lime zinc	68.0	2.0		
	Corning	8371	Lens element	Lead-potash-soda	44.9	1.0		
	Corning	8395	Dense flint lens	Potash-lead	31.1			
Reflecting	Corning	7940	Teles. mirror	100% silica	99.9			
	Corning	7971	Teles. mirror	Titanium sili.	92.6			
	Owens-Illinois	Cer-vit C-101	Teles. mirror	β-Spodumene, s.s.	66.4	21.4		

Table 2–1 (continued)
COMPOSITIONS OF GLASSES AND GLASS-CERAMICS

Type	Na_2O	K_2O/Li_2O	CaO	BaO/PbO	ZnO/CdO	Optical TiO_2	ZrO_2	As_2O_3/Sb_2O_3	Other
Transmitting	5.5				12.7/1.0			/1.1	0.7 Se; 0.2 S
	0.9	/0.2		/23.3	4.2/				1.3 FeO; 2.7 SnO; 1.4 Cl; 0.1 SO_3
	6.0	3.5/	0.5						0.1 CoO
				/82.0				/0.2	
	2.2	0.1/0.7	0.5						1.1F
			5.2	17.8/					2.5 Cl; 0.9 NiO; 1.8 CoO
Refracting	1.6	7.6/		/50.8		3.8	1.0	0.3/	
		/2.6		7.1/4.7			2.4		
	9.3	8.4/	0.1	0.9/		0.2		0.9/	
	8.0		5.0	20.0/14.2		3.0	7.3	0.1/0.1	0.2 Ag; 0.2 Cl; 0.2 Br; 0.2 F; 0.04 CuO
	7.9	9.4/	8.2	/45.6	3.5/	0.4		0.1/0.2	
	3.0	5.5/		/65.9				/0.1	
	1.0	2.0/						/0.1	
Reflecting	0.5	0.1/3.9	3.6			7.4	1.9	/0.4	
						1.8			

Table 2–1 (continued)
COMPOSITIONS OF GLASSES AND GLASS-CERAMICS

Industrial — Lab/Pharmaceutical

Manufacturer	Manufacturing code	Major use	Description	SiO_2	Al_2O_3	B_2O_3
Corning	1720·	Ignition tube	Lime-magnesia-aluminosilicate	60.7	17.3	5.0
Corning	7331	Gauge	Soda-alumino-borosilicate	77.6	5.3	8.8
Corning	7740	Lab, process	Soda-borosili-cate	80.3	2.3	13.3
Corning	7800	Pharmaceutical	Soda-barium borosilicate	74.5	6.4	9.0
Corning	7913	Lab, gauge	96% silica	96.5	0.5	3.0
Owens-Illinois	R-6	Lab	Soda-lime			
Owens-Illinois	KG-33	Lab	Soda-borosilicate			
Owens-Illinois	N-51A	Pharmaceutical	Soda-barium-borosilicate			

Manufacturer	Na_2O	K_2O	Li_2O	MgO	CaO	BaO	As_2O_3	Sb_2O_3	Other
Corning	1.0	0.2		7.4	8.6		0.5		
Corning	5.3	0.4	0.4				1.0	0.9	$0.1\ TiO_2$; $0.1\ ZrO_2$
Corning	4.0								
Corning	6.2	0.8			0.5	2.2			0.1 F; 0.1 Cl
Corning									
Owens-Illinois									
Owens-Illinois									
Owens-Illinois									

Table 2−1 (continued)
COMPOSITIONS OF GLASSES AND GLASS-CERAMICS

Speciality

Type	Manufacturer	Manufacturing code	Major use	Description	SiO$_2$	Al$_2$O$_3$
Fiber glass	Johns Manville	E		Lime aluminosilicate	54.5	14.6
	Johns Manville	753		Soda-lime borosilicate	63.5	5.5
	Johns Manville	475		Borosilicate	58.0	5.8
	Owens-Corning	E	Reinforcement	Lime aluminoborosilicate	54.0	14.0
	Owens-Corning	T	Insulation	Soda-lime	59.	4.5
	Owens-Corning	C	Acid resistant	Soda-lime borosilicate	65.	4.
	Owens-Corning	SF	Insulation	Soda-titania-zirconia	59.5	5.
Glass-ceramic	Corning	0336	Architectural	β-Spodumene, s.s.	64.6	20.0
	Corning	9455	Heat exchanger	β-Spodumene, s.s.	71.8	22.9
	Corning	9606	Radome	Cordierite-quartz-rutile	56.0	19.8

Type	B$_2$O$_3$	Na$_2$O	K$_2$O	Li$_2$O	MgO	CaO	BaO	ZnO	TiO$_2$	ZrO$_2$	Fe$_2$O$_3$	Other
Fiber glass	6.8	0.8		0.7		21.2			0.6		0.4	0.4 F
	5.5	14.6	1.0		3.0	6.0					0.1	0.7 F; 0.2 SO$_3$
	10.6	10.0	3.0		0.5	2.5	5.0	4.0			0.1	0.6 F
	10.0				4.5	17.5						
	3.5	11.	0.5		5.5	16.						
	5.5	8.	0.5		3.	14.						
	7	14.5							8.	4.		1.9 F
Glass-ceramic	2.0	0.6	0.2	3.5	1.8			2.2	4.4			0.8 As$_2$O$_3$
				5.1					0.1			
				14.8					9.0			0.3 As$_2$O$_3$

Housewares

Type	Manufacturer	Manufacturing code	Major use	Description	SiO$_2$
Cooking ware	Corning	0281	Lids	Soda-lime	72.7
	Corning	1710	Top-of-stove	Lime-magnesia-aluminosilicate	60.2
	Corning	7740	Ovenware	Soda borosilicate	80.3
	Corning	9608	Cookware	β-Spodumene, s.s.	69.7
Tableware	Corning	6720	Table and ovenware	Soda-zinc-lime-aluminosilicate	59.4
Appliance	Corning	9617	Cooktop	β-Spodumene, s.s.	67.4

Type	Al$_2$O$_3$	B$_2$O$_3$	Na$_2$O	K$_2$O	Li$_2$O	MgO	CaO	ZnO	TiO$_2$	ZrO$_2$	As$_2$O$_3$/Sb$_2$O$_3$	Other
Cooking ware	1.5	0.7	14.5	0.3			4.1	5.7			/0.2	0.2 SO$_3$
	18.0	4.4	1.6			9.0	6.7					0.1 ZrO$_2$
	2.3	13.3	4.0									
	17.1		0.4	0.1	2.5	2.8		1.0	4.8	0.1	0.5/	
Tableware	10.1	1.4	8.5	2.2				4.7	9.8		0.3/	3.5 F
Appliance	20.4		0.3	0.1	3.5	1.6		1.2	4.8		0.4/	0.2 F

Table 2–1 (continued)

COMPOSITIONS OF GLASSES AND GLASS-CERAMICS

Flat

Type	Manufacturer	Description	Major use	SiO_2	Al_2O_3	Na_2O	K_2O	MgO	CaO	Fe_2O_3	SO_3
Sheet	Libbey-Owens Ford	Soda-lime	Window	72.6	1.1	13.3	0.1	3.8	8.6	0.1	0.3
Plate	Pilkington	Soda-lime	Window	72.7	1.4	12.8	0.7	3.8	8.2	0.14	0.2
	Libbey-Owens Ford	Soda-lime	Architectural	72.2	0.1	13.9	–	2.1	11.2	0.1	0.4
Float	Libbey-Owens Ford	Soda-lime	Windshields	72.9	0.1	13.9	–	4.0	8.6	0.1	0.3
	Pilkington	Soda-lime	Windshields	72.7	1.0	13.0	0.6	3.9	8.3	0.095	0.22
Combination sheet-plate-float	Glaverbel	Soda-lime	Window-windshield	72.2	1	13.6	0.3	4.2	8.5	0.1	0.3

Container

Type	Manufacturer	Manufacturing code	Description	Major use	SiO_2	Al_2O_3	B_2O_3	Na_2O	K_2O	MgO	CaO	BaO	Fe_2O_3	TiO_2	As_2O_3/Sb_2O_3	SO_3
	Thatcher	A-3004	Soda-lime flint	Bottles	71.93	1.58	0.46	14.84	0.31	1.36	8.90	0.27	0.079	0.031	0.019/	0.22
	Owens-Illinois		Soda-lime	Containers												

Lighting

Type	Manufacturer	Manufacturing code	Description	Major use	SiO_2	Al_2O_3	B_2O_3	Na_2O	K_2O	MgO	CaO	BaO	PbO	Fe_2O_3	As_2O_3/Sb_2O_3	SO_3	Cl/F
Bulb	Corning	0081	Soda-lime	Lamp	73.4	1.4		16.2	0.4	3.4	5.0			0.05		0.1	
	General Electric	X-4	Soda-lime	Lamp	72.5	1.3		15.9		3.0	6.5					0.3	
Tube	Corning	0010	Potash-soda-lead	Exhaust, flare	61.8	2.2		7.0	7.3				21.5		0.2/		
	Corning	0080	Soda-lime	Fluorescent tube	73.6	1.4		16.2	0.4	3.4	4.8					0.2	
	Corning	0088		Tubing	70.5	2.0	2.6	12.2	5.3	3.0	4.2					0.2	
	Pilkington	PWM	Soda-lime	Fluorescent tube	71.4	2.2		15.0	1.7	4.0	4.6	0.8					
	Owens-Illinois	KG-1	Potash-soda lead	Exhaust, flare													
Sealed beam	Corning	7251	Soda-boro-silicate	Auto head lamps	78.1	2.0	14.9	4.9									0.1/

Table compiled by D. E. Campbell and H. E. Hagy.

Table 2–2
PROPERTIES OF GLASSES AND GLASS-CERAMICS

	Code	Softening point, °C	Annealing point, °C	Strain point, °C	Working point, °C	0–300 expansion, $10^{-7}/°C$	Young's modulus, 10 psi
Corning glasses	0010	626	432	392	983	93.5	9.0
	0080	696	514	473	1,005	93.5	10.2
	0081	696	514	473	1,013	93.5	10.3
	0088	700	521	480	1,017	92.	10.6
	0120	630	435	395	985	89.5	8.6
	0137	661	478	436	977	97.	–
	0138	670	494	451	970	99.	10.3
	0281	714	532	491	1,024	86.	10.4
	1710	915	713	669	1,189	42.	12.5
	1720	915	712	667	1,202	42.	12.7
	1723	908	710	665	1,168	46.	12.5
	1990	500	370	340	756	124.	8.4
	2403	–	–	–	–	–	–
	3320	780	540	493	1,171	40.	9.4
	4602	760	560	519	1,033	54.	10.3
	5543	673	459	417	1,076	71.	–
	6720	780	540	505	1,023	78.5	10.2
	7040[a]	702	490	449	1,080	47.5	8.6
	7050[a]	703	501	461	1,027	46.	8.7
	7052[a]	712	480	436	1,128	46.	8.2
	7056[a]	718	512	472	1,058	51.5	9.2
	7059	844	639	593	1,160	46.	9.8
	7070[a]	–	496	456	1,068	32.	7.4
	7251[a]	780	544	500	1,167	36.7	9.3
	7331	800	555	511	1,232	43.	11.0
	7570	440	363	342	–	84[b]	8.0
	7581	–	–	–	–	–	–
	7720[a]	755	523	484	1,146	36.	9.0
	7740	821	560	510	1,252	32.5	9.1
	7800	795	576	533	1,189	50.	10.2
	7910	1,500	910	820	–	7.5	9.6
	7913	1,530	1,020	890	–	7.5	9.6
	7940	~1,550	1,050	910	–	5.6	10.5
	7971	1,500	1,000	890	–	0.3	9.8
	8039	617	475	440	866	87.	–
	8097	675	511	473	947	51.	10.4
	8260	713	548	511	–	82.	12.3
	8316	686	553	520	886	91.	–
	8361	726	543	500	1,032	94.	10.5
	8371	598	429	391	905	88.	8.4
	8395	547	413	386	760	87.	8.0
	8463	377	316	300	–	104.	7.5
	8871	527	384	350	783	102.	8.4
	9008	646	446	408	1,002	89.	10.2
	9068	688	503	462	1,005	99.	10.1
	9362	627	445	405	958	91.5	–
	9741	705	450	408	1,161	39.5	7.2
	9863	596	482	453	–	97.	–

Table 2–2 (continued)
PROPERTIES OF GLASSES AND GLASS-CERAMICS

	Code	Softening point, °C	Annealing point, °C	Strain point, °C	Working point, °C	0–300 expansion, $10^{-7}/°C$	Young's modulus, 10 psi
Corning	0336	–	–	–	–	–	–
glass	9455	–	–	–	–	–	–
ceramics	9606	–	–	–	–	57.	17.2
	9608	–	–	–	–	4–20	12.5
	9617	–	–	–	–	–	–
General	X4	710	520	–	–	93.	–
Electric							
glasses							
Glaverbel	Flat	724	545	517	1,032	80.1[c]	–
glasses	glass						
Johns-	E	846	671	627	1,071	–	–
Manville	753	677	527	485	941	–	–
	475	679	529	491	924	–	–
Owens-	C	750	–	–	–	72.	–
Corning	E	830	–	–	–	60.	10.5
glasses	T	715	–	–	–	80.	–
	SF	675	–	–	–	75.	–
Owens-	A-3004	732	548	501	1,045	84.	–
Illinois	ES-1	735	476	430	1,095	33.	–
glasses	EN-1	716	482	437	1,115	47.	–
	EG-19	665	484	440	990	95.	–
	EZ-1	915	715	670	1,200	42.	–
	K-704	713	485	443	1,065	49.	–
	K-705	715	503	464	1,010	47.	–
	K-772	755	520	478	1,120	35.	–
	KG-1	626	435	394	970	94.	–
	KG-12	632	438	400	980	90.	–
	KG-33	827	565	513	1,240	32.	–
	N51-A	798	580	538	1,190	50.	–
	R-6	700	525	486	985	93.	–
	TH-6	670	488	447	995	98.5	–
	TL-10	691	503	461	1,015	99.5	–
	TM-SK	654	451	410	1,025	90.	–
Owens-	C-101	–	–	–	–		13.4
Illinois							
glass-							
ceramics							
Pilkington	Float	734	551	522	1,047	80.	10.7
glasses	sheet						
	PMW	707	524	495	1,023	92.	–
Thatcher	Flint	709	539	502	986	–	–

Table 2–2 (continued)
PROPERTIES OF GLASSES AND GLASS-CERAMICS

	Log$_{10}$ dc volume resistivity (ohm-centimeter)		Loss factor at 20°C	Dielectric constant, 20°C	Refractive index, D line	Density, grams per cubic centimeter	Corrosion resistance		
	250°C	350°C					Acid	Water	Weather
Corning glasses	8.9	7.0	0.16	6.7	1.540	2.86	2	2	2
	6.4	5.1	0.9	7.2	1.510	2.47	2	2	3
	6.4	5.1	2.1	5.0	1.510	2.47	2	2	3
	7.5	5.9	0.42	7.1	1.512	2.47	1–2	2	2
	10.1	8.0	0.12	6.7	1.560	3.05	2	2	2
	10.1	8.4	0.09	8.6	1.574	3.18	4	3	2–3
	–	–	–	–	–	3.02	4	3	2–3
	6.5	5.2	–	–	1.515	2.48	2	2	2–3
	10.8	9.0	0.37	6.3	–	2.52	3	1	1
	10.8	9.0	0.38	7.2	1.53	2.54	3	1	1
	13.5	11.3	0.16	6.3	1.545	2.64	3	1	1–2
	10.1	–	–	–	–	3.50	4	2–3	2–3
	–	–	–	–	–	–	2?	2?	1?
	8.6	7.1	0.28	4.0	1.48	2.27	1?	1	1
	–	–	–	–	1.51	2.52	4	3	3
	–	–	–	–	1.53	2.84	2?	2?	2?
	–	–	–	–	1.51	2.58	2	1	1
	9.4	7.7	0.20	4.8	1.48	2.24	4?	2	2
	8.4	6.8	0.33	4.9	1.479	2.24	4	2	2
	9.2	7.4	0.15	5.1	1.484	2.27	3	1	2
	10.3	8.4	0.27	5.7	1.487	2.29	2	2	2
	13.1	11.0	0.1	5.9	1.53	2.76	4	1	1
	11.2	9.1	0.06	4.1	1.47	2.13	2?	2	2?
	8.1	6.6	0.45	4.9	1.476	2.25	2	1	1
	7.1	5.8	–	–	1.486	2.32	2	1	1
	10.6	–	0.22	15.	–	5.42	4	2	2
	–	–	–	–	–	–	4	2–3	2–3
	8.8	7.3	0.23	4.6	1.487	2.35	2	2	1
	8.1	6.6	0.40	4.6	1.474	2.23	1	1	1
	7.0	5.7	–	–	1.491	2.36	1	1	1
	–	–	–	–	1.458	2.18	1	1	1
	9.7	8.1	0.04	3.8	1.458	2.18	1	1	1
	12.4	10.7	0.001	3.8	1.458	2.20	1	1	1
	12.0	10.3	<0.002	4.0	1.484	2.21	1	1	1
	–	–	0.16	10.0	1.700	4.02	4	2	2
	–	–	–	–	1.523	2.54	2–3	2	2
	–	–	–	–	1.517	2.51	2	2	2?
	–	–	–	–	1.653	3.52	3–4	2	2
	8.7	6.9	0.24	7.2	1.523	2.54	2	2	2
	–	–	–	–	1.621	3.61	3–4	1–2	1–2
	11.1	9.2	0.10	11.7	1.751	4.73	4	1–2	1–2
	–	–	–	–	1.97	6.22	4	1	2–3

Table 2–2 (continued)
PROPERTIES OF GLASSES AND GLASS-CERAMICS

	Log$_{10}$ dc volume resistivity (ohm-centimeter)		Loss factor at 20°C	Dielectric constant, 20°C	Refractive index, D line	Density, grams per cubic centimeter	Corrosion resistance		
	250°C	350°C					Acid	Water	Weather
Corning glasses (cont.)	10.9	–	0.05	8.4	1.656	3.84	4	1–2	2
	9.1	7.2	0.18	6.6	1.513	2.66	2	2	2
	–	–	–	–	–	2.685	2	2	2
	9.4	7.7	–	–	–	3.12	2–3	2	2–3
	9.4	7.6	–	–	1.47	2.16	3	2	2
	–	–	–	–	–	2.97	4	3	3
Corning glass-ceramics	–	–	–	–	–	–	3	1	1
	–	–	–	–	–	–	4	1	1
	10.	8.7	0.30	5.6	–	2.6	4	1–2	1
	8.1	6.8	0.34	6.9	–	2.5	2	1	1
	–	–	–	–	–	–	2	1	1
General Electric glasses	–	–	–	–	–	–	–	–	–
Glaverbel glasses	–	–	–	–	–	–	–	–	–
Johns-Manville	–	–	–	–	–	2.61	–	–	–
	–	–	–	–	–	2.52	–	–	–
	–	–	–	–	–	2.61	–	–	–
Owens-Corning glasses	–	–	–	7.8	1.549	2.61	–	–	–
	–	–	–	6.4	1.548	2.60	–	–	–
	–	–	–	7.3	1.541	2.54	–	–	–
	–	–	–	8.3	1.537	2.57	–	–	–
Owens Illinois glasses	–	–	–	–	1.51	2.48	2	2	2
	12.6	10.2	0.4	4.0	1.47	2.15	–	–	–
	9.0	7.2	1.3	5.1	1.49	2.27	–	–	–
	10.3	8.3	–	–	1.59	3.23	–	–	–
	10.7	8.9	2.3	6.3	1.53	2.52	–	–	–
	9.2	7.4	1.5	5.0	1.48	2.24	–	–	–
	8.3	6.9	1.7	4.8	1.48	2.25	–	–	–
	9.1	7.5	0.9	4.6	1.48	2.35	–	–	–
	8.7	6.8	–	–	1.54	2.85	–	–	–
	9.9	7.8	1.0	6.7	1.56	3.05	–	–	–
	8.1	6.6	2.1	4.6	1.47	2.23	1	1	1
	6.8	5.4	5.7	5.9	1.49	2.36	1	1	1
	6.6	5.2	6.1	7.2	1.52	2.53	1–2	1–2	2
	9.8	7.8	0.6	7.8	1.56	2.96	–	–	–
	8.7	6.9	1.4	7.2	1.52	2.63	–	–	–
	8.9	7.0	–	–	1.51	2.67	–	–	–
Owens-Illinois glass ceramics	–	–	–	–	–	–	–	–	–

Table 2–2 (continued)
PROPERTIES OF GLASSES AND GLASS-CERAMICS

	Log_{10} dc volume resistivity (ohm-centimeter)		Loss factor at 20°C	Dielectric constant, 20°C	Refractive index, D line	Density, grams per cubic centimeter	Corrosion resistance		
	250°C	350°C					Acid	Water	Weather
Pilkington glasses	5.4	4.3	–	7.5	1.518	2.497	–	–	–
	6.0	4.4	–	–	1.511	2.49	–	–	–
Thatcher	–	–	–	–	–	2.50	2	2	2

[a]Prolonged heating in the annealing temperature range may result in serious degradation of the corrosion resistance owing to appearance of a highly soluble second phase.

[b]0 to 200°C range.

[c]20 to 400°C range.

Table compiled by D. E. Campbell and H. E. Hagy.

For sealing purposes, the average coefficient from room temperature to the vicinity of the strain point is of primary interest. In general, this coefficient is about 10% higher than the 0 to 300°C value. However, for high reliability in seal design, complete expansion curves should be consulted with expert guidance.

The ASTM measurement procedure adopted for glass and glass-ceramics is designation E-228.

3. Young's Modulus

Room temperature values are listed. The change with temperature is small, generally being less than 10% over a very wide temperature range. Measurements are made in accordance with ASTM Designation C-623.

4. Electrical Properties

Glasses are used in the electrical industry for insulators, lamp and electronic tube enclosures, substrates under conductive coatings, sealing beads, and fuse bodies. The desirable properties of electrical glasses include:

High dielectric strength
High volume resistivity
Low power and loss factors
Transparency
Ability to hold a vacuum.

For a better acquaintance with these terms, a brief discussion of each is presented.

Dielectric strength – Researchers in the dielectric strength field believe there are two types of dielectric failure – one thermal, the other intrinsic. In thermal breakdown, the sample temperature increases, thereby lowering the sample's electrical resistance. This lower electrical resistance causes greater sample heating until dielectric failure occurs. In disruptive dielectric failure, the sample temperature does not increase. This type of failure is usually associated with voids and defects in the material.

Measurements are made according to ASTM Designation D-149, and the values are very high compared to most other materials.

Volume resistivity – The dependence of volume resistivity on temperature is expressed as

$$\log \rho = A + B/T, \tag{1}$$

where A and B are constants and T is in degrees Kelvin.

Measurements are made according to ASTM Designation C-657, and as with the dielectric strength dc ρ values are high.

Dielectric properties – Glass is a dielectric which passes a displacement current when ac fields are applied. If the dielectric is ideal, this current is 90 degrees ahead of the applied voltage. Actually, the current through the capacitor is

$$I = V \left(\frac{\omega \epsilon_0 A}{t} \right) (K'' + jK'). \tag{2}$$

In this equation, V is the voltage, ω is the angular frequency, $+ j$ designates a phase 90 degrees ahead of the applied voltage, ϵ_0 is the dielectric constant of air, A is the capacitor area, t is the capacitor thickness, K' is the relative dielectric constant, and K'' is the relative loss factor.

Thus, a good dielectric has a low K''/K' ratio. This ratio is called the loss tangent ($\tan \delta$). The power consumed is equal to the product of voltage and current of the real part of the equation

$$W = VI = V^2 \left(\frac{\omega \epsilon_0}{t} \right) (K'' + jK'). \tag{3}$$

Table 2–2 lists some of the electrical properties of glasses.

The dielectric properties were measured according to ASTM Designation D-150.

In determining a particular material for a dielectric application, it may be worthwhile to observe that, in addition to the obvious effect of high dielectric constant on the capacitance of the circuit element, the dielectric strength may be more important. According to Equation 3, the amount of energy a capacitor can store varies as the first power of the dielectric constant and the second power of the voltage. Thus, a material with twice the dielectric strength is as effective as a material with four times the dielectric constant.

CHEMICAL PROPERTIES – CORROSION RESISTANCE

More commonly referred to as chemical durability, this property is defined by ASTM as "the lasting quality (both physical and chemical) of a glass surface. It is frequently evaluated, after prolonged weathering in storage, in terms of chemical and physical changes in the glass surface, or in terms of change in contents of a vessel."

Chemical durability is strongly dependent upon composition. In addition, previous thermal history and mode of forming can have pronounced effects. For glass-ceramics, the amount and composition of the crystalline and glass phases, as well as the microstructure, have effects that may well substantially alter the intrinsic durability expected on the basis of the overall bulk composition.

Generally, with but a few exceptions, silicate glasses and glass-ceramics have superior resistance to chemical attack compared to most materials. Indeed, these materials are relatively inert to the oxidizing action of all the gaseous nonmetals, except fluorine, with which they will react. Neutral or acid salts, e.g., NaCl, K_2SO_4, $Na_2S_2O_7$, KNO_3, etc., in their molten or aqueous dissolved state are usually without detrimental effects. On the other hand, under the reducing conditions of some molten metals, e.g., aluminum, alkalies, and even hydrogen at elevated temperatures, glasses and glass-ceramics may be seriously degraded. With respect to liquids, these materials are resistant to a wide variety of solutions and solvents. As a rough guide, the degree of attack of glasses and glass-ceramics by the following reagents decreases in this order: hydrofluoric acid, strong alkalies, weak alkalies, chelates, acid, water, neutral salts, organic solvents.

Attack of glass and glass-ceramic surfaces may be evidenced in a number of ways, including hazing, iridescence, etching or pitting of a surface, surface electrical leakage, change in wettability, and contamination of stored solutions. In most instances, such degradation, whether by atmospheric gases or by solution attack, is associated with the presence of water. Thus, tests for evaluating chemical durability usually involve water and include acid, base, water, and high humidity (weathering).

Acid attack by most acids except hydrofluoric involves selective extraction of the alkali (fluxing or network modifying) elements, principally by an exchange of ions in the glass for hydrogen ions in the solution. Since the protons are hydrated, some hydration of the surface layer is concomitant. Ideally, the process is diffusion-controlled as a result of buildup of the surface reaction layer so that the amount of material extracted decreases with the square root of time. In practice, the buildup of constituents in the attacking solution attenuates the square root of time dependence. The temperature effect is small; the rate of attack increases only about one and a half times for each $10°C$. The effect of pH and, aside from hydrofluoric and phosphoric acids, the kind of acid is largely without effect. In the case of hydrofluoric acid[a] and, to a considerably lesser extent, phosphoric acid,[b] the attack involves complete disintegration of the glass structure. In these instances, the rate of attack increases linearly with time, and the temperature effect may increase the attack by as much as twofold per $10°C$ temperature rise.

Strong alkali attack for most silicate glasses and glass-ceramics is quite severe. Most glasses and glass-ceramics will lose between 75 and 330 $\times$ 10^{-4} mm/day when exposed to 5 wt/wt % NaOH at $95°C$. Since the process involves destruction of the basic silica network much like the action of hydrofluoric acid, the attack rate is quite dependent on alkali concentration (pH) and is linear with time. The reaction rate approximately doubles for each unit pH rise during the attack of a durable chemically resistant borosilicate. The temperature effect is about a twofold increase in rate for each $10°C$ temperature rise. Weak alkalies such as ammonium hydroxide do not have nearly so drastic an effect, although the attack is much more severe than by acids.

Water attack is not nearly so harsh as alkaline attack, but it can approach the severity of acid degradation. The mechanism is thought to consist initially of leaching, similar to that of acids, followed by alkaline attack as the water becomes alkaline with buildup of alkaline leach products. The temperature effect on the rate of water attack will vary accordingly, corresponding to that of acid at pH levels of less than 7 and approaching that of alkaline attack, as the pH trends higher. In the vapor state, water, i.e., steam, when properly handled, is usually without effect on chemically resistant compositions. Wet steam (untrapped), on the other hand, can be severely corrosive.

Weathering refers to corrosion by atmospheric agents such as water vapor and other reactive gases

[a]Chemically resistant borosilicate glass dissolves in 50% (wt/wt) hydrofluoric acid at a rate of about 2.5 cm/day at room temperature.

[b]Hot 80% (wt/wt) phosphoric acid dissolves chemically resistant borosilicate glass at the rate of about 127 $\times$ 10^{-4} mm/day at $180°C$.

such as sulfur dioxide. The mechanism initially involves selective removal of alkaline components, which, unless neutralized by acidic atmospheric components such as SO_2, accumulate on the surface and become the focus of an alkaline attack. Cleaning will often delay any appearance of visible degradation.

To be meaningful, tests for the chemical durability of glass and glass-ceramics should be performance related. If the material is to be used as a container, the release of glass components into the contents of the vessel is important; if it is to be used architecturally, appearance is important; if employed as a camera lens, the conditions for testing would be expected to be mild. Thus, evaluation of all glasses and glass-ceramics using the same test is not practical. Nevertheless, some rough universal guide is essential to the materials scientist. In the following tabulation, the durability ratings of the materials listed are defined:

Definitions of Chemical Durability Ratings

Class	Acid 95°C-5 w/w % HCl–24 hr	Water 90°C–4 hr	Weathering 50°C–98% RH-3 Mo
1	$< 40 \times 10^{-7}$ cm	$< 5 \times 10^{-6}$	No discernible surface change
2	25 to 250×10^{-7}	5 to 30×10^{-6} cm	Slight discernible surface change
3	25 to 250×10^{-6}	$> 30 \times 10^{-6}$ cm	Serious surface change
4	$> 25 \times 10^{-5}$ cm	–	–

One must qualify the above effort to provide in one tabulation unified classifications of glasses of such diverse composition as estimates and, inasmuch as possible, upper limits. That is, most glasses will very likely perform better than shown. This is particularly true for the estimates for water attack, wherein the table shows the maximum depth of layer affected in 4 hr at 90°C. Since water attack, to a large extent, proceeds as the square root of time, most of the attack will be evident in the early stages of exposure. All tests upon which the estimates have been based are static tests. Except for the water test, which utilizes 40- to 50-mesh powdered glass, the tests presume exposure of surfaces as formed. The weathering classification relates to the appearance of the surface immediately upon removal from the test chamber.

For quantitative techniques for evaluating durability, one is advised to consult the ASTM Standards, C-225. The United States Pharmacopoeia and the National Formulary contain adapted versions of the same tests which are primarily oriented to container applications. Other than these and other like tests promulgated by corresponding agencies in other countries, there are no standardized test procedures. However, many specialized tests are described in the literature, and those interested are advised to consult the bibliography published by the International Glass Commission (ICG) in 1965 and 1972–73 (Parts 1 and 2).

Acknowledgments

The authors gratefully acknowledge the assistance of several of their colleagues, including T. S. Magliocca and G. B. Hares – Compositions; W. H. Barney – Electrical Properties; P. B. Adams – Corrosion Resistance; and G. W. McLellan for his suggestions relative to format and content. Mrs. E. L. Cross is also to be thanked for her arduous efforts, particularly in transcribing the data into tabular form.

Table 2–3
HEAT OF FORMATION OF INORGANIC OXIDES[a]

The ΔH_O values are given in gram calories per mole. The a, b, and I values listed here make it possible for one to calculate the ΔF and ΔS values by use of the following equations:

$$\Delta F_t = \Delta H_O + 2.303 aT \log T + b \times 10^{-3} T^2 + c \times 10^5 T^{-1} + IT$$
$$\Delta S_t = -a - 2.303a \log T - 2b \times 10^{-3} T + c \times 10^5 T^{-2} - I$$

Coefficients in Free-energy Equations

Reaction	Temperature range of validity	ΔH_O	2.303a	b	c	I
2 Ac(c) + 3/2 O₂(g) = Ac₂O₃ (c)	298.16–1,000 K	-446,090	-16.12	—	—	+109.89
2 Al(c) + 1/2 O₂ (g) = Al₂O(g)	298.16–931.7 K	-31,660	+14.97	—	—	-72.74
2 Al(l) + 1/2 O₂(g) = Al₂O(g)	931.7–2,000 K	-38,670	+10.36	—	—	-51.53
Al(c) + 1/2 O₂ (g) = AlO(g)	298.16–931.7 K	+10,740	+5.76	—	—	-37.61
Al(l) + 1/2 O₂ (g) = AlO(g)	931.7–2,000 K	+8,170	+5.76	—	—	-34.85
2 Al(c) + 3/2 O₂ (g) = Al₂O₃ (corundum)	298.16–931.7 K	-404,080	-15.68	+2.18	+3.935	+123.64
2 Al(l) + 3/2 O₂ (g) = Al₂O₃ (corundum)	931.7–2,000 K	-407,950	-6.19	-0.78	+3.935	+102.37
2 Am(c) + 3/2 O₂ (g) = Am₂O₃ (c)	298.16–1,000 K	-422,090	-16.12	—	—	+107.89
Am(c) + O₂ (g) = AmO₂ (c)	298.16–1,000 K	-240,600	-4.61	—	—	+55.91
2 Sb(c) + 3/2 O₂ (g) = Sb₂O₃ (cubic)	298.16–842 K	-169,450	+6.12	-6.01	-0.30	+52.21
2 Sb(c) + 3/2 O₂ (g) = Sb₂O₃ (orthorhombic)	298.16–903 K	-168,060	+6.12	-6.01	-0.30	+50.56
2 Sb(l) + 3/2 O₂ (g) = Sb₂O₃ (orthorhombic)	903–928 K	-175,370	+15.29	-7.75	-0.30	+33.12
2 Sb(l) + 3/2 O₂ (g) = Sb₂O₃ (l)	928–1,698 K	-173,940	-32.84	+0.75	-0.30	+166.52
2 Sb(l) + 3/2 O₂ (g) = 1/2 Sb₄O₆ (g)	1,698–1,713 K	-132,760	+10.91	+0.75	-0.30	+0.96
2 Sb(g) + 3/2 O₂ (g) = 1/2 Sb₄O₆ (g)	1,713–2,000 K	-234,760	-0.74	+0.75	-0.30	+98.17
2 Sb(c) + 2 O₂ (g) = Sb₂O₄ (c)	298.16–903 K	-208,310	+6.31	-5.36	-0.40	+73.02
2 Sb(l) + 2 O₂ (g) = Sb₂O₄ (c)	903–1,500 K	-215,610	+15.47	-7.10	-0.40	+55.61
6 Sb(c) + 13/2 O₂ (g) = Sb₆O₁₃ (c)	298.16–903 K	-649,160	+38.46	-25.13	-1.30	+192.54
6 Sb(c) + 13/2 O₂ (g) = Sb₆O₁₃ (c)	903–1,500 K	-691,370	+14.13	-30.35	-1.30	+315.93
2 Sb(c) + 5/2 O₂ (g) = Sb₂O₅ (c)	298.16–903 K	-226,060	+37.12	-22.66	-0.50	+18.61
2 Sb(l) + 5/2 O₂ (g) = Sb₂O₅ (c)	903–1,500 K	-240,130	+29.01	-24.40	-0.50	+59.74
2 As(c) + 3/2 O₂ (g) = As₂O₃ (orthorhombic)	298.16–542 K	-154,870	+29.54	-21.33	-0.30	-8.83
2 As(c) + 3/2 O₂ (g) = As₂O₃ (monoclinic)	298.16–586 K	-150,760	+29.54	-21.33	-0.30	-16.95
2 As(c) + 3/2 O₂ (g) = As₂O₃ (l)	542–730.3 K	-156,260	-43.29	+2.97	-0.30	+180.95
2 As(c) + 3/2 O₂ (g) = 1/2 As₄O₆ (g)	730.3–883 K	-135,930	+0.46	+2.97	-0.30	+26.88

[a]This table is included here because there is no separate section for oxides.

Table 2–3 (continued)
HEAT OF FORMATION OF INORGANIC OXIDES

Reaction	Temperature range of validity	ΔH_o	2.303a	b	c	I
$1/2\ As_4(g) + 3/2\ O_2(g) = 1/2\ As_4O_6(g)$	883–2,000 K	−154,450	−2.90	+0.75	−0.30	+59.71
$2\ As(c) + 2\ O_2(g) = As_2O_4(c)$	298.16–883 K	−173,690	+21.52	−13.42	−0.40	+34.38
$1/2\ As_4(g) + 2\ O_2(g) = As_2O_4(c)$	883–1,500 K	−192,210	+18.15	−15.64	−0.40	+67.22
$2\ As(c) + 5/2\ O_2(g) = As_2O_5(c)$	298.16–883 K	−217,080	+12.32	−4.65	−0.50	+80.50
$1/2\ As_4(g) + 5/2\ O_2(g) = As_2O_5(c)$	883–2,000 K	−235,600	+8.96	−6.87	−0.50	+113.33
$Ba(\alpha) + 1/2\ O_2(g) = BaO(c)$	298.16–648 K	−134,590	−7.60	+0.87	+0.42	+45.76
$Ba(\beta) + 1/2\ O_2(g) = BaO(c)$	648–977 K	−134,140	−3.34	−0.56	+0.42	+34.01
$Ba(l) + 1/2\ O_2(g) = BaO(c)$	977–1,911 K	−135,900	−2.19	−0.56	+0.42	+32.37
$Ba(g) + 1/2\ O_2(g) = BaO(c)$	1,911–2,000 K	−176,400	−8.01	−0.56	+0.42	+72.66
$Ba(\alpha) + O_2(g) = BaO_2(c)$	298.16–648 K	−154,830	−11.05	+0.87	+0.42	+74.48
$Ba(\beta) + O_2(g) = BaO_2(c)$	648–977 K	−154,380	−6.79	−0.56	+0.42	+62.73
$Ba(l) + O_2(g) = BaO_2(c)$	977–1,500 K	−156,140	−5.64	−0.56	+0.42	+61.09
$Be(c) + 1/2\ O_2(g) = BeO(c)$	298.16–1,556 K	−144,220	−1.91	−0.46	+1.24	+30.64
$Be(l) + 1/2\ O_2(g) = BeO(c)$	1,556–2,000 K	−144,300	+6.06	−1.75	+1.485	+7.25
$Bi(c) + 1/2\ O_2(g) = BiO(c)$	298.16–544 K	−50,450	−4.61	—	—	+35.51
$Bi(l) + 1/2\ O_2(g) = BiO(c)$	544–1,600 K	−52,920	−4.61	—	—	+40.05
$2\ Bi(c) + 3/2\ O_2(g) = Bi_2O_3(c)$	298.16–544 K	−139,000	−11.56	+2.15	−0.30	+96.52
$2\ Bi(l) + 3/2\ O_2(g) = Bi_2O_3(c)$	544–1,090 K	−142,270	+2.30	−3.25	−0.30	+67.55
$2\ Bi(l) + 3/2\ O_2(g) = Bi_2O_3(l)$	1,090–1,600 K	−147,350	−32.84	+0.75	−0.30	+174.59
$2\ Bi(c) + 3/2\ O_2(g) = B_2O(c)$	298.16–723 K	−304,690	+11.72	−7.55	+0.355	+34.25
$2\ B(c) + 3/2\ O_2(g) = B_2O_3(gl)$	298.16–723 K	−298,670	+26.57	−15.90	−0.30	−10.40
$2\ B(c) + 3/2\ O_2(g) = B_2O_3(l)$	723–2,000 K	−308,100	−38.41	+5.15	−0.30	+173.24
$Cd(c) + 1/2\ O_2(g) = CdO(c)$	298.16–594 K	−62,330	−2.05	+0.71	−0.10	+29.17
$Cd(l) + 1/2\ O_2(g) = CdO(c)$	594–1,038 K	−63,240	+2.07	−0.76	−0.10	+20.14
$Cd(g) + 1/2\ O_2(g) = CdO(c)$	1,038–2,000 K	−89,320	−2.83	−0.76	−0.10	+60.05
$Ca(\alpha) + 1/2\ O_2(g) = CaO(c)$	298.16–673 K	−151,850	−6.56	+1.46	+0.68	+43.93
$Ca(\beta) + 1/2\ O_2(g) = CaO(c)$	673–1,124 K	−151,730	−4.14	+0.41	+0.68	+37.63
$Ca(l) + 1/2\ O_2(g) = CaO(c)$	1,124–1,760 K	−153,480	−1.36	−0.29	+0.68	+31.49
$Ca(g) + 1/2\ O_2(g) = CaO(c)$	1,760–2,000 K	−194,670	−7.18	−0.29	+0.68	+73.84
$Ca(\alpha) + O_2(g) = CaO_2(c)$	298.16–500 K	−158,230	−12.32	+1.46	+0.68	+78.28

Table 2–3 (continued)
HEAT OF FORMATION OF INORGANIC OXIDES

Reaction	Temperature range of validity	ΔH_o	$2.303a$	b	c	I
C(graphite) + 1/2 O$_2$ (g) = CO(g)	298.16–2,000 K	−25,400	+2.05	+0.27	−1.095	−28.79
C(graphite) + O$_2$ (g) = CO$_2$ (g)	298.16–2,000 K	−93,690	+1.63	−0.7	−0.23	−5.64
2 Ce(c) + 3/2 O$_2$ (g) = Ce$_2$O$_3$ (c)	298.16–1,048 K	−435,600	−4.60	–	–	+92.84
2 Ce(l) + 3/2 O$_2$ (g) = Ce$_2$O$_3$ (c)	1,048–1,900 K	−440,400	−4.60	+2.34	−0.20	+97.42
Ce(c) + O$_2$ (g) = CeO$_2$ (c)	298.16–1,048 K	−245,490	−6.42	–	–	+67.79
Ce(l) + O$_2$ (g) = CeO$_2$ (c)	1,048–2,000 K	−247,930	+0.71	−0.66	−0.20	+51.73
2 Cs(c) + 1/2 O$_2$ (g) = Cs$_2$O(c)	298.16–301.5 K	−75,900	–	–	–	+36.60
2 Cs(l) + 1/2 O$_2$ (g) = Cs$_2$O(c)	301.5–763 K	−76,900	–	–	–	+39.92
2 Cs(l) + 1/2 O$_2$ (g) = Cs$_2$O(l)	763–963 K	−75,370	−9.21	–	–	+64.47
2 Cs(g) + 1/2 O$_2$ (g) = Cs$_2$O(l)	963–1,500 K	−113,790	−23.03	–	–	+145.60
2 Cs(c) + O$_2$ (g) = Cs$_2$O$_2$ (c)	298.16–301.5 K	−96,500	−2.30	–	–	+62.30
2 Cs(l) + O$_2$ (g) = Cs$_2$O$_2$ (c)	301.5–870 K	−97,800	−4.61	–	–	+72.34
2 Cs(l) + O$_2$ (g) = Cs$_2$O$_2$ (l)	870–963 K	−96,060	−18.42	–	–	+110.94
2 Cs(g) + O$_2$ (g) = Cs$_2$O$_2$ (l)	963–1,500 K	−134,000	−31.08	–	–	+188.11
2 Cs(c) + 3/2 O$_2$ (g) = Cs$_2$O$_3$ (c)	298.16–301.5 K	−112,690	−11.51	–	–	+110.10
2 Cs(l) + 3/2 O$_2$ (g) = Cs$_2$O$_3$ (c)	301.5–775 K	−113,840	−12.66	–	–	+116.77
2 Cs(l) + 3/2 O$_2$ (g) = Cs$_2$O$_3$ (l)	775–963 K	−110,740	−26.48	–	–	+152.70
2 Cs(g) + 3/2 O$_2$ (g) = Cs$_2$O$_3$ (l)	963–1,500 K	−148,680	−39.14	–	–	+229.87
Cs(c) + O$_2$ (g) = CsO$_2$ (c)	298.16–301.5 K	−63,590	−11.51	–	–	+72.29
Cs(l) + O$_2$ (g) = CsO$_2$ (c)	301.5–705 K	−64,240	−12.66	–	–	+77.30
Cs(l) + O$_2$ (g) = CsO$_2$ (l)	705–963 K	−61,770	−18.42	–	–	+90.20
Cs(g) + O$_2$ (g) = CsO$_2$ (l)	963–1,500 K	−80,500	−24.18	–	–	+126.83
Cl$_2$ (g) + 1/2 O$_2$ (g) = Cl$_2$O(g)	298.16–2,000 K	+17,770	−0.71	−0.12	+0.49	+16.81
1/2 Cl$_2$ (g) + 1/2 O$_2$ (g) = ClO(g)	298.16–1,000 K	+33,000	−0.76	–	–	−0.24
1/2 Cl$_2$ (g) + O$_2$ (g) = ClO$_2$ (g)	298.16–2,000 K	+24,150	−0.76	−0.105	−0.665	+19.08
1/2 Cl$_2$ (g) + 3/2 O$_2$ (g) = ClO$_3$ (g)	298.16–500 K	+37,740	+5.76	–	–	+21.42
Cl$_2$ (g) + 7/2 O$_2$ (g) = Cl$_2$O$_7$ (g)	298.16–500 K	+65,040	+12.66	–	–	+78.01
2 Cr(c) + 3/2 O$_2$ (g) = Cr$_2$O$_3$ (β)	298.16–1,823 K	−274,670	−14.07	+2.01	+0.69	+105.65
2 Cr(l) + 3/2 O$_2$ (g) = Cr$_2$O$_3$ (β)	1,823–2,000 K	−278,030	+2.33	−0.35	+1.57	+58.29
Cr(c) + O$_2$ (g) = CrO$_2$ (c)	298.16–1,000 K	−142,500	–	–	–	+42.00

Table 2–3 (continued)
HEAT OF FORMATION OF INORGANIC OXIDES

Reaction	Temperature range of validity	ΔH_0	2.303a	b	c	I
$Cr(c) + 3/2\ O_2(g) = CrO_3(c)$	298.16–471 K	-141,590	-13.82	—	—	+103.90
$Cr(c) + 3/2\ O_2(g) = CrO_3(l)$	471–600 K	-141,580	-32.24	—	—	+153.14
$Co(\alpha, \beta) + 1/2\ O_2(g) = CoO(c)$	298.16–1,400 K	-56,910	+0.69	—	—	+16.03
$Co(\gamma) + 1/2\ O_2(g) = CoO(c)$	1,400–1,763 K	-58,160	-1.15	—	—	+22.71
$Co(l) + 1/2\ O_2(g) = CoO(c)$	1,763–2,000 K	-65,680	-6.22	—	—	+43.43
$3\ Co(\alpha, \beta, \gamma) + 2\ O_2(g) = Co_3O_4(c)$	298.16–1,500 K	-207,300	-2.30	—	—	+90.56
$2\ Cu(c) + 1/2\ O_2(g) = Cu_2O(c)$	298.16–1,357 K	-40,550	-1.15	-1.10	-0.10	+21.92
$2\ Cu(l) + 1/2\ O_2(g) = Cu_2O(c)$	1,357–1,502 K	-43,880	+8.47	-2.60	-0.10	-3.72
$2\ Cu(l) + 1/2\ O_2(g) = Cu_2O(l)$	1,502–2,000 K	-37,710	-12.48	+0.25	-0.10	+54.44
$Cu(c) + 1/2\ O_2(g) = CuO(c)$	298.16–1,357 K	-37,740	-0.64	-1.40	-0.10	+24.87
$Cu(l) + 1/2\ O_2(g) = CuO(c)$	1,357–1,720 K	-39,410	+4.17	-2.15	-0.10	+12.05
$Cu(l) + 1/2\ O_2(g) = CuO(l)$	1,720–2,000 K	-41,060	-11.35	+0.25	-0.10	+59.09
$F_2(g) + 1/2\ O_2(g) = F_2O(g)$	298.16–2,000 K	+5,070	-0.41	-0.15	+0.535	+16.04
$2\ Ga(c) + 1/2\ O_2(g) = Ga_2O(c)$	298.16–302.7 K	-81,110	+10.32	-5.75	-0.10	-3.66
$2\ Ga(l) + 1/2\ O_2(g) = Ga_2O(c)$	302.7–1,000 K	-83,360	+13.49	-5.75	-0.10	-4.08
$2\ Ga(c) + 3/2\ O_2(g) = Ga_2O_3(c)$	298.16–302.7 K	-256,240	+14.64	-3.75	-0.30	+32.23
$2\ Ga(l) + 3/2\ O_2(g) = Ga_2O_3(c)$	302.7–2,000 K	-258,490	+17.82	-3.75	-0.30	+31.79
$Ge(c) + 1/2\ O_2(g) = GeO(c)$	298.16–1,200 K	-60,900	+1.27	-1.49	-0.10	+17.19
$Ge(c) + 1/2\ O_2(g) = GeO(g)$	298.16–1,200 K	-21,870	+6.72	-0.075	-0.10	-41.25
$Ge(c) + O_2(g) = GeO_2(gl)$	298.16–1,200 K	-127,830	+4.28	-2.52	-0.20	+30.54
$2\ Au(c) + 3/2\ O_2(g) = Au_2O_3(c)$	298.16–500 K	-2,160	-10.36	—	—	+95.14
$Hf(c) + O_2(g) = HfO_2\ (monoclinic)$	298.16–2,000 K	-268,380	-9.74	-0.28	+1.54	+78.16
$H_2(g) + 1/2\ O_2(g) = H_2O(l)$	298.16–373.16 K	-70,600	-18.26	+0.64	-0.04	+91.67
$H_2(g) + 1/2\ O_2(g) = H_2O(g)$	298.16–2,000 K	-56,930	+6.75	-0.64	-0.08	-8.74
$D_2(g) + 1/2\ O_2(g) = D_2O(l)$	298.16–374.5 K	-72,760	-18.10	—	—	+93.59
$D_2(g) + 1/2\ O_2(g) = D_2O(g)$	298.16–2,000 K	-58,970	+5.50	-0.75	+0.085	-3.74
$1/2\ H_2(g) + 1/2\ O_2(g) = OH(g)$	298.16–2,000 K	+10,350	+0.90	+0.005	-0.26	-6.69
$H_2(g) + O_2(g) = H_2O_2(l)$	298.16–425 K	-47,140	-13.52	-7.13	—	+99.39
$H_2(g) + O_2(g) = H_2O_2(g)$	298.16–1,500 K	-32,570	+4.77	-0.96	+0.97	+13.84
$2\ In(c) + 3/2\ O_2(g) = In_2O_3(c)$	298.16–429.6 K	-220,410	+5.43	-0.50	-0.30	+59.49

Table 2–3 (continued)
HEAT OF FORMATION OF INORGANIC OXIDES

Reaction	Temperature range of validity	ΔH_o	$2.303a$	b	c	I
2 In(l) + 3/2 O$_2$(g) = In$_2$O$_3$ (c)	429.6–2,000 K	−220,970	+13.22	−3.00	−0.30	+41.36
I$_2$(c) + 5/2 O$_2$(g) = I$_2$O$_5$(c)	298.16–386.8 K	−42,040	+2.30	—	—	+113.71
I$_2$(l) + 5/2 O$_2$(g) = I$_2$O$_5$(c)	386.8–456 K	−43,490	+16.12	—	—	+81.70
I$_2$(g) + 5/2 O$_2$(g) = I$_2$O$_5$(c)	456–500 K	−58,020	−6.91	—	—	+174.79
Ir(c) + O$_2$(g) = IrO$_2$(c)	298.16–1,300 K	−39,480	+8.17	−6.39	−0.20	+20.33
0.947 Fe(α) + 1/2 O$_2$(g) = Fe$_{0.947}$O(c)	298.16–1,033 K	−65,320	−11.26	+2.61	+0.44	+48.60
0.947 Fe(β) + 1/2 O$_2$(g) = Fe$_{0.947}$O(c)	1,033–1,179 K	−62,380	+4.08	−0.75	+0.235	+3.00
0.947 Fe(γ) + 1/2 O$_2$(g) = Fe$_{0.947}$O(c)	1,179–1,650 K	−66,750	−8.04	+0.67	−0.10	+42.28
0.947 Fe(γ) + 1/2 O$_2$(g) = Fe$_{0.947}$O(l)	1,650–1,674 K	−64,200	−18.72	+1.67	−0.10	+73.45
0.947 Fe(δ) + 1/2 O$_2$(g) = Fe$_{0.947}$O(l)	1,674–1,803 K	−59,650	−6.84	+0.25	−0.10	+34.81
0.947 Fe(l) + 1/2 O$_2$(g) = Fe$_{0.947}$O(l)	1,803–2,000 K	−63,660	−7.48	+0.25	−0.10	+39.12
3 Fe(α) + 2 O$_2$(g) = Fe$_3$O$_4$ (magnetite)	298.16–900 K	−268,310	+5.87	−12.45	+0.245	+73.11
3 Fe(α) + 2 O$_2$(g) = Fe$_3$O$_4$ (β)	900–1,033 K	−272,300	−54.27	+11.65	+0.245	+233.52
3 Fe(β) + 2 O$_2$(g) = Fe$_3$O$_4$ (β)	1,033–1,179 K	−262,990	−5.71	+1.00	−0.40	+89.19
3 Fe(γ) + 2 O$_2$(g) = Fe$_3$O$_4$ (β)	1,179–1,674 K	−276,990	−44.05	+5.50	−0.40	+213.52
3 Fe(δ) + 2 O$_2$(g) = Fe$_3$O$_4$ (β)	1,674–1,803 K	−262,560	−6.40	+1.00	−0.40	+91.05
3 Fe(l) + 2 O$_2$(g) = Fe$_3$O$_4$ (β)	1,803–1,874 K	−275,280	−8.74	+1.00	−0.40	+104.84
3 Fe(l) + 2 O$_2$(g) = Fe$_3$O$_4$ (l)	1,874–2,000 K	−257,240	−26.89	+1.00	−0.40	+155.46
2 Fe(α) + 3/2 O$_2$(g) = Fe$_2$O$_3$ (hematite)	298.16–950 K	−200,000	−13.84	−1.45	+1.905	+108.26
2 Fe(α) + 3/2 O$_2$(g) = Fe$_2$O$_3$ (β)	950–1,033 K	−202,960	−42.64	+7.85	+0.13	+188.48
2 Fe(β) + 3/2 O$_2$(g) = Fe$_2$O$_3$ (β)	1,033–1,050 K	−196,740	−10.27	+0.75	−0.30	+92.26
2 Fe(β) + 3/2 O$_2$(g) = Fe$_2$O$_3$ (γ)	1,050–1,179 K	−193,200	−0.39	−0.13	−0.30	+59.96
2 Fe(γ) + 3/2 O$_2$(g) = Fe$_2$O$_3$ (γ)	1,179–1,674 K	−202,540	−25.95	+2.87	−0.30	+142.85
2 Fe(δ) + 3/2 O$_2$(g) = Fe$_2$O$_3$ (γ)	1,674–1,800 K	−192,920	−0.85	−0.13	−0.30	+61.21
2 La(c) + 3/2 O$_2$(g) = La$_2$O$_3$ (c)	298.16–1,153 K	−431,120	−13.31	+0.80	+1.34	+112.36
2 La(l) + 3/2 O$_2$(g) = La$_2$O$_3$ (c)	1,153–2,000 K	−434,330	−4.88	−0.80	+1.34	+91.17
Pb(c) + 1/2 O$_2$(g) = PbO (red)	298.16–600.5 K	−52,800	−2.76	−0.80	−0.10	+32.49
Pb(l) + 1/2 O$_2$(g) = PbO (red)	600.5–762 K	−53,780	−0.51	−1.75	−0.10	+28.44
Pb(c) + 1/2 O$_2$(g) = PbO (yellow)	298.16–600.5 K	−52,040	+0.81	−2.00	−0.10	+22.13
Pb(l) + 1/2 O$_2$(g) = PbO (yellow)	600.5–1,159 K	−53,020	+3.06	−2.95	−0.10	+18.08

Table 2–3 (continued)
HEAT OF FORMATION OF INORGANIC OXIDES

Reaction	Temperature range of validity	ΔH_o	2.303a	b	c	I
$Pb(l) + 1/2\ O_2(g) = PbO(l)$	1,159–1,745 K	-53,980	-12.94	+0.25	-0.10	+64.22
$Pb(c) + 1/2\ O_2(g) = PbO(g)$	298.16–600.5 K	+10,270	+1.91	+1.08	+0.295	-23.21
$Pb(l) + 1/2\ O_2(g) = PbO(g)$	600.5–2,000 K	+9,300	+4.17	+0.13	+0.295	-27.29
$3\ Pb(c) + 2\ O_2(g) = Pb_3O_4(c)$	298.16–600.5 K	-174,920	+8.82	-8.20	-0.40	+72.78
$3\ Pb(l) + 2\ O_2(g) = Pb_3O_4(c)$	600.5–1,000 K	-177,860	+15.59	-11.05	-0.40	+60.57
$Pb(c) + O_2(g) = PbO_2(c)$	298.16–600.5 K	-66,120	+0.64	-2.45	-0.20	+45.58
$Pb(l) + O_2(g) = PbO_2(c)$	600.5–1,100 K	-67,100	+2.90	-3.40	-0.20	+41.50
$2\ Li(c) + 1/2\ O_2(g) = Li_2O(c)$	298.16–452 K	-142,220	-3.06	+5.77	-0.10	+34.19
$2\ Li(l) + 1/2\ O_2(g) = Li_2O(c)$	452–1,600 K	-141,380	+16.97	-2.63	-0.10	-17.05
$2\ Li(c) + O_2(g) = Li_2O_2(c)$	298.16–452 K	-151,880	-1.38	—	—	+54.83
$2\ Li(l) + O_2(g) = Li_2O_2(c)$	452–500 K	-153,260	-1.38	—	—	+57.88
$Mg(c) + 1/2\ O_2(g) = MgO$ (periclase)	298.16–923 K	-144,090	-1.06	+0.13	+0.25	+29.16
$Mg(l) + 1/2\ O_2(g) = MgO$ (periclase)	923–1,393 K	-145,810	+1.84	-0.62	+0.64	+23.07
$Mg(g) + 1/2\ O_2(g) = MgO$ (periclase)	1,393–2,000 K	-180,700	-3.75	-0.62	+0.64	+65.69
$Mg(c) + O_2(g) = MgO_2(c)$	298.16–500 K	-150,230	-9.12	+0.13	+0.25	+70.84
$Mn(\alpha) + 1/2\ O_2(g) = MnO(c)$	298.16–1,000 K	-92,600	-4.21	+0.97	+0.155	+29.66
$Mn(\beta) + 1/2\ O_2(g) = MnO(c)$	1,000–1,374 K	-91,900	+1.84	-0.39	+0.34	+12.15
$Mn(\gamma) + 1/2\ O_2(g) = MnO(c)$	1,374–1,410 K	-89,810	+7.30	-0.72	+0.34	-6.05
$Mn(\delta) + 1/2\ O_2(g) = MnO(c)$	1,410–1,517 K	-89,390	+8.68	-0.72	+0.34	-10.70
$Mn(l) + 1/2\ O_2(g) = MnO(c)$	1,517–2,000 K	-93,350	+7.99	-0.72	+0.34	-5.90
$3\ Mn(\alpha) + 2\ O_2(g) = Mn_3O_4(\alpha)$	298.16–1,000 K	-332,400	-7.41	+0.66	+0.145	+106.62
$3\ Mn(\beta) + 2\ O_2(g) = Mn_3O_4(\alpha)$	1,000–1,374 K	-330,310	+10.75	-3.42	+0.70	+54.07
$3\ Mn(\gamma) + 2\ O_2(g) = Mn_3O_4(\alpha)$	1,374–1,410 K	-324,050	+27.12	-4.41	+0.70	-0.50
$3\ Mn(\delta) + 2\ O_2(g) = Mn_3O_4(\alpha)$	1,410–1,445 K	-322,800	+31.27	-4.41	+0.70	-14.46
$3\ Mn(\delta) + 2\ O_2(g) = Mn_3O_4(\beta)$	1,445–1,517 K	-328,870	-4.56	+1.00	-0.40	+95.20
$3\ Mn(l) + 2\ O_2(g) = Mn_3O_4(\beta)$	1,517–1,800 K	-340,730	-6.63	+1.00	-0.40	+109.60
$2\ Mn(\alpha) + 3/2\ O_2(g) = Mn_2O_3(c)$	298.16–1,000 K	-230,610	-5.96	-0.06	+0.945	+80.74
$2\ Mn(\beta) + 3/2\ O_2(g) = Mn_2O_3(c)$	1,000–1,374 K	-229,210	+6.15	-2.78	+1.315	+45.70
$2\ Mn(\gamma) + 3/2\ O_2(g) = Mn_2O_3(c)$	1,374–1,410 K	-225,030	+17.06	-3.44	+1.315	+9.33
$2\ Mn(\delta) + 3/2\ O_2(g) = Mn_2O_3(c)$	1,410–1,517 K	-224,200	+19.82	-3.44	+1.315	+0.05

Table 2–3 (continued)
HEAT OF FORMATION OF INORGANIC OXIDES

Reaction	Temperature range of validity	ΔH_o	2.303a	b	c	I
2 Mn(l) + 3/2 O$_2$ (g) = Mn$_2$O$_3$ (c)	1,517–1,700 K	−232,100	+18.44	−3.44	+1.315	+9.65
Mn(α) + O$_2$ (g) = MnO$_2$ (c)	298.16–1,000 K	−126,400	−8.61	+0.97	+1.555	+70.14
2 Hg(l) + 1/2 O$_2$ (g) = Hg$_2$O(c)	298.16–629.88 K	−22,400	−4.61	–	–	+43.29
2 Hg(g) + 1/2 O$_2$ (g) = Hg$_2$O(c)	629.88–1,000 K	−53,800	−16.12	–	–	+125.36
Hg(l) + 1/2 O$_2$ (g) = HgO (red)	298.16–629.88 K	−21,760	+0.85	−2.47	−0.10	+24.81
Hg(g) + 1/2 O$_2$ (g) = HgO (red)	629.88–1,500 K	−36,920	−2.92	−2.47	−0.10	+59.42
Mo(c) + O$_2$ (g) = MoO$_2$ (c)	298.16–2,000 K	−132,910	−3.91	–	–	+47.42
Mo(c) + 3/2 O$_2$ (g) = MoO$_3$ (c)	298.16–1,068 K	−182,650	−8.86	−1.55	+1.54	+90.07
Mo(c) + 3/2 O$_2$ (g) = MoO$_3$ (l)	1,068–1,500 K	−179,770	−36.34	+1.40	−0.30	+167.61
2 Nd(c) + 3/2 O$_2$ (g) = Nd$_2$O$_3$ (hexagonal)	298.16–1,113 K	−435,150	−16.19	+3.21	+1.78	+125.68
2 Nd(l) + 3/2 O$_2$ (g) = Nd$_2$O$_3$ (hexagonal)	1,113–1,500 K	−437,090	+4.03	−2.13	+1.78	+71.77
Np(c) + O$_2$ (g) = NpO$_2$ (c)	298.16–913 K	−246,450	−3.45	–	–	+52.44
Np(l) + O$_2$ (g) = NpO$_2$ (c)	913–1,500 K	−249,010	−4.61	–	–	+58.68
Ni(α) + 1/2 O$_2$ (g) = NiO(c)	298.16–633 K	−57,640	−4.61	+2.16	−0.10	+34.41
Ni(β) + 1/2 O$_2$ (g) = NiO(c)	633–1,725 K	−57,460	−0.14	−0.46	−0.10	+23.27
Ni(l) + 1/2 O$_2$ (g) = NiO(c)	1,725–2,000 K	−58,830	+7.23	−1.36	−0.10	+1.76
2 Nb(c) + 2 O$_2$ (g) = Nb$_2$O$_4$ (c)	298.16–2,000 K	−382,050	−9.67	–	–	+116.23
2 Nb(c) + 5/2 O$_2$ (g) = Nb$_2$O$_5$ (c)	298.16–1,785 K	−458,640	−16.14	−0.56	+1.94	+157.66
2 Nb(c) + 5/2 O$_2$ (g) = Nb$_2$O$_5$ (l)	1,785–2,000 K	−463,630	−66.04	+2.21	−0.50	+317.84
N$_2$ (g) + 1/2 O$_2$ (g) = N$_2$O(g)	298.16–2,000 K	+18,650	−1.57	−0.27	+0.92	+23.47
1/2 N$_2$ (g) + 1/2 O$_2$ (g) = NO(g)	298.16–2,000 K	+21,590	−0.28	+0.45	−0.03	−2.20
N$_2$ (g) + 3/2 O$_2$ (g) = N$_2$O$_3$ (g)	298.16–500 K	+17,390	−0.35	–	–	+54.30
1/2 N$_2$ (g) + O$_2$ (g) = NO$_2$ (g)	298.16–2,000 K	+7,730	+0.53	−0.265	+0.605	+13.74
N$_2$ (g) + 2 O$_2$ (g) = N$_2$O$_4$ (g)	298.16–1,000 K	+1,370	+2.14	−3.24	+1.38	+68.34
N$_2$ (g) + 5/2 O$_2$ (g) = N$_2$O$_5$ (c)	298.16–305 K	−10,200	–	–	–	+131.70
N$_2$ (g) + 5/2 O$_2$ (g) = N$_2$O$_5$ (g)	298.16–500 K	+3,600	+14.97	–	–	+86.50
Os(c) + 2 O$_2$ (g) = OsO$_4$ (yellow)	298.16–329 K	−92,260	+14.97	–	–	+32.45
Os(c) + 2 O$_2$ (g) = OsO$_4$ (white)	298.16–315 K	−90,560	+9.67	–	–	+27.60
Os(c) + 2 O$_2$ (g) = OsO$_4$ (l)	315–403 K	−88,970	+8.17	–	–	+35.78
Os(c) + 2 O$_2$ (g) = OsO$_4$ (g)	298.16–1,000 K	−81,200		−2.86	+1.94	+20.34

Table 2–3 (continued)
HEAT OF FORMATION OF INORGANIC OXIDES

Reaction	Temperature range of validity	ΔH_0	2.303a	b	c	I
$3/2\ O_2\ (g) = O_3\ (g)$	298.16–2,000 K	+33,980	+2.03	−0.48	+0.36	+11.45
P (white) + $1/2\ O_2\ (g) = PO(g)$	298.16–317.4 K	−9,370	+2.53	—	—	−25.40
$P(l) + 1/2\ O_2\ (g) = PO(g)$	317.4–553 K	−9,390	+3.45	—	—	−27.63
$1/4\ P_4\ (g) + 1/2\ O_2\ (g) = PO(g)$	553–1,500 K	−12,640	+2.30	—	—	−18.61
4 P (white) + $5\ O_2\ (g) = P_4O_{10}$ (hexagonal)	298.16–317.4 K	−711,520	+95.67	−51.50	−1.00	−28.24
$4\ P(l) + 5\ O_2\ (g) = P_4O_{10}$ (hexagonal)	317.4–553 K	−711,800	+97.98	−51.50	−1.00	−33.13
$P_4\ (g) + 5\ O_2\ (g) = P_4O_{10}$ (hexagonal)	553–631 K	−725,560	+87.45	−51.07	−2.405	+20.87
$P_4\ (g) + 5\ O_2\ (g) = P_4O_{10}\ (g)$	631–1,500 K	−722,330	−43.45	+2.93	−2.405	+348.20
$Pu(c) + O_2\ (g) = PuO_2\ (c)$	298.16–1,500 K	−246,450	−3.45	—	—	+52.48
$Po(c) + O_2\ (g) = PoO_2\ (c)$	298.16–900 K	−61,510	−9.21	—	—	+72.80
$2\ K(c) + 1/2\ O_2\ (g) = K_2O(c)$	298.16–336.4 K	−86,400	—	—	—	+33.90
$2\ K(l) + 1/2\ O_2\ (g) = K_2O(c)$	336.4–1,049 K	−87,380	+1.15	—	—	+33.90
$2\ K(g) + 1/2\ O_2\ (g) = K_2O(c)$	1,049–1,500 K	−133,090	−16.12	—	—	+129.64
$2\ K(c) + O_2\ (g) = K_2\ O_2\ (c)$	298.16–336.4 K	−118,300	−2.30	—	—	+59.60
$2\ K(l) + O_2\ (g) = K_2\ O_2\ (c)$	336.4–763 K	−119,780	−4.61	—	—	+69.85
$2\ K(l) + O_2\ (g) = K_2\ O_2\ (l)$	763–1,049 K	−118,250	−18.42	—	—	+107.66
$2\ K(g) + O_2\ (g) = K_2\ O_2\ (l)$	1,049–1,500 K	−161,870	−31.08	—	—	+187.49
$2\ K(c) + 3/2\ O_2\ (g) = K_2\ O_3\ (c)$	298.16–336.4 K	−126,640	−12.66	—	—	+111.75
$2\ K(l) + 3/2\ O_2\ (g) = K_2\ O_3\ (c)$	336.4–703 K	−127,790	−12.66	—	—	+115.16
$2\ K(l) + 3/2\ O_2\ (g) = K_2\ O_3\ (l)$	703–1,000 K	−125,330	−27.63	—	—	+154.28
$K(c) + O_2\ (g) = KO_2\ (c)$	298.16–336.4 K	−68,940	−10.36	—	—	+66.45
$K(l) + O_2\ (g) = KO_2\ (c)$	336.4–653 K	−69,510	−10.36	—	—	+68.15
$K(l) + O_2\ (g) = KO_2\ (l)$	653–1,000 K	−67,880	−18.42	—	—	+88.34
$K(c) + 3/2\ O_2\ (g) = KO_3\ (c)$	298.16–336.4 K	−63,340	−10.36	—	—	+85.85
$K(l) + 3/2\ O_2\ (g) = KO_3\ (c)$	336.4–500 K	−63,910	−10.36	—	—	+87.55
$2\ Pr(c) + 3/2\ O_2\ (g) = Pr_2\ O_3$ (c, C-type)	298.16–1,205 K	−440,600	−4.60	—	—	+78.38
$2\ Pr(l) + 3/2\ O_2\ (g) = Pr_2\ O_3$ (c, C-type)	1,205–2,000 K	−446,100	−4.60	—	—	+82.94
$6\ Pr(c) + 11/2\ O_2\ (g) = Pr_6\ O_{11}\ (c)$	298.16–1,205 K	−1,374,000	—	—	—	+241.04
$6\ Pr(l) + 11/2\ O_2\ (g) = Pr_6\ O_{11}\ (c)$	1,205–1,500 K	−1,390,500	—	—	—	+254.73
$Pr(c) + O_2\ (g) = PrO_2\ (c)$	298.16–1,200 K	−230,990	−6.42	+2.34	−0.20	+61.07

Table 2–3 (continued)
HEAT OF FORMATION OF INORGANIC OXIDES

Reaction	Temperature range of validity	ΔH_0	2.303a	b	c	I
Ra(c) + 1/2 O$_2$ (g) = RaO(c)	298.16–1,000 K	–130,000	–	–	–	+23.50
Re(c) + 3/2 O$_2$ (g) = ReO$_3$ (c)	298.16–433 K	–149,090	–16.12	–	–	+110.49
Re(c) + 3/2 O$_2$ (g) = ReO$_3$ (l)	433–1,000 K	–146,750	–31.32	–	–	+145.16
2 Re(c) + 7/2 O$_7$ (g) = Re$_2$O$_7$ (c)	298.16–569 K	–301,470	–34.54	–	–	+250.57
2 Re(c) + 7/2 O$_7$ (g) = Re$_2$O$_7$ (l)	569–635.5 K	–295,810	–73.68	–	–	+348.45
2 Re(c) + 7/2 O$_2$ (g) = Re$_2$O$_7$ (g)	635.5–1,500 K	–256,460	+3.45	–	–	+70.33
2 Re(c) + 4 O$_2$ (g) = Re$_2$O$_8$ (c)	298.16–420 K	–313,870	–41.45	–	–	+293.57
2 Re(c) + 4 O$_2$ (g) = Re$_2$O$_8$ (l)	420–600 K	–318,470	–87.50	–	–	+425.32
2 Rh(c) + 1/2 O$_2$ (g) = Rh$_2$O(c)	298.16–2,000 K	–23,740	–8.06	–	–	+35.64
Rh(c) + 1/2 O$_2$ (g) = RhO(c)	298.16–1,500 K	–22,650	–7.37	–	–	+40.54
2 Rh(c) + 3/2 O$_2$ (g) = Rh$_2$O$_3$ (c)	298.16–1,500 K	–70,060	–13.58	–	–	+101.72
2 Rb(c) + 1/2 O$_2$ (g) = Rb$_2$O(c)	298.16–312.2 K	–78,900	–	–	–	+32.20
2 Rb(l) + 1/2 O$_2$ (g) = Rb$_2$O(c)	312.2–750 K	–79,950	–	–	–	+35.56
2 Rb(l) + 1/2 O$_2$ (g) = Rb$_2$O(l)	750–952 K	–78,830	–10.36	–	–	+63.85
2 Rb(g) + 1/2 O$_2$ (g) = Rb$_2$O(l)	952–1,500 K	–120,290	–23.03	–	–	+145.14
2 Rb(c) + O$_2$ (g) = Rb$_2$O$_2$ (c)	298.16–312.2 K	–102,000	–2.30	–	–	+57.40
2 Rb(l) + O$_2$ (g) = Rb$_2$O$_2$ (c)	312.2–840 K	–103,360	–4.61	–	–	+67.52
2 Rb(l) + O$_2$ (g) = Rb$_2$O$_2$ (l)	840–952 K	–101,680	–18.42	–	–	+105.91
2 Rb(g) + O$_2$ (g) = Rb$_2$O$_2$ (l)	952–1,500 K	–143,130	–31.08	–	–	+187.16
2 Rb(c) + 3/2 O$_2$ (g) = Rb$_2$O$_3$ (c)	298.16–312.2 K	–118,190	–11.51	–	–	+104.70
2 Rb(l) + 3/2 O$_2$ (g) = Rb$_2$O$_3$ (c)	312.2–760 K	–119,400	–12.66	–	–	+111.43
2 Rb(l) + 3/2 O$_2$ (g) = Rb$_2$O$_3$ (l)	760–952 K	–116,740	–27.63	–	–	+151.06
2 Rb(g) + 3/2 O$_2$ (g) = Rb$_2$O$_3$ (l)	952–1,500 K	–157,720	–39.14	–	–	+228.39
Rb(c) + O$_2$ (g) = RbO$_2$ (c)	298.16–312.2 K	–52,330	–10.36	–	–	+66.25
Rb(l) + O$_2$ (g) = RbO$_2$ (c)	312.2–685 K	–65,120	–11.51	–	–	+71.30
Rb(l) + O$_2$ (g) = RbO$_2$ (l)	685–952 K	–63,070	–18.42	–	–	+87.89
Rb(g) + O$_2$ (g) = RbO$_2$ (l)	952–1,500 K	–83,560	–24.18	–	–	+126.57
Ru(α, β γ) + O$_2$ (g) = RuO$_2$ (c)	298.16–1,500 K	–57,290	–6.91	–	–	+62.01
2 Sm(c) + 3/2 O$_2$ (g) = Sm$_2$O$_3$ (c)	298.16–1,623 K	–430,600	–4.60	–	–	+78.38
2 Sm(l) + 3/2 O$_2$ (g) = Sm$_2$O$_3$ (c)	1,623–2,000 K	–438,000	–4.60	–	–	+82.94

Table 2–3 (continued)
HEAT OF FORMATION OF INORGANIC OXIDES

Reaction	Temperature range of validity	ΔH_o	2.303a	b	c	I
$2\,Sc(c) + 3/2\,O_2\,(g) = Sc_2O_3\,(c)$	298.16–1,673 K	−409,960	+7.78	−1.84	−0.30	+52.73
$2\,Sc(l) + 3/2\,O_2\,(g) = Sc_2O_3\,(c)$	1,673–2,000 K	−412,950	+18.47	−2.93	−0.30	+21.88
$Se(c) + 1/2\,O_2\,(g) = SeO(g)$	298.16–490 K	+9,280	−3.04	+4.40	+0.30	−14.78
$Se(l) + 1/2\,O_2\,(g) = SeO(g)$	490–1,027 K	+9,420	+8.70	—	+0.30	−44.50
$1/2\,Se_2\,(g) + 1/2\,O_2\,(g) = SeO(g)$	1,027–2,000 K	−7,400	−0.37	—	+0.19	−0.80
$Se(c) + O_2\,(g) = SeO_2\,(c)$	298.16–490 K	−53,770	+14.94	−9.41	−0.20	+6.94
$Se(l) + O_2\,(g) = SeO_2\,(c)$	490–595 K	−53,640	+27.59	−14.31	—	−25.05
$Se(l) + O_2\,(g) = SeO_2\,(g)$	595–1,027 K	−32,840	+6.79	—	—	−10.80
$1/2\,Se_2\,(g) + O_2\,(g) = SeO_2\,(g)$	1,027–2,000 K	−49,000	−0.74	—	−0.295	+27.61
$Si(c) + 1/2\,O_2\,(g) = SiO(g)$	298.16–1,683 K	−21,090	+3.84	−0.16	—	−33.14
$Si(l) + 1/2\,O_2\,(g) = SiO(g)$	1,683–2,000 K	−30,170	−7.78	−0.12	+0.25	−40.01
$Si(c) + O_2\,(g) = SiO_2\,(\alpha\text{-quartz})$	298.16–848 K	−210,070	+3.98	−3.32	+0.605	+34.59
$Si(c) + O_2\,(g) = SiO_2\,(\beta\text{-quartz})$	848–1,683 K	−209,920	−3.36	−0.19	−0.745	+53.44
$Si(l) + O_2\,(g) = SiO_2\,(\beta\text{-quartz})$	1,683–1,883 K	−219,000	+0.58	−0.47	−0.20	+46.58
$Si(l) + O_2\,(g) = SiO_2\,(l)$	1,883–2,000 K	−228,590	−15.66	—	—	+103.97
$Si(c) + O_2\,(g) = SiO_2\,(\alpha\text{-cristobalite})$	298.16–523 K	−207,330	+19.96	−9.75	−0.745	−9.78
$Si(c) + O_2\,(g) = SiO_2\,(\beta\text{-cristobalite})$	523–1,683 K	−209,820	−3.34	−0.24	−0.745	+53.35
$Si(l) + O_2\,(g) = SiO_2\,(\beta\text{-cristobalite})$	1,683–2,000 K	−218,900	+0.60	−0.52	−0.20	+46.49
$Si(c) + O_2\,(g) = SiO_2\,(\alpha\text{-tridymite})$	298.16–390 K	−207,030	+22.29	−11.62	−0.745	−15.64
$Si(c) + O_2\,(g) = SiO_2\,(\beta\text{-tridymite})$	390–1,683 K	−209,350	−1.59	−0.54	−0.745	+47.86
$Si(l) + O_2\,(g) = SiO_2\,(\beta\text{-tridymite})$	1,683–1,953 K	−218,430	+2.35	−0.82	−0.20	+41.00
$2\,Ag(c) + 1/2\,O_2\,(g) = Ag_2O(c)$	298.16–1,000 K	−7,740	−4.14	—	—	+27.84
$2\,Ag(c) + O_2\,(g) = Ag_2O_2\,(c)$	298.16–500 K	−6,620	−3.22	—	—	+52.17
$2\,Na(c) + 1/2\,O_2\,(g) = Na_2O(c)$	298.16–371 K	−99,820	−7.51	+5.47	−0.10	+50.43
$2\,Na(l) + 1/2\,O_2\,(g) = Na_2O(c)$	371–1,187 K	−100,150	+4.97	−2.45	−0.10	+22.19
$2\,Na(g) + 1/2\,O_2\,(g) = Na_2O(c)$	1,187–1,190 K	−156,200	−20.72	—	—	+145.48
$2\,Na(g) + 1/2\,O_2\,(g) = Na_2O(l)$	1,190–2,000 K	−150,250	−23.03	—	—	+147.58
$2\,Na(c) + O_2\,(g) = Na_2O_2\,(c)$	298.16–371 K	−122,500	−2.30	—	—	+57.51
$2\,Na(l) + O_2\,(g) = Na_2O_2\,(c)$	371–733 K	−124,320	−5.76	—	—	+71.30
$2\,Na(l) + O_2\,(g) = Na_2O_2\,(l)$	733–1,187 K	−123,220	−20.72	—	—	+112.66

Table 2–3 (continued)

HEAT OF FORMATION OF INORGANIC OXIDES

Reaction	Temperature range of validity	ΔH_0	2.303a	b	c	I
2 Na(g) + O₂(g) = Na₂O₂(l)	1,187–1,500 K	−174,800	−31.08	—	—	+187.97
Na(c) + O₂(g) = NaO₂(c)	298.16–371 K	−63,040	−8.06	—	—	+56.98
Na(l) + O₂(g) = NaO₂(c)	371–1,000 K	−64,220	−11.51	—	—	+69.04
Sr(c) + 1/2 O₂(g) = SrO(c)	298.16–1,043 K	−142,410	−6.79	+0.305	+0.675	+44.33
Sr(l) + 1/2 O₂(g) = SrO(c)	1,043–1,657 K	−143,370	−2.42	−0.38	+0.675	+32.77
Sr(g) + 1/2 O₂(g) = SrO(c)	1,657–2,000 K	−181,180	−8.24	−0.38	+0.675	+74.32
Sr(c) + O₂(g) = SrO₂(c)	298.16–1,000 K	−155,540	−11.40	+0.305	+0.675	+75.44
S(rhombohedral) + 1/2 O₂(g) = SO(g)	298.16–368.6 K	+19,250	−1.24	+2.95	+0.225	−18.84
S(monoclinic) + 1/2 O₂(g) = SO(g)	368.6–392 K	+19,200	−1.29	+3.31	+0.225	−18.72
S(λ,μ) + 1/2 O₂(g) = SO(g)	392–718 K	+20,320	+10.22	−0.17	+0.225	−50.05
1/2 S₂(g) + 1/2 O₂(g) = SO(g)	298.16–2,000 K	+3,890	+0.07	—	—	−1.50
S(rhombohedral) + O₂(g) = SO₂(g)	298.16–368.6 K	−70,980	+0.83	+2.35	+0.51	−5.85
S(monoclinic) + O₂(g) = SO₂(g)	368.6–392 K	−71,020	+0.78	+2.71	+0.51	−5.74
S(λ,μ) + O₂(g) = SO₂(g)	392–718 K	−69,900	+12.30	−0.77	+0.51	−37.10
1/2 S₂(g) + O₂(g) = SO₂(g)	298.16–2,000 K	−86,330	+2.42	−0.70	+0.31	+10.71
S(rhombohedral) + 3/2 O₂(g) = SO₃(c–I)	298.16–335.4 K	−111,370	−6.45	—	—	+88.32
S(rhombohedral) + 3/2 O₂(g) = SO₃(c–II)	298.16–305.7 K	−108,680	−11.97	—	—	+94.95
S(rhombohedral) + 3/2 O₂(g) = SO₃(l)	298.16–335.4 K	−107,430	−21.18	—	—	+113.76
S(rhombohedral) + 3/2 O₂(g) = SO₃(g)	298.16–368.6 K	−95,070	+1.43	+0.66	+1.26	+16.81
S(monoclinic) + 3/2 O₂(g) = SO₃(g)	368.6–392 K	−95,120	+1.38	+1.02	+1.26	+16.93
S(λ,μ) + 3/2 O₂(g) = SO₃(g)	392–718 K	−94,010	+12.89	−2.46	+1.26	−14.40
1/2 S₂(g) + 3/2 O₂(g) = SO₃(g)	298.16–1,500 K	−110,420	+3.02	−2.39	+1.06	+33.41
2 Ta(c) + 5/2 O₂(g) = Ta₂O₅(c)	298.16–2,000 K	−492,790	−17.18	−1.25	+2.46	+161.68
Tc(c) + O₂(g) = TcO₂(c)	298.16–500 K	−103,400	—	—	—	+41.00
Tc(c) + 3/2 O₂(g) = TcO₃(c)	298.16–500 K	−129,000	—	—	—	+64.50
2 Tc(c) + 7/2 O₂(g) = Tc₂O₇(c)	298.16–392.7 K	−266,000	—	—	—	+147.00
2 Tc(c) + 7/2 O₂(g) = Tc₂O₇(l)	392.7–500 K	−258,930	—	—	—	+129.00
Te(c) + 1/2 O₂(g) = TeO(g)	298.16–723 K	+43,110	+1.91	+0.84	+0.315	−27.22
Te(l) + 1/2 O₂(g) = TeO(g)	723–1,360 K	+39,750	+6.08	+0.09	+0.315	−33.94
1/2 Te₂(g) + 1/2 O₂(g) = TeO(g)	1,360–2,000 K	+23,730	−0.90	+0.09	+0.315	−0.29

Table 2–3 (continued)
HEAT OF FORMATION OF INORGANIC OXIDES

Reaction	Temperature range of validity	ΔH_o	2.303a	b	c	I
$Te(c) + O_2(g) = TeO_2(c)$	298.16–723 K	−78,090	−2.10	−2.35	−0.20	+51.27
$Te(l) + O_2(g) = TeO_2(c)$	723–1,006 K	−81,530	+1.84	−3.10	−0.20	+45.30
$Te(l) + O_2(g) = TeO_2(l)$	1,006–1,300 K	−82,090	−21.74	+0.50	−0.20	+113.04
$2\,Tl(\alpha) + O_2(g) = Tl_2O(c)$	298.16–505.5 K	−44,110	−6.91	—	—	+42.30
$2\,Tl(\beta) + O_2(g) = Tl_2O(c)$	505.5–573 K	−44,260	−6.91	—	—	+42.60
$2\,Tl(\beta) + 1/2\,O_2(g) = Tl_2O(l)$	573–576 K	−40,880	−13.82	—	—	+55.76
$2\,Tl(l) + 1/2\,O_2(g) = Tl_2O(l)$	576–773 K	−42,320	−11.51	—	—	+51.89
$2\,Tl(l) + 1/2\,O_2(g) = Tl_2O(g)$	773–1,730 K	−18,400	+11.51	—	—	−45.55
$2\,Tl(g) + 1/2\,O_2(g) = Tl_2O(g)$	1,730–2,000 K	−104,670	−16.12	—	—	+41.59
$2\,Tl(\alpha) + 3/2\,O_2(g) = Tl_2O_3(c)$	298.16–505.5 K	−99,410	−16.12	—	—	+119.09
$2\,Tl(\beta) + 3/2\,O_2(g) = Tl_2O_3(c)$	505.5–576 K	−99,560	−16.12	—	—	+119.39
$2\,Tl(l) + 3/2\,O_2(g) = Tl_2O_3(c)$	576–990 K	−101,010	−13.82	—	—	+115.55
$2\,Tl(l) + 3/2\,O_2(g) = Tl_2O_3(l)$	990–1,500 K	−94,550	−27.63	—	—	+150.39
$2\,Tl(\alpha) + 2\,O_2(g) = Tl_2O_4(c)$	298.16–505.5 K	−117,680	−23.03	—	—	+161.19
$2\,Tl(\beta) + 2\,O_2(g) = Tl_2O_4(c)$	505.5–576 K	−117,830	−23.03	—	—	+161.49
$2\,Tl(l) + 2\,O_2(g) = Tl_2O_4(c)$	576–1,000 K	−119,270	−20.72	—	—	+157.63
$Th(c) + O_2(g) = ThO_2(c)$	298.16–2,000 K	−294,350	−5.25	+0.59	+0.775	+62.81
$Sn(c) + 1/2\,O_2(g) = SnO(c)$	298.16–505 K	−68,600	−3.57	+1.65	−0.10	+32.59
$Sn(l) + 1/2\,O_2(g) = SnO(c)$	505–1,300 K	−69,670	+3.06	−1.50	−0.10	+18.39
$Sn(c) + 1/2\,O_2(g) = SnO(g)$	298.16–505 K	−1,000	−0.97	+3.24	+0.32	−17.41
$Sn(l) + 1/2\,O_2(g) = SnO(g)$	505–2,000 K	−2,070	+5.66	+0.09	+0.32	−31.62
$Sn(c) + O_2(g) = SnO_2(c)$	298.16–505 K	−142,010	−14.00	+2.45	+2.38	+90.74
$Sn(l) + O_2(g) = SnO_2(c)$	505–1,898 K	−143,080	−7.37	−0.70	+2.38	+76.53
$Sn(l) + O_2(g) = SnO_2(l)$	1,898–2,000 K	−139,130	−21.97	+0.50	−0.20	+120.11
$Ti(\alpha) + 1/2\,O_2(g) = TiO(\alpha)$	298.16–1,150 K	−125,010	−4.01	−0.29	+0.83	+36.28
$Ti(\beta) + 1/2\,O_2(g) = TiO(\alpha)$	1,150–1,264 K	−125,040	+1.17	−1.55	+0.83	+21.90
$Ti(\beta) + 1/2\,O_2(g) = TiO(\beta)$	1,264–2,000 K	−125,210	−1.77	−1.25	−0.10	+30.83
$Ti(\alpha) + 1/2\,O_2(g) = TiO(g)$	298.16–1,150 K	+11,710	+3.71	+1.07	−0.10	−35.50
$Ti(\beta) + 1/2\,O_2(g) = TiO(g)$	1,150–2,000 K	+11,680	+8.89	−0.19	−0.10	−49.88
$2\,Ti(\alpha) + 3/2\,O_2(g) = Ti_2O_3(\alpha)$	298.16–473 K	−360,660	+32.08	−23.49	−0.30	−10.66

Table 2–3 (continued)
HEAT OF FORMATION OF INORGANIC OXIDES

Reaction	Temperature range of validity	ΔH_0	2.303a	b	c	I
2 Ti(α) + 3/2 O₂ (g) = Ti₂O₃ (β)	473–1,150 K	−369,710	−30.95	+2.62	+4.80	+162.79
2 Ti(β) + 3/2 O₂ (g) = Ti₂O₃ (β)	1,150–2,000 K	−369,760	−20.59	+0.10	+4.80	+134.03
3 Ti(α) + 5/2 O₂ (g) = Ti₃O₅ (α)	298.16–450 K	−587,980	−4.19	−9.72	−0.50	+131.05
3 Ti(α) + 5/2 O₂ (g) = Ti₃O₅ (β)	450–1,150 K	−586,330	−18.31	+1.03	−0.50	+159.98
3 Ti(β) + 5/2 O₂ (g) = Ti₃O₅ (β)	1,150–2,000 K	−586,420	−2.76	−2.75	−0.50	+116.81
Ti(α) + O₂ (g) = TiO₂ (rutile)	298.16–1,150 K	−228,360	−12.80	+1.62	+1.975	+82.81
Ti(β) + O₂ (g) = TiO₂ (rutile)	1,150–2,000 K	−228,380	−7.62	+0.36	+1.975	+68.43
W(c) + O₂(g) = WO₂(c)	298.16–1,500 K	−137,180	−1.38	–	–	+45.56
4 W(c) + 11/2 O₂ (g) = W₄O₁₁ (c)	298.16–1,700 K	−745,730	−32.70	–	–	+321.84
W(c) + 3/2 O₂ (g) = WO₃ (c)	298.16–1,743 K	−201,180	−2.92	−1.81	−0.30	+70.89
W(c) + 3/2 O₂ (g) = WO₃ (l)	1,743–2,000 K	−203,140	−35.74	+1.13	−0.30	+173.27
U(α) + O₂ (g) = UO₂ (c)	298.16–935 K	−262,880	−19.92	+3.70	+2.13	+100.54
U(β) + O₂ (g) = UO₂ (c)	935–1,045 K	−260,660	−4.28	−0.31	+1.78	+55.50
U(γ) + O₂ (g) = UO₂ (c)	1,045–1,405 K	−262,830	−6.54	−0.31	+1.78	+64.41
U(l) + O₂ (g) = UO₂ (c)	1,405–1,500 K	−264,790	−5.92	–	–	+63.50
3 U(α) + 4 O₂ (g) = U₃O₈ (c)	298.16–935 K	−863,370	−56.57	+10.68	+5.20	+330.19
3 U(β) + 4 O₂ (g) = U₃O₈ (c)	935–1,045 K	−856,720	−9.67	−1.35	+4.15	+195.12
3 U(γ) + 4 O₂ (g) = U₃O₈ (c)	1,045–1,405 K	−863,230	−16.44	−1.35	+4.15	+221.79
3 U(l) + 4 O₂ (g) = U₃O₈ (c)	1,405–1,500 K	−869,460	−10.91	−1.35	+4.15	+208.82
U(α) + 3/2 O₂ (g) = UO₃ (hexagonal)	298.16–935 K	−294,090	−18.33	+3.49	+1.535	+114.94
U(β) + 3/2 O₂ (g) = UO₃ (hexagonal)	935–1,045 K	−291,870	−2.69	−0.52	+1.185	+69.90
U(γ) + 3/2 O₂ (g) = UO₃ (hexagonal)	1,045–1,400 K	−294,040	−4.95	−0.52	+1.185	+78.80
V(c) + 1/2 O₂ (g) = VO(c)	298.16–2,000 K	−101,090	−5.39	−0.36	+0.53	+38.69
V(c) + 1/2 O₂ (g) = VO(g)	298.16–2,000 K	+52,090	+1.80	+1.04	+0.35	−28.42
2 V(α) + 3/2 O₂ (g) = V₂O₃ (c)	298.16–2,000 K	−299,910	−17.98	+0.37	+2.41	+118.83
2 V(c) + 2 O₂ (g) = V₂O₄ (α)	298.16–345 K	−342,890	−11.03	+3.00	−0.40	+117.38
2 V(c) + 2 O₂ (g) = V₂O₄ (β)	345–1,818 K	−345,330	−24.36	+1.30	+3.545	+155.55
2 V(c) + 2 O₂ (g) = V₂O₄ (l)	1,818–2,000 K	−339,880	−59.59	+3.00	−0.40	+264.42
6 V(c) + 13/2 O₂ (g) = V₆O₁₃ (c)	298.16–1,000 K	−1,076,340	−95.33	–	–	+557.61
2 V(c) + 5/2 O₂ (g) = V₂O₅ (c)	298.16–943 K	−381,960	−41.08	+5.20	+6.11	+228.50

Table 2–3 (continued)
HEAT OF FORMATION OF INORGANIC OXIDES

Reaction	Temperature range of validity	ΔH_O	2.303a	b	c	I
2 V(c) + 5/2 O_2 (g) = V_2O_5 (l)	943–2,000 K	–365,840	–38.91	+3.25	–0.50	+207.54
2 Y(c) + 3/2 O_2 (g) = Y_2O_3 (c)	298.16–1,773 K	–419,600	+2.76	–1.73	–0.30	+66.36
2 Y(l) + 3/2 O_2 (g) = Y_2O_3 (c)	1,773–2,000 K	–422,850	+13.36	–2.75	–0.30	+35.56
Zn(c) + 1/2 O_2 (g) = ZnO(c)	298.16–692.7 K	–84,670	–6.40	+0.84	+0.99	+43.25
Zn(l) + 1/2 O_2 (g) = ZnO(c)	692.7–1,180 K	–85,520	–1.45	–0.36	+0.99	+31.25
Zn(g) + 1/2 O_2 (g) = ZnO(c)	1,180–2,000 K	–115,940	–7.28	–0.36	+0.99	+74.94
Zr(α) + O_2 (g) = ZrO_2 (α)	298.16–1,135 K	–262,980	–6.10	+0.16	+1.045	+65.00
Zr(β) + O_2 (g) = ZrO_2 (α)	1,135–1,478 K	–264,190	–5.09	–0.40	+1.48	+63.58
Zr(β) + O_2 (g) = ZrO_2 (β)	1,478–2,000 K	–262,290	–7.76	+0.50	–0.20	+69.50

From *Contributions to the Data on Theoretical Metallurgy*, Bulletin 542, U.S. Bureau of Mines, 1954, 60.

Alumina and Other Refractory Materials

ALUMINA AND OTHER REFRACTORY MATERIALS

R. N. Kleiner
GTE Sylvania, Inc.

Ceramics have been an important class of materials because of their inertness and heat-resistant properties. Although these materials have been used for many years, ceramics are in a very active stage of development. Applications based on unique properties of ceramic materials range from refractory bricks and crucibles for molten metals to miniature electronic devices; from nuclear fuels and control elements to magnetic memory units. The field of application of ceramic materials is diverse; the materials are numerous.

Vast amounts of data have been generated on properties of ceramic materials, and volumes of literature have been written on the subject. In this section, data on the properties of several ceramic materials are presented in a condensed tabulation. It is beyond the scope of this work to reference all of the work done on the materials discussed. The ceramics in this section were selected to show some of the properties of commercially available ceramic materials that make them unique for many applications.

The ceramic materials described in this section are alumina, beryllia, zirconia, mullite, cordierite, silicon carbide, and silicon nitride. The properties tabulated are crystal chemical, thermodynamic, physical, mechanical, thermal, and electrical properties.

The crystal chemical and thermodynamic properties are independent of processing conditions and are listed separately from the other properties. The physical, mechanical, thermal, and electrical properties are a function of the processing conditions. These properties can be modified through process changes to meet specifications for different applications. To show these effects, data on the physical, mechanical, thermal, and electrical properties for some of the materials are given for different levels of density and purity.

ALUMINA (Al_2O_3)

Good mechanical strength, inertness, refractoriness, and availability make alumina one of the most widely used of the ceramic materials. Some applications for alumina are electrical insulators, abrasives, cutting tools, radomes, and wear-resistant parts. Alumina is used in cements, refractories, glasses, coatings, cermets, and seals. The ability to densify this material with little or no glass phase present is advantageous in developing properties approaching the theoretical limit for this material. This has resulted in new applications for alumina such as envelopes for high-pressure sodium vapor lamps and microwave windows.

The crystal chemical and thermodynamic properties for alpha alumina are given in Table 3–1. Table 3–2 lists the physical, mechanical, thermal, and electrical properties. In Table 3–2, properties also are included on single crystal alumina. Properties are listed for polycrystalline alumina available from some of the alumina component manufacturers. The effects that density and purity levels have on these properties can be determined from the table.

BERYLLIA (BeO)

Beryllia is characterized by a higher thermal conductivity than any other ceramic. The thermal conductivity of beryllia is about ten times that of alumina and is approximately the same as aluminum. A good mechanical strength in combination with the high thermal conductivity gives beryllia good thermal shock-resistant characteristics.

Beryllia is used in electronic components as an electrical insulator and a heat sink. It is used as crucibles for melting uranium, thorium, and beryllium. Other applications are in the nuclear industry as moderators for fast neutrons, a matrix for fuel elements, shielding, and control rod assemblies.

Beryllia is toxic in the powder form, and special precautions must be taken in fabrication from the powder and in grinding the finished parts.

The crystal chemical and thermodynamic properties of beryllia are given in Table 3–3, and the physical, mechanical, thermal, and electrical properties are given in Table 3–4.

ZIRCONIA (ZrO_2)

Zirconia has a melting point of 2,700°C in a

neutral or oxidizing atmosphere, and it is not wet by most steel alloys and noble metals. The high melting point and chemical inertness, combined with a low thermal conductivity, make zirconia a good material for refractory applications. It is also used as an opacifier for glazes and enamels and is a major constituent of lead-zirconate-titanate piezoelectrics.

Zirconia is monoclinic at room temperature and transforms to the denser tetragonal form at about 1,100 to 1,200°C, undergoing a disruptive volume change of about 9%. Zirconia is often stabilized with CaO, Y_2O_3, or MgO additions, which, when fired, results in a material having a cubic structure. The cubic form differs only slightly from the monoclinic and tetragonal forms and is stable above and below the transformation temperature.

The crystal chemical and thermodynamic properties of unstabilized zirconia are given in Table 3–5. The physical, mechanical, thermal, and electrical properties of stabilized zirconia are given in Table 3–6.

MULLITE ($3Al_2O_3 \cdot 2SiO_2$)

Mullite is a refractory, mixed-oxide, ceramic material which has good thermal shock resistance properties. Mullite is used for refractory liners and laboratory ware. The silica in mullite is not present as free silica and usually does not pose a contamination problem. Mullite often appears as a separate phase in classical, triaxial porcelain whiteware ceramics. The needlelike habit of the mullite crystals adds strength to the body.

The physical, mechanical, thermal, and electrical properties of mullite are given in Table 3–7. The effect of different density levels on the properties of mullite can be determined from the table.

CORDIERITE ($2MgO \cdot 2Al_2O_3 \cdot 5SiO_2$)

Cordierite is a ceramic material that has low dielectric losses and is used in high frequency insulators. Cordierite also has a low coefficient of expansion, good thermal shock resistance, and is used for heating element supports and burner tips.

Pure cordierite bodies in general have a short vitrification range so additions are often made to extend the range. Reactions to form cordierite are often sluggish, and precautions must be taken to fire the body high enough to complete the reactions.

The physical, mechanical, thermal, and electrical properties of cordierite are given in Table 3–8.

SILICON CARBIDE (SiC)

Silicon carbide has a high melting point and hardness. It can be used to 1,650°C in an oxidizing atmosphere and to 2,300°C in an inert atmosphere. Its high thermal conductivity gives it high thermal shock resistance, which makes it useful in refractory furnace parts. Silicon carbide is also used as an abrasive and is currently being explored as a material for gas turbine components because of its high temperature modulus of rupture and corrosion resistance.

Silicon carbide occurs in alpha and beta forms. The alpha phase is hexagonal or rhombohedral and the beta phase is cubic. The beta phase undergoes an irreversible transformation to the alpha phase above 1,650°C. The base of the hexagonal unit cell is $\sim$ 3.08 Å and the c-axis is a multiple of $\sim$ 2.52 Å. Numerous stacking sequences can occur in the alpha form, giving rise to many polytypes. Some of these polytypes given in Ramsdell notation are 2H, 4H, 6H, 15R, 21R, and 24R (etc.). In this notation, the number refers to the number of layers in the unit cell and the letter refers to either hexagonal or rhombohedral symmetry.

Silicon carbide had not been successfully fired to near theoretical density without hot pressing in the past. The problem has been that silicon carbide is unreactive up to the temperature at which it decomposes. Recently silicon carbide has been successfully densified to greater than 96% theoretical density by using boron as a sintering aid.[a] This development will promote other applications for silicon carbide which were previously limited by the restrictive geometry and cost associated with hot pressed materials. The properties of silicon carbide are given in Tables 3–9 and 3–10.

[a]Prochazka, S., Investigation of Ceramics for High-temperature Turbine Vanes, General Electric Co., Quarterly Progress Report SRD-73-145, USN Contract N62269-73-C0356, June 20–September 19, 1973, General Electric Co., Schenectady, N.Y.

SILICON NITRIDE (Si₃N₄)

Silicon nitride has high hardness, wear resistance, high corrosion resistance, high thermal shock resistance, and good high temperature strength. The material is currently used for applications such as furnace refractories and supports, bearings and seals, nozzles, thermocouple sheaths, and crucibles. Silicon nitride is currently being explored as a material for gas turbine components because of its high temperature fracture energy, corrosion resistance, and high thermal shock resistance.

Silicon nitride occurs in alpha and beta forms. These phases are hexagonal and differ only slightly in lattice parameters. The alpha phase undergoes an irreversible transformation to the beta phase above 1,500 to 1,600°C.

This material, as with silicon carbide, has not been successfully fired to near theoretical density without hot pressing. A great deal of research effort is currently being given to silicon nitride because of its potential high temperature corrosion and wear resistance properties. If developments are made with respect to densification without hot pressing, as with silicon carbide, the number of applications for this material will be greatly increased. Complex components are currently fabricated using the reaction bonded approach. This consists of fabricating parts from silicon and nitriding by firing the parts in nitrogen. The bodies formed by this process are limited in that they exhibit porosities in the order of 30%. Theoretically dense bodies have been made by hot pressing with MgO as a sintering aid.

The crystal chemical and thermodynamic properties for silicon nitride are given in Table 3–11. The physical, mechanical, thermal, and electrical properties are given in Table 3–12.

Table 3–1
CRYSTAL CHEMICAL AND THERMODYNAMIC
PROPERTIES OF ALUMINA

System		α-Al$_2$O$_3$[1] Hexagonal
Lattice constants		
a	nm	0.4758
b	nm	
c	nm	1.2991
β		
c/a		2.72
Density	(kg/cm³) × 10⁻³	3.97
Crystal lattice energy	(J/kg-mol) × 10⁻⁶	15,520.468
Standard heat of formation, $-\Delta H^\circ_{298}$	(J/kg-mol) × 10⁻⁶	1,675.557
Entropy, S°_T at 298.15 K	(J/[kg-mol · deg]) × 10⁻³	51.020
Free energy of formation of oxides, $-\Delta F$ at 298 K	(J/kg-mol) × 10⁻⁶	1,577.461
Specific heat capacity, C_p at 298.16 K	(J/kg · deg)	774.977
Heat of fusion, ΔH	(J/kg-mol) × 10⁻⁶	108.86
Melting point	K	2,319.7 ± 8
Boiling point	K	3,253

Table compiled by R. N. Kleiner.

Table 3–2

PHYSICAL, MECHANICAL, THERMAL, AND ELECTRICAL PROPERTIES OF ALUMINA

Property		Single² crystal	~99%³	99.9%⁴	99.9%⁵	99.9%⁶
Physical						
Phase			α-Al$_2$O$_3$	α-Al$_2$O$_3$	α-Al$_2$O$_3$	α-Al$_2$O$_3$
Crystal structure			Hexagonal			
Density	g/cm³		3.98	>3.97	3.99	3.96
Melting point	°C		2,050	2,040		
Water absorption	%			0	0	0
Gas permeability				0	0	0
Grain size	μm			>30	15–45	1–6
Surface finish	μin. (authentic average)				25	20
Color				Translucent white	Translucent white	Ivory
Mechanical						
Modulus of elasticity	psi	63×10^6	59×10^6	57×10^6	57×10^6	56×10^6
Modulus of rigidity	psi		22×10^3	23×10^6	23.5×10^6	23×10^6
Bulk modulus	psi				34×10^6	33×10^6
Poisson's ratio			0.27	0.23	0.22	0.22
Flexural strength 25°C	psi	92,000	67,000	40,000	41,000	80,000
Flexural strength 1,000°C	psi	60,000	50,000		25,000	60,000
Compressive strength 25°C	psi		420,000	325,000	370,000	550,000
Compressive strength 1,000°C	psi		100,000		70,000	280,000
Tensile strength 25°C	psi		38,000	32,000	30,000	45,000
Tensile strength 1,000°C	psi				15,000	32,000
Transverse sonic velocity	m/sec		32,000		9.9×10^3	9.9×10^3
Hardness	R45N				85	90
Impact resistance	Charpy in.-lb					
Thermal						
Coefficient of expansion	10⁻⁶/°C					
−200– 25°C						
25– 200					3.4	3.4
25– 300						
25– 400					6.5	6.5
25– 500						
25– 600					7.4	7.4

Table 3–2 (continued)
PHYSICAL, MECHANICAL, THERMAL, AND ELECTRICAL PROPERTIES OF ALUMINA

	99.5%[7]	99.5%[8]	96%[9]	96%[10]	90%[11]	85%[12]
Physical						
Phase	α-Al$_2$O$_3$	α-Al$_2$O$_3$	α-Al$_2$O$_3$	α-Al$_2$O$_3$	α-Al$_2$O$_3$	α-Al$_2$O$_3$
Crystal structure / Density	3.87	3.87	3.70	3.72	3.60	3.41
Melting point	0	0	0	0	0	0
Water absorption	0	0	0	0	0	0
Gas permeability						
Grain size		5–50		2–20	2–10	2–12
Surface finish		35		65	65	65
Color	White	Ivory	White	White	White	White
Mechanical						
Modulus of elasticity		54×10^6	47×10^6	44×10^6	40×10^6	32×10^6
Modulus of rigidity		22×10^6	19×10^6	18×10^6	17×10^6	14×10^6
Bulk modulus		33×10^6		25×10^6	23×10^6	20×10^6
Poisson's ratio		0.22	0.22	0.21	0.22	0.22
Flexural strength	45,000	55,000	46,000	52,000	49,000	43,000
Compressive strength	>300,000	380,000	375,000	300,000	360,000	280,000
Tensile strength		38,000	25,000	28,000	32,000	22,500
Transverse sonic velocity		9.8×10^3		14,000 9.1×10^3	15,000 8.8×10^3	8.2×10^3
Hardness	81	83	78	78	79	75
Impact resistance						
Thermal						
Coefficient of expansion						
−200– 25°C		3.4		3.4	3.4	3.4
25– 200	6.9	7.1		6.0	6.1	5.3
25– 300			6.4			
25– 400		7.6		7.4	7.0	6.2
25– 500			7.5			
25– 600						

Table 3–2 (continued)
PHYSICAL, MECHANICAL, THERMAL, AND ELECTRICAL PROPERTIES OF ALUMINA

	Single[2] crystal	~99%[3]	99.9%[4]	99.9%[5]	99.9%[6]
Thermal					
Coefficient of expansion (cont.)					
25– 700					
25– 800					
25– 900					
25–1,000				7.8	7.8
25–1,100				8.0	8.0
25–1,200				8.3	8.3
25–1,500					
Conductivity	cal/(sec) (cm²) (°C/cm)				
20°C	.103	.079		0.095	0.093
100				0.068	0.066
300	.047	.039			
400				0.032	0.032
500	.029	.029			
800				0.015	0.015
1,200					
Electrical					
Dielectric constant at 25° C					
1 kHz				10.1	9.9
1 MHz				10.1	9.8
10 MHz				10.1	
100 MHz					
1 GHz					
10 GHz				10.1	9.8
50 GHz					
Dissipation factor at 25° C					
1 kHz				0.00050	0.0020
1 MHz				0.00004	0.0002
10 MHz					
100 MHz				0.00006	
1 GHz					
10 GHz				0.00009	0.0050
50 GHz					

Table 3–2 (continued)

PHYSICAL, MECHANICAL, THERMAL, AND ELECTRICAL PROPERTIES OF ALUMINA

	99.5%[7]	99.5%[8]	96%[9]	96%[10]	90%[11]	85%[12]
Thermal						
Coefficient of expansion (cont.)						
25– 700						
25– 800		8.0		8.0	7.7	6.9
25– 900			7.9			
25–1,000		8.3		8.2	8.1	7.2
25–1,100						
25–1,200						
25–1,500				8.4	8.4	7.5
Conductivity						
20°C		0.085	0.084	0.059	0.040	0.035
100		0.062	0.041	0.045	0.032	0.029
300						
400		0.029	0.026	0.024	0.019	0.016
500			0.020			
800		0.015		0.013	0.012	0.010
1,200						
Electrical						
Dielectric constant at 25°C						
1 kHz		9.8		9.0	8.8	8.2
1 MHz		9.7	9.3	9.0	8.8	8.2
10 MHz	9.58					
100 MHz				9.0	8.7	8.2
1 GHz	9.3	9.7	9.3	8.9		8.2
10 GHz			9.2	8.9	8.7	8.2
50 GHz				8.7		
Dissipation factor at 25°C						
1 kHz		0.0002		0.0011	0.0006	0.0014
1 MHz		0.0003	0.0003	0.0001	0.0004	0.0009
10 MHz	0.00003					
100 MHz	0.00014			0.0002	0.0004	0.0009
1 GHz		0.0002	0.0003	0.0001		0.0014
10 GHz			0.0009	0.0006	0.0009	0.0019
50 GHz				0.0068		

Table 3–2 (continued)

PHYSICAL, MECHANICAL, THERMAL, AND ELECTRICAL PROPERTIES OF ALUMINA

		Single[2] crystal	∿99%[3]	99.9%[4]	99.9%[5]	99.9%[6]
Loss factor at 25°C						
1 kHz					0.0050	0.020
1 MHz					0.0004	0.002
10 MHz						
100 MHz					0.0006	
1 GHz					0.0010	0.005
10 GHz						
50 GHz						
Dielectric strength	AC volts/mil					
0.250 in. thick	(average RMS values					240
0.125	at 60 Hz AC)				230	325
0.050					340	460
0.025					510	590
0.010					650	800
Volume resistivity	(ohm-cm²/cm)					
25°C					>10[15]	
100						1.0×10^{15}
300						3.3×10^{12}
500						9.0×10^{9}
700						
900						
1,000						1.1×10^{7}
1,500						
2,000						
Te value	°C					1,170

Table 3-2 (continued)
PHYSICAL, MECHANICAL, THERMAL, AND ELECTRICAL PROPERTIES OF ALUMINA

	99.5%[7]	99.5%[8]	96%[9]	96%[10]	90%[11]	85%[12]
Loss factor at 25°C						
1 kHz		0.002		0.010	0.005	0.011
1 MHz		0.003	0.0028	0.001	0.004	0.007
10 MHz	0.00029					
100 MHz	0.00130	0.002	0.0028	0.002	0.004	0.007
1 GHz			0.0082	0.001	0.008	0.011
10 GHz				0.005		0.016
50 GHz				0.059		
Dielectric strength						
0.250 in. thick		220	210	210	235	240
0.125		290		275	320	340
0.050		430		370	450	440
0.025		580		450	580	550
0.010		840		580	760	720
Volume resistivity						
25°C	$>10^{14}$	$>10^{14}$	$>10^{14}$	$>10^{14}$	$>10^{14}$	$>10^{14}$
100						
300	2.0×10^{11}		2.0×10^{13}	3.1×10^{11}	1.4×10^{11}	4.6×10^{10}
500			1.1×10^{10}	4.0×10^{9}	2.8×10^{8}	4.0×10^{8}
700			7.3×10^{7}	1.0×10^{8}	7.0×10^{6}	7.0×10^{6}
900	2.5×10^{6}		3.5×10^{6}			
1,000			6.8×10^{5}	1.0×10^{6}	8.6×10^{5}	
1,500						
2,000						
Te value	>975		840	1,000	960	850

Table compiled by R. N. Kleiner.

Table 3–3
CRYSTAL CHEMICAL AND THERMODYNAMIC PROPERTIES OF BERYLLIA

System		BeO[1] Hexagonal
Lattice constants		
a	nm	0.269
b	nm	
c	nm	0.437
β		
c/a		1.621
Density	$(kg/cm^3) \times 10^{-3}$	3.03
Crystal lattice energy	$(J/kg\text{-mol}) \times 10^{-6}$	4,521.744
Standard heat of formation, $-\Delta H^{\circ}_{298}$	$(J/kg\text{-mol}) \times 10^{-6}$	599.131
Entropy, S°_T at 298 K	$(J/[kg\text{-mol} \cdot deg]) \times 10^{-3}$	14.109
Free energy of formation of oxides, $-\Delta F$ at 298 K	$(J/kg\text{-mol}) \times 10^{-6}$	581.965
Specific heat capacity, C_p at 298 K	$(J/kg \cdot deg)$	1,017.392
Heat of fusion, ΔH	$(J/kg\text{-mol}) \times 10^{-6}$	71.217
Melting point	K	$2,843 \pm 30$
Boiling point	K	4,123

Table compiled by R. N. Kleiner.

Table 3–4

PHYSICAL, MECHANICAL, THERMAL, AND ELECTRICAL PROPERTIES OF BERYLLIA

		~99%[13]	99.5%[14]	96%[15]
Physical				
Phase		α-BeO	α-BeO	α-BeO
Crystal structure		Hexagonal		
Density	g/cm³	3.008	2.90	2.85
Melting point	°C	2,570		
Water absorption	%		0	0
Gas permeability			0	0
Grain size	μm		10–40	15–140
Surface finish	μin. (authentic average)		22	
Color			White	Blue
Mechanical				
Modulus of elasticity	psi	56×10^6	51×10^6	44×10^6
Modulus of rigidity	psi	21.5×10^6	20×10^6	17×10^6
Bulk modulus	psi	54.5×10^6	35×10^6	31×10^6
Poisson's ratio		0.34	0.26	0.30
Flexural strength	psi { 25°C / 1,000°C	38,000	40,000	25,000
Compressive strength	psi { 25°C / 1,000°C	300,000	310,000 / 40,000	9,000 / 225,000
Tensile strength	psi { 25°C / 1,000°C	14,000	20,000 / 5,000	
Transverse sonic velocity	m/sec		11.1×10^3	10.7×10^3
Hardness	R45N		67	64
Impact resistance	Charpy D256 in.-lb			
Thermal				
Coefficient of expansion	10^{-6}/°C			
−200– 25°C				
25– 200			2.4	2.4
25– 300			6.4	6.3
25– 400		7.4		
25– 500			7.7	7.5
25– 600				
25– 700				
25– 800			8.5	8.4

Table 3–4 (continued)
PHYSICAL, MECHANICAL, THERMAL, AND ELECTRICAL PROPERTIES OF BERYLLIA

	~99%[13]	99.5%[14]	96%[15]
Thermal			
Coefficient of expansion (cont.)			
25– 900			
25–1,000		8.9	8.9
25–1,100			
25–1,200		9.4	9.2
25–1,500			
Conductivity cal/(sec)(cm²)(°C/cm)			
20°C	.47	0.67	0.38
100	.34	0.48	0.32
300			
400	.14	0.20	0.16
500			
800	.07	0.07	0.06
1,200			
Electrical			
Dielectric constant at 25°C			
1 kHz		6.7	
1 MHz		6.7	
10 MHz			
100 MHz		6.6	
1 GHz		6.6	
10 GHz		6.6	
50 GHz			
Dissipation factor at 25°C			
1 kHz		0.0010	0.0005
1 MHz		0.0002	0.0001
10 MHz			
100 MHz		0.0002	
1 GHz		0.0002	0.0005
10 GHz		0.0004	0.0004
50 GHz			

Table 3–4 (continued)
PHYSICAL, MECHANICAL, THERMAL, AND ELECTRICAL PROPERTIES OF BERYLLIA

		~99%[13]	99.5%[14]	96%[15]
Loss factor at 25°C				
1 kHz			0.0070	0.0030
1 MHz			0.0010	0.0007
10 MHz				
100 MHz			0.0010	0.0030
1 GHz			0.0010	0.0030
10 GHz			0.0030	
50 GHz				
Dielectric strength	AC volts/mil			
0.250 in. thick	(average RMS values at		260	240
0.125	60 Hz AC)		340	340
0.050			490	400
0.025			610	
0.010			800	
Volume resistivity	(ohm-cm^2/cm)			
25°C		$>10^{17}$	$>10^{17}$	
100				
300		$>10^{15}$	$>10^{15}$	
500			5×10^{13}	1×10^{13}
700			1.5×10^{10}	2×10^{10}
900				
1,000			7×10^7	4×10^7
1,500				
2,000				
Te value	°C		1,240	1,170

Table compiled by R. N. Kleiner.

Table 3–5

CRYSTAL CHEMICAL AND THERMODYNAMIC PROPERTIES OF ZIRCONIA

System		ZrO_2 [1] Monoclinic
Lattice constants		
a	nm	0.517
b	nm	0.526
c	nm	0.530
β		80° 10[1]
c/a		
Density	$(kg/cm^3) \times 10^{-3}$	5.56
Crystal lattice energy	$(J/kg\text{-}mol) \times 10^{-6}$	11,195.503
Standard heat of formation, $-\Delta H^\circ_{298}$	$(J/kg\text{-}mol) \times 10^{-6}$	1,094.848
Entropy, S°_T at 298 K	$(J/[kg\text{-}mol \cdot deg]) \times 10^{-3}$	50.367
Free energy of formation of oxides, $-\Delta F$ at 298 K	$(J/kg \cdot deg)\ 10^{-6}$	1,037.070
Specific heat capacity, C_p at 298 K	$(J/kg \cdot deg)$	
Heat of fusion, ΔH	$(J/kg\text{-}mol) \times 10^{-6}$	87.085
Melting point	K	2,963
Boiling point	K	4,573

Table compiled by R. N. Kleiner.

Table 3–6
PHYSICAL, MECHANICAL, THERMAL, AND
ELECTRICAL PROPERTIES OF ZIRCONIA

		~5–10% CaO[16]	Reference 17
Physical			
Phase		CaO stabilized	MgO stabilized
Crystal structure		Cubic	Cubic
Density	g/cm³	5.5	5.43
Melting point	°C	2,500	
Water absorption	%		
Gas permeability			0.5
Grain size	μm		
Surface finish	μin. (authentic average)		
Color			
Mechanical			
Modulus of elasticity	psi	22.6 × 10⁶	15 × 10⁶
Modulus of rigidity	psi	8.5 × 10⁶	
Bulk modulus	psi	13.85 × 10⁶	
Poisson's ratio		0.324	
Flexural strength	psi { 25°C / 1,000°C	20–35,000	30,000
Compressive strength	psi { 25°C / 1,000°C	85–190,000	
Tensile	psi { 25°C / 1,000°C	21,000	
Transverse sonic velocity	m/sec		
Hardness	R45N		
Impact resistance	Charpy in.-lb		
Thermal			
Coefficient of expansion	10^{-6}/°C		
−200–25°C			
25–200			
25–300			
25–400			
25–500			
25–600			
25–700			
25–800			
25–900			
25–1,000			
25–1,100			
25–1,200			
25–1,500			
Conductivity	cal/(sec) (cm²) (°C/cm)		
20°C			
100			
300			
400		0.0045	
500			
800		0.0049	
1,200		0.0057	
Electrical			
Dielectric constant at 25°C			
1 kHz			
1 MHz			
10 MHz			

Table 3—6 (continued)
PHYSICAL, MECHANICAL, THERMAL, AND
ELECTRICAL PROPERTIES OF ZIRCONIA

~5—10% CaO[16]   Reference 17

Electrical
  Dielectric constant at 25°C (cont.)
    100 MHz
      1 GHz
     10 GHz
     50 GHz

  Dissipation factor at 25°C
      1 kHz
      1 MHz
     10 MHz
    100 MHz
      1 GHz
     10 GHz
     50 GHz

  Loss factor at 25°C
      1 kHz
      1 MHz
     10 MHz
    100 MHz
      1 GHz
     10 GHz
     50 GHz

  Dielectric strength   AC volt/mil (average
    0.250 in. thick   RMS values at 60 Hz AC)
    0.125
    0.050
    0.025
    0.010

  Volume resistivity   (ohm-cm^2/cm)
       25°C
      100
      300
      500
      700
      900
    1,000
    1,500
    2,000

  Te value   °C

Table compiled by R. N. Kleiner.

Table 3-7

PHYSICAL, MECHANICAL, THERMAL, AND ELECTRICAL PROPERTIES OF MULLITE

		Reference 18	Reference 19	Reference 20	Reference 21
Physical					
Phase		$3Al_2O_3 \cdot 2SiO_2$	$3Al_2O_3 \cdot 2SiO_2$	$3Al_2O_3 \cdot 2SiO_2$	$3Al_2O_3 \cdot 2SiO_2$
Crystal structure		Orthorhombic			
Density	g/cm³	3.13–3.26	2.6	2.8	3.1
Melting point	°C	1,850			
Water absorption	%			0	
Gas permeability					
Grain size	μm				
Surface finish	μin. (authentic average)				
Color				Tan	
Mechanical					
Modulus of elasticity	psi	21×10^6		23×10^6	
Modulus of rigidity	psi				
Bulk modulus	psi				
Poisson's ratio					
Flexural strength	psi {25°C / 1,000°C}	25,000	20,000	27,000	6,000
Compressive strength	psi {25°C / 1,000°C}	100–190,000		80,000	
Tensile strength	psi {25°C / 1,000°C}	16,000			
Transverse sonic velocity	m/sec				
Hardness	R45N			71	
Impact resistance	Charpy in.-lb				
Thermal					
Coefficient of expansion	10⁻⁶/°C				
−200–25°C					
25–200					
25–300					
25–400			4.6		2.3
25–500					
25–600				3.7	
25–700					
25–800			4.8		4.3
25–900				5.0	
25–1,000					

Table 3–7 (continued)

PHYSICAL, MECHANICAL, THERMAL, AND ELECTRICAL PROPERTIES OF MULLITE

	Reference 18	Reference 19	Reference 20	Reference 21
Thermal				
Coefficient of expansion (cont.)				
25–1,100				
25–1,200				
25–1,500				
Conductivity cal/(sec) (cm²) (°C/cm)				
20°C				
100	.143	0.005		0.003
300				
400	.102			
500				
800	.090			
1,200				
Electrical				
Dielectric constant at 25°C				
1 kHz				
1 MHz				
10 MHz				
100 MHz				
1 GHz				
10 GHz				
50 GHz				
Dissipation factor at 25°C				
1 kHz				
1 MHz				
10 MHz				
100 MHz				
1 GHz				
10 GHz				
50 GHz				
Loss factor at 25°C				
1 kHz				
1 MHz				
10 MHz				

Table 3–7 (continued)

PHYSICAL, MECHANICAL, THERMAL, AND ELECTRICAL PROPERTIES OF MULLITE

		Reference 18	Reference 19	Reference 20	Reference 21
Loss factor at 25°C (cont.)					
100 MHz					
1 GHz					
10 GHz					
50 GHz					
Dielectric strength	AC volts/mil (average				
0.250 in. thick	RMS values at 60 Hz AC)				
0.125					
0.050					
0.025					
0.010					
Volume resistivity	(ohm-cm^2/cm)				
25°C			$>10^{14}$		
100					
300			10^{10}		
500			10^{8}		
700					
900					
1,000					
1,500					
2,000					
Te value	°C		670		

Table compiled by R. N. Kleiner.

Table 3–8

PHYSICAL, MECHANICAL, THERMAL, AND ELECTRICAL PROPERTIES OF CORDIERITE

		Reference 22	Reference 23	Reference 24	Reference 25
Physical					
Phase		$2MgO \cdot 2Al_2O_3 \cdot 5SiO_2$	$2MgO \cdot 2Al_2O_3 \cdot 5SiO_2$	$2MgO \cdot 2Al_2O_3 \cdot 5SiO_2$	$2MgO \cdot 2Al_2O_3 \cdot 5SiO_2$
Crystal structure		Orthorhombic			
Density	g/cm³	2.51	2.3	2.1	1.8
Melting point	°C	1,471			
Water absorption	%		0–1	10–15	
Gas permeability					
Grain size	μm				14–17
Surface finish	μin. (authentic average)				
Color					
Mechanical					
Modulus of elasticity	psi		17×10^6	8.8×10^6	
Modulus of rigidity	psi		7×10^6	3.8×10^6	
Bulk modulus	psi				
Poisson's ratio			0.21	0.17	
Flexural strength	psi { 25°C / 1,000°C }	16,000	15,000	8,000	3,400
Compressive strength	psi { 25°C / 1,000°C }	50,000	50,000	30,000	18,500
Tensile strength	psi { 25°C / 1,000°C }	7,800		3,500	2,500
Transverse sonic velocity	m/sec				
Hardness	R45N				
Impact resistance	Charpy in.-lb.	4.3	4.0	2.5	2.5
Thermal					
Coefficient of expansion	$10^{-6}/°C$				
−200–25°C					
25–200					
25–300					
25–400			2.4	2.2	0.6
25–500					
25–600					
25–700			3.3	2.8	1.5
25–800					
25–900					1.7
25–1,000			3.7	2.8	

Table 3–8 (continued)

PHYSICAL, MECHANICAL, THERMAL, AND ELECTRICAL PROPERTIES OF CORDIERITE

	Reference 22	Reference 23	Reference 24	Reference 25
Thermal				
Coefficient of expansion (cont.)				
25–1,100				
25–1,200				
25–1,500				
Conductivity	cal/(sec) (cm²) (°C/cm)			
20°C	2.7			
100		0.0077	0.0043	
300		0.0062	0.0041	
400				
500		0.0055	0.0040	
800		0.0055	0.0038	
1,200				
Electrical				
Dielectric constant at 25°C				
1 kHz		5.3	5.0	4.1
1 MHz			4.9	
10 MHz				
100 MHz				
1 GHz				
10 GHz				
50 GHz				
Dissipation factor at 25°C				
1 kHz		0.0047	0.004	0.012
1 MHz			0.003	
10 MHz				
100 MHz				
1 GHz				
10 GHz				
50 GHz				
Loss factor at 25°C				
1 kHz		0.025	0.020	0.048
1 MHz			0.015	
10 MHz				

Table 3–8 (continued)

PHYSICAL, MECHANICAL, THERMAL, AND ELECTRICAL PROPERTIES OF CORDIERITE

		Reference 22	Reference 23	Reference 24	Reference 25
Loss factor at 25°C (cont.)					
100 MHz					
1 GHz					
10 GHz					
50 GHz					
Dielectric strength	AC volts/mil (average RMS values at 60 Hz AC)				
0.250 in. thick			200	100	60
0.125					
0.050					
0.025					
0.010					
Volume resistivity	(ohm-cm^2/cm)				
25°C			1.0×10^{14}	$>10^{14}$	1.0×10^{14}
100			2.5×10^{11}	3.0×10^{13}	1.0×10^{13}
300			3.3×10^{7}	2.0×10^{10}	3.0×10^{9}
500			7.7×10^{5}	9.0×10^{7}	4.9×10^{7}
700			8.0×10^{4}	3.0×10^{6}	4.7×10^{6}
900			1.9×10^{4}	3.5×10^{5}	7.0×10^{5}
1,000					
1,500					
2,000					
Te value	°C		485	780	850

Table compiled by R. N. Kleiner.

Table 3—9
CRYSTAL CHEMICAL AND THERMODYNAMIC PROPERTIES
OF SILICON CARBIDE

		α-SiC	β-SiC
System		Hexagonal-6H	Cubic
Lattice constants			
a	nm	3.073[26]	4.358[27]
b	nm		
c	nm	15.08[26]	
c/a			
Density	(kg/cm^3) $\times$ 10^{-3}	3.218[26]	
Crystal lattice energy	(J/kg-mol) $\times$ 10^{-6}		
Standard heat of formation, $-\Delta H^{\circ}_{298}$	(J/kg-mol) $\times$ 10^{-6}		
Entropy, S°_T at 298 K	(J/[kg-mol·deg]) $\times$ 10^{-3}		
Free energy of formation, $-\Delta F$ at 298 K	(J/kg-mol) $\times$ 10^{-6}		
Specific heat capacity, C_p at 298 K	(J/kg·deg)		
Heat of fusion, ΔH	(J/kg-mol) $\times$ 10^{-6}		
Melting point	K		
Boiling point	K		

Table compiled by R. N. Kleiner.

Table 3–10

PHYSICAL, MECHANICAL, THERMAL, AND ELECTRICAL PROPERTIES OF SILICON CARBIDE

		Reference 28		Reference 29	Reference 30	Reference 31
Physical						
Phase		α-SiC	β-SiC			
Crystal structure		Hexagonal	Cubic			
Density	g/cm³	3.208	3.21	3.1	2.6	3.10
Melting point	°C					
Water absorption	%			0		
Gas permeability				0		
Grain size	μm					
Surface finish	μin. (authentic average)				150	
Color						
Mechanical						
Modulus of elasticity	psi	70 × 10⁶		69 × 10⁶	30 × 10⁶	60 × 10⁶
Modulus of rigidity	psi	24 × 10⁶				
Bulk modulus	psi	14 × 10⁶				
Poisson's ratio		0.19				0.24
Flexural strength	psi {25°C / 1,000°C}	25,000		25,000	14–18,000	76,000
Compressive strength	psi {25°C / 1,000°C}	200,000		200,000		
Tensile strength	psi {25°C / 1,000°C}	5–20,000				
Transverse sonic velocity	m/sec					
Hardness	R45N					
Impact resistance	Charpy in.-lb	0.80				
Thermal						
Coefficient of expansion	10⁻⁶/°C					
−200–25°C						
25–200						
25–300						
25–400		4.34				
25–500						
25–600						
25–700						
25–800						
25–900				3.8		
25–1,000						4.3

Table 3–10 (continued)

PHYSICAL, MECHANICAL, THERMAL, AND ELECTRICAL PROPERTIES OF SILICON CARBIDE

	Reference 28	Reference 29	Reference 30	Reference 31
Thermal				
Coefficient of expansion (cont.)				
25–1,100				
25–1,200				
25–1,500				
Conductivity cal/(sec) (cm²) (°C/cm)				
20°C		4.0	4.8	
100				
300	3.26			
400	2.04			
500			.612	
800	1.19		.560	0.2
1,200			.493	0.093
Electrical				
Dielectric constant at 25°C				
1 kHz				
1 MHz				
10 MHz				
100 MHz				
1 GHz				
10 GHz				
50 GHz				
Dissipation factor at 25°C				
1 kHz				
1 MHz				
10 MHz				
100 MHz				
1 GHz				
10 GHz				
50 GHz				
Loss factor at 25°C				
1 kHz				
1 MHz				
10 MHz				

Table 3–10 (continued)

PHYSICAL, MECHANICAL, THERMAL, AND ELECTRICAL PROPERTIES OF SILICON CARBIDE

	Reference 28	Reference 29	Reference 30	Reference 31
Loss factor at 25°C (cont.)				
100 MHz				
1 GHz				
10 GHz				
50 GHz				
Dielectric strength	AC volt/mil (average			
0.250 in. thick	RMS values at 60 Hz AC)			
0.125				
0.050				
0.025				
0.010				
Volume resistivity	(ohm-cm²/cm)			
25°C				
100				
300				
500				
700				
900				
1,000				
1,500				
2,000				
Te value	°C			

Table compiled by R. N. Kleiner.

Table 3–11
CRYSTAL CHEMICAL AND THERMODYNAMIC PROPERTIES
OF SILICON NITRIDE

		α-Si$_3$N$_4$	β-Si$_3$N$_4$
System		Hexagonal[32]	Hexagonal[33]
Lattice constants			
a	nm	7.758[32]	7.608[33]
b	nm		
c	nm	5.623[32]	2.911[33]
β			
c/a			
Density	(kg/cm^3) $\times$ 10^{-3}		
Crystal lattice energy	(J/kg-mol) $\times$ 10^{-6}		
Standard heat of formation, $-\Delta H^\circ_{298}$	(J/kg-mol) $\times$ 10^{-6}		
Entropy, S°_T at 298 K	(J/[kg-mol$\cdot$deg]) $\times$ 10^{-3}		
Free energy of formation, $-\Delta F$ at 298 K	(J/kg-mol) $\times$ 10^{-6}		
Specific heat capacity, C_p at 298 K	(J/kg$\cdot$deg)		
Heat of fusion, ΔH	(J/kg-mol) $\times$ 10^{-6}		
Melting point	K		
Boiling point	K		

Table compiled by R. N. Kleiner.

Table 3–12
PHYSICAL, MECHANICAL, THERMAL, AND ELECTRICAL PROPERTIES OF SILICON NITRIDE

	Units	Reference 34	Reference 35	Reference 36	Reference 37	Reference 38	Reference 39
Physical							
Phase		α-Si$_3$N$_4$					
Crystal structure		Hexagonal		α 0–30% β 100–70%			
Density	g/cm^3		3.2	3.12–3.18	2.6	2.5–2.6	2.0–2.7
Melting point	°C	1,870					
Water absorption	%			0–0.1			
Gas permeability							
Grain size	μm						
Surface finish	μin. (authentic average)						
Color							
Mechanical							
Modulus of elasticity	psi	8–31 × 10^6	35 × 10^6	31.5 × 10^6	32 × 10^6	24 × 10^6	13.9–31.6 × 10^6
Modulus of rigidity	psi						
Bulk modulus	psi						
Poisson's ratio			0.27		0.27		0.257
Flexural strength	psi {25°C / 1,000°C	10–100,000	100,000	80–100,000	35,000	35,400	10–30,000 / 10–30,000
Compressive strength	psi {25°C / 1,000°C	72–90,000			35,000		77–110,000
Tensile strength	psi {25°C / 1,000°C						
Transverse sonic velocity	m/sec						
Hardness	R45N						
Impact resistance	Charpy in.-lb						
Thermal							
Coefficient of expansion	10^{-6}/°C						
–200–25°C							
25–200							
25–300							
25–400							
25–500							
25–600							
25–700							
25–800							
25–900							

Table 3–12 (continued)

PHYSICAL, MECHANICAL, THERMAL, AND ELECTRICAL PROPERTIES OF SILICON NITRIDE

		Reference 34	Reference 35	Reference 36	Reference 37	Reference 38	Reference 39
Thermal							
Coefficient of expansion (cont.)							
25–1,000		2.85					
25–1,100			3.2		3.2		3.2
25–1,200							
25–1,500							
Conductivity	cal/(sec) (cm²) (°C/cm)						
20°C							
100							
300							
400							
500			0.042		0.036		
800							
1,200			0.033		0.034		
Electrical							
Dielectric constant at 25°C							
1 kHz							
1 MHz							
10 MHz							
100 MHz							
1 GHz							
10 GHz							
50 GHz							
Dissipation factor at 25°C							
1 kHz							
1 MHz							
10 MHz							
100 MHz							
1 GHz							
10 GHz							
50 GHz							

Table 3–12 (continued)

PHYSICAL, MECHANICAL, THERMAL, AND ELECTRICAL PROPERTIES OF SILICON NITRIDE

	Reference 34	Reference 35	Reference 36	Reference 37	Reference 38	Reference 39
Loss factor at 25°C						
1 kHz						
1 MHz						
10 MHz						
100 MHz						
1 GHz						
10 GHz						
50 GHz						
Dielectric strength AC volt/mil (average						
0.250 in. thick RMS values at 60 Hz AC)						
0.125						
0.050						
0.025						
0.010						
Volume resistivity (ohm-cm²/cm)						
25°C						
100						
300						
500						
700						
900						
1,000						
1,500						
2,000						
Te value °C						

Table compiled by R. N. Kleiner.

Table 3–12 (continued)
PHYSICAL, MECHANICAL, THERMAL, AND ELECTRICAL PROPERTIES OF SILICON NITRIDE

REFERENCES

1. **Samsonov, G. V., Ed.,** *The Oxide Handbook,* IFI/Plenum, Arlington, Va., 1972.
2. *Engineering Properties of Selected Ceramic Materials,* American Ceramic Society, Columbus, Oh., 1966, 5.4.1–1.
3. *Engineering Properties of Selected Ceramic Materials,* American Ceramic Society, Columbus, Oh., 1966, 5.4.1–1.
4. General Electric Bulletin L-4R – 9-70.
5. *Coors Ceramic Handbook,* Bulletin 952, revised January 1972, Coors Porcelain Co., Golden, Colo. (Vistol).
6. *Coors Ceramic Handbook,* Bulletin 952, revised January 1972, Coors Porcelain Co., Golden, Colo.(AD-999).
7. *Alumina Ceramics,* Wesgo Catalog No. C118, Ceramic Div., Western Gold and Platinum Co., Belmont, Cal.(AL-995).
8. *Coors Ceramic Handbook,* Bulletin 952, revised January 1972, Coors Porcelain Co., Golden, Colo.(AD-995).
9. American Lava Corp. Chart No. 711 (Alsimag 614).
10. *Coors Ceramic Handbook,* Bulletin 952, revised January 1972, Coors Porcelain Co., Golden, Colo.(AD-96).
11. *Coors Ceramic Handbook,* Bulletin 952, revised January 1972, Coors Porcelain Co., Golden, Colo.(AD-90).
12. *Coors Ceramic Handbook,* Bulletin 952, revised January 1972, Coors Porcelain Co., Golden, Colo.(AD-85).
13. *Engineering Properties of Selected Ceramic Materials,* American Ceramic Society, Columbus, Oh., 1966, 5.4.2–1.
14. *Coors Ceramic Handbook,* Bulletin 952, revised January 1972, Coors Porcelain Co., Golden, Colo.(BD-995-2).
15. *Coors Ceramic Handbook,* Bulletin 952, revised January 1972, Coors Porcelain Co., Golden, Colo.(BD-96).
16. *Engineering Properties of Selected Ceramic Materials,* American Ceramic Society, Columbus, Oh., 1966, 5.4.5–1.
17. **King, A. G.,** *Zircoa News Focus,* November 1968 (Zircoa 1706).
18. *Engineering Properties of Selected Ceramic Materials,* American Ceramic Society, Columbus, Oh., 1966, 5.5.1–1.
19. Kyoto Ceramic Co. Bulletin (K-635).
20. Norton Company Bulletin 1-CTM-P3 (DHP mulnorite).
21. Kyoto Ceramic Co. Bulletin(K-692).
22. *Engineering Properties of Selected Ceramic Materials,* American Ceramic Society, Columbus, Oh., 1966, 5.4.1–1.
23. American Lava Chart No. 711 (Alsimag 701).
24. American Lava Chart No. 711 (Alsimag 202).
25. American Lava Chart No. 711 (Alsimag 447).
26. JCPDS X-ray Powder Data File, Card No. 22-173, Joint Committee on Powder Diffraction Standards, Swarthmore, Pa.
27. **Pearson, W. B.,** *Handbook of Lattice Spacings and Structures of Metals,* Pergamon Press, Elmsford, N.Y., 1958, 956.
28. *Engineering Properties of Selected Ceramic Materials,* American Ceramic Society, Columbus, Oh., 1966, 5.2.3–1.
29. Carborundum Co. Bulletin ("KT" silicon carbide).
30. Norton Co. Bulletin (Crystar).
31. United Kingdom Atomic Energy Authority (UKAEA) Bulletin, September 1973 (Refel silicon carbide).
32. JCPDS X-ray Powder Data File, Card No. 4-250, Joint Committee on Powder Diffraction Standards, Swarthmore, Pa.
33. JCPDS X-ray Powder Data File, Card No. 9-259, Joint Committee on Powder Diffraction Standards, Swarthmore, Pa.
34. *Engineering Properties of Selected Ceramic Materials,* American Ceramic Society, Columbus, Oh., 1966, 5.3.3–1.
35. United Kingdom Atomic Energy Authority (UKAEA) Bulletin (hot pressed silicon nitride).
36. **Deeby, G. G. et al.,** Dense silicon nitride, *Powder Metallurgy,* 8, 145, 1961.
37. United Kingdom Atomic Energy Authority (UKAEA) Bulletin (reaction-bonded silicon nitride).
38. Advanced Materials Engineering Ltd. Data Sheet TD/P/74-1 (reaction bonded silicon nitride).
39. Advanced Materials Engineering Ltd. Bulletin 5/71/3 (reaction bonded silicon nitride).

Composites

4.1 FIBERS

F. S. Galasso
United Aircraft Research Laboratories

Table 4.1–1 includes experimental and commercial fibers. The filaments are continuous while the whiskers are shorter in length.

The Al_2O_3 fibers are single crystal sapphire produced by Tyco. The $Al_2O_{3F.P.}$ fiber has been flame polished.

Boron fiber is produced by Hamilton Standard and Avco; the designation B/W indicates that the fiber is produced by chemical vapor deposition on a tungsten wire substrate. The B_4C and BN fibers are produced by the Carborundum Company.

Carbon fibers T-25 through T-400 are formed by Union Carbide and are yarns. The diameter given is that of an individual filament in the yarn. Test results on these fibers are made in a strand test and the gauge length is 10 in. For strength measurements on other filaments in this table, a common length is 1 in., and they are conducted mostly on single filaments.

HMG 25-50, Type A-C, Types 1 and 2, and 2T-6T are also bundles of carbon filaments and are produced by Hitco, Hercules-Courtaulds, Morganite-Whittaker, and Great Lakes Carbon, respectively.

The fiber SiC/B/W is produced by Hamilton Standard and sold under the trade name BORSIC.® It consists of a silicon carbide coated boron fiber which in turn was formed by chemical vapor deposition of boron on tungsten wire.

Other data included in Table 4.1–1 are the density in g/cm^3 and $lb/in.^3$, the ultimate tensile strength (UTS), the specific tensile strength (which is the UTS divided by the density), the tensile modulus, and the specific tensile modulus.

Table 4.1–1
PROPERTIES OF EXPERIMENTAL AND COMMERCIAL FIBERS

Fibers	Diameter (mils)	Density (g/cm^3)	Density ($lb/in.^3$)	UTS (10^3 psi)	Specific UTS (10^6 in.)	Modulus (10^6 psi)	Specific modulus (10^6 in.)
Filaments							
Glass							
"D"		2.2	0.078	350	4.5	8	103
"C"		2.5	0.090	450	5.0	10	111
"E"	0.4	2.6	0.092	500	5.4	12	130
"M"		2.9	0.103	500	4.7	16	155
SiO_2	1.4	2.2	0.078	850	10.8	10	133
"S"	0.4	2.5	0.090	650	7.2	13	140
970 s/66		2.5	0.090	661	7.3	15	161
X2285		2.4	0.088	676	7.7	14	159
2124		2.4	0.088	600	7.0	14	159
UARL 344		3.3	0.119	772	6.5	19	157
UARL 417		3.1	0.112			18	157
PRD-49		1.5	0.053	373	7.0	20	388
Metal							
Al	6.0	2.7	0.092	13	0.1	10	109
Fe_{ss}	6.0	7.9	0.282	347	1.2	30	106
Nb	6.0	8.6	0.308	50	0.2	15	49
Mo	10.0	10.2	0.364	270	0.7	47	129
Ta	6.0	16.8	0.594	48	0.1	27	45
Ti	6.0	4.5	0.161	78	0.5	17	106
W2% ThO_2	5.0	19.3	0.690	350	0.5	50	72
W3% Re	5.0	19.3	0.690	470	0.7	50	72

Table 4.1–1 (continued)
PROPERTIES OF EXPERIMENTAL AND COMMERCIAL FIBERS

Fibers	Diameter (mils)	Density (g/cm^3)	Density (lb/in.^{3})	UTS (10^3 psi)	Specific UTS (10^6 in.)	Modulus (10^6 psi)	Specific modulus (10^6 in.)
Other							
Al$_2$O$_3$	10	3.97	0.143	400	2.8	67	470
Al$_2$O$_3$ F.P.	10	3.97	0.143	525	3.7	67	470
B/W	4.0	2.7	0.096	450	5.0	57	610
	5.6	2.6	0.092	450	5.2	59	650
B$_4$C	0.47	2.3	0.085	207	2.4	56	660
B$_4$C/W	2.5	2.7	0.095	390	4.1	62	650
BN	0.3	2.0	0.072	210	2.9	12	167
Be	5.0	1.8	0.066	185	2.8	35	530
C-T25	0.3	1.5	0.054	180	3.2	25	464
-T50	0.3	1.7	0.060	315	5.2	57	940
-T50S	0.2	1.7	0.060	285	4.7	57	940
-T75	0.2	1.8	0.064	380	5.8	79	1,200
-T75S	0.2	1.8	0.064	345	5.2	79	1,200
-T400	0.3	1.8	0.064	425	6.6	30	470
C-HMG-25		1.5	0.054	175	3.2	25	460
-HMG-40		1.7	0.061	218	3.6	43	705
-HMG-50		1.8	0.064	308	4.8	56	875
C-Type A		1.9	0.068	300	4.4	31	456
-Type B		2.0	0.071	275	3.9	55	775
-Type C		1.9	0.068	350	5.2	38	559
C-Type 1		2.0	0.071	250	3.5	60	845
-Type 2		1.8	0.064	400	6.3	38	595
C-Kureha		1.7	0.061	75–250	1.2–4.1	4–38	66–625
Nippon-Kayaku		1.8	0.064	85	1.3		
C-2T		1.8	0.064	157	2.4	18	281
-3T		1.8	0.064	299	4.7	31	484
-4T		1.8	0.064	344	5.4	36	563
-5T		1.8	0.064	393	6.2	48	750
-6T		1.9	0.068	411	6.1	58	855
SiC/B/W	4.3	2.8	0.100	430	4.3	55	550
	5.7	2.7	0.096	450	5.0	57	610
SiC/W	4.0	3.5	0.125	300	2.4	65	520
TiB$_2$/W				150		70	
ZrO$_2$		4.9	0.175	300	1.7	50	286
Whiskers							
Al$_2$O$_3$	<0.1–0.3	4.0	0.143	1,000	7.0	70	489
BeO		2.9	0.103	1,900	18.4	50	486
B$_4$C		2.5	0.091	2,000	22.0	70	770
C		1.7	0.060	2,850	47.6	102	1,700
SiC	0.1	3.2	0.114	3,000	26.1	70	608
Si$_3$N$_4$	<0.1–0.4	3.2	0.114	700	6.1	40	350
Cr		7.6	0.27	1,290	5.0	35	130
Cu		9.0	0.32	430	1.0	18	60
Fe		6.4	0.23	1,900	8.0	29	130

Table 4.1–1 (continued)
PROPERTIES OF EXPERIMENTAL AND COMMERCIAL FIBERS

REFERENCES

Bacon, J., United Aircraft Corporation, Research Laboratories, East Hartford, Conn.
Bacon, R., Union Carbide Corporation, Parma, O.
Benn, W., Great Lakes Carbon, New York, N.Y.
Galasso, F. S., *High Modulus Fibers and Composites,* Gordon and Breach, New York, 1969.
Hurley, G., Tyco Corporate Technology Center, Waltham, Mass.
Lin, R.-Y., The Carborundum Company, Niagara Falls, N.Y.
Rauch, H., General Electric Company, Philadelphia, Pa.

Table compiled by F. S. Galasso.

Table 4.2
METAL MATRIX COMPOSITES

F. S. Galasso
United Aircraft Research Laboratories

Table 4.2–1 lists data for metal matrix composites. Most of the data are for boron fiber and BORSIC® (silicon carbide coated boron fiber) reinforced composites, so they are listed first.

The first column contains the fiber used to reinforce the composite and the next column lists the diameter of the fiber. In the case of carbon fiber, the diameter is that of a single filament in a yarn or tow.

The volume percent fiber in the composite and the matrix metal alloy are then given. The fabrication process is listed in the next column. In diffusion bonding the matrix containing the fibers is hot pressed; in braze bonding the matrix is partially molten during pressing; in the diffusion bonding of plasma sprayed monolayer tapes (Diff. B. Pl. Sp. Tape) made by plasma spraying the matrix onto the fibers are hot pressed to form a composite. The eutectic (Eut.) process is a proprietary McDonnell method for forming composites.

The condition of the composite is either as fabricated (As Fab.) or heat treated (Ht. Trt.). In the next columns, the tensile strength ($\sigma \, UTS_{11}$), tensile modulus (E_{11}), and elongation (ϵf_{11}) along the fiber axis are given. Then the tensile strength ($\sigma \, UTS_{22}$), modulus (E_{22}), and elongation (ϵf_{22}) perpendicular to the fiber axes are listed in the next columns.

Table 4.2–1
DATA FOR METAL MATRIX COMPOSITES

Fiber	Diameter (mils)	Vol % fiber	Matrix	Fabrication process	Condition of the composite	UTS_{11} (10^3 psi)	E_{11} (10^6 psi)	ϵf_{11} (%)	UTS_{22} (10^3 psi)	E_{22} (10^6 psi)	ϵf_{22} (%)	Source
Boron	4.0	42	6061Al	Diff. B. Pl. Sp. Tape	Ht. Trt.				34.0			1
		45	6061Al	Diff. B. Pl. Sp. Tape	As Fab.				15.0	18.0	0.30	1
		50	6061Al	Diff. B. Pl. Sp. Tape	As Fab.	200.0	33.0	0.7				1
		50	6061Al	Diff. B.	As Fab.	170.0	33.0	0.6	15.0	22.0	0.30	1
		50	6061Al	Diff. B.	Ht. Trt.	160.0	33.0	0.6	22.0	20.0	0.30	1
		50	6061Al	Diff. B.	As Fab.	170.0	33.0	0.6	15.0	22.0	0.30	2
		50	6061Al	Diff. B.	Ht. Trt.	160.0	33.0	0.6	22.0	20.0	0.30	2
		40	6061Al	Diff. B.	As Fab.	175.0	29.0	0.7	15.5	16.0	0.30	3
		40	6061Al	Diff. B.	Ht. Trt.	190.0			25.0			3
		50	6061Al	Diff. B.	As Fab.	215.0	34.0	0.7	15.5	20.0	0.30	3
		50	6061Al	Diff. B.	Ht. Trt.	220.0			22.3			3
		60	6061Al	Diff. B.	As Fab.	260.0	38.0	0.7	15.6	23.0	0.30	3
		60	6061Al	Diff. B.	Ht. Trt.	250.0			21.2			3
		44	2024Al	Diff. B. Pl. Sp. Tape	As Fab.	194.0			16.0			1
		40	2024Al	Diff. B.	As Fab.	150	30.0	0.5	25.0	18.0	0.25	4
						160 (260°C)	28 (260°C)		20 (260°C)	15 (260°C)		
		50	7075Al	Diff. B.	As Fab.	198.0	34.0	0.6	14.3			3
		50	7075Al	Diff. B.	Ht. Trt.	225.0	32.0	0.7	33.5			3
	5.6	50	6061Al	Diff. B.	As Fab.	200.0	32.0	0.6	20.0	20.0	0.30	1
		50	6061Al	Diff. B.	As Fab.	200.0	32.0	0.6	12.0	20.0	0.30	2
		40	6061Al	Diff. B.	As Fab.	187.0			22.7	16.0	0.40	3
		40	6061Al	Diff. B.	Ht. Trt.	203.0			34.0	17.0	0.50	3
		50	6061Al	Diff. B.	As Fab.				22.0	20.0	0.60	3
		50	6061Al	Diff. B.	Ht. Trt.				43.5	20.0	0.60	3
		60	6061Al	Diff. B.	As Fab.	261.0			27.3	23.0	0.30	3
		60	6061Al	Diff. B.	Ht. Trt.	255.0			46.8	24.0	0.40	3
		50	1100Al	Eut.		209.0	30.8					5
	8.0	50	6061Al	Diff. B.	As Fab.	225.0	33.0	0.8	19.2	20.0	0.7	3
		50	6061Al	Diff. B.	Ht. Trt.	238.0	34.0	0.8	36.8	21.0	0.2	3

Table 4.2–1 (continued)
DATA FOR METAL MATRIX COMPOSITES

Fiber	Diameter (mils)	Vol % fiber	Matrix	Fabrication process	Condition of the composite	UTS_{11} (10^3 psi)	E_{11} (10^6 psi)	ϵf_{11} (%)	UTS_{22} (10^3 psi)	E_{22} (10^6 psi)	ϵf_{22} (%)	Source[a]
BORSIC®	4.2	50	1100Al	Diff. B. Pl. Sp. Tape	As Fab.	172.0	30.9	0.6	20.7	18.5	0.47	5
		40	1100Al	Eut.	As Fab.	−311.0	29.6	−1.4	−30.6	22.4	−0.43	1
		50	2024Al	Diff. B. Pl. Sp. Tape	As Fab.				13.0 / 12 (320°C)			1
		50	2024Al	Diff. B. Pl. Sp. Tape	Ht. Trt.				22.0 / 13 (320°C)			1
		50	6061Al	Diff. B. Pl. Sp. Tape	As Fab.	160.0	33.0	0.6	13.0 / 6 (320°C)	20.0	0.15	1
		50	6061Al	Diff. B. Pl. Sp. Tape	Ht. Trt.	90 (500°C)			21.0 / 6 (320°C)			1
	5.7	54	1100Al	Diff. B. Pl. Sp. Tape	As Fab.	207.0	30.2	0.7	13.1	22.0	0.60	1
		50	1100Al	Eut.	As Fab.	200.0	39.0	0.6	20.0	24.0	0.30	5
		60	6061Al	Diff. B. Pl. Sp. Tape	As Fab.				40.0	24.0	0.25	1
		60	6061Al	Diff. B. Pl. Sp. Tape	Ht. Trt.							1
		50	6061Al	Braze B.	Ht. Trt.	136.0	29.6	0.5				5
	4.2	32	Beta III	Diff. B.	Ht. Trt.	163–173	29.0	0.58–0.70				6
		27	6/4 Ti	Diff. B.	As Fab.	154.0	26.0	0.6				6
		50	6/4 Ti	Diff. B.	As Fab.	175.0	34.0	0.6	40.0			1
		30	6/4 Ti	Diff. B.	As Fab.	155.0	28.0					3
		40	6/4 Ti	Diff. B.	As Fab.	160.0						3
		50	6/4 Ti	Diff. B.	As Fab.	164.0	38.0	0.5	40.0	29.0	0.4	3
		60	6/4 Ti	Diff. B.	As Fab.	170.0						3
	5.7	30	6/4 Ti	Diff. B.	As Fab.	179.0	29.0	0.7	85.0	26.0	2.7	3
		50	6/4 Ti	Diff. B.	As Fab.	195.0	38.0	0.6	65.0	31.0	0.8	3
Be	5	50	Al	Diff. B.	As Fab.	65.0	25.0		50		5	7
Be	5	50	6/4 Ti	Diff. B.	As Fab.	135.0	32.0	4.0				8
Be	3	73	Ti	Diff. B.	As Fab.	83.0	42.0	2.5				9

Table 4.2–1 (continued)

DATA FOR METAL MATRIX COMPOSITES

Fiber	Diameter (mils)	Vol % fiber	Matrix	Fabrication process	Condition of the composite	UTS_{11} (10³ psi)	E_{11} (10⁶ psi)	ef_{11} (%)	UTS_{22} (10³ psi)	E_{22} (10⁶ psi)	ef_{22} (%)	Source[a]
Be	20	50	Ti	Diff. B.		150.0	34.0	4.0	50	26		10
Be		43	Ti	Diff. B.	As Fab.	140.0	26.0	>2.0	50		>1	11
C-Thornel 50	0.3 Yarn	28	Al-1170 Si	Liq. Infil.	As Fab.	106.0 / 90 (500°C)		0.5				12
Morg. I	0.3 Tow	50	Ni	Diff. B.	As Fab.	105.0 / 35 (1050°C)						1
Morg. I	0.3 Tow	45	Ni₃Al	Diff. B.	As Fab.	85.0 / 35 (710°C)						1
UARL monofilament	3.4	45	Al	Diff. B.	As Fab.	98.0	24.0					1
Mo TZM	10	15	Cb	Expl.	As Fab.	127.0		3.0				13
Mo TZM	10	36	Hastelloy X	Diff. B.	As Fab.	124.0 / 42 (1100°C)		6.0				14
SiC	4	40	Ni	Electrodep.		130.0	40.0		80			15
SiC			Ti	Diff. B.		140.0						3
Steel	3	30	Ag	Diff. B.		39.0						16
Steel	2	40	Ag	Hot extrusion		60.0						17
Steel	2	20	Al	Liq. Infil.		50.0						
W	2	50	Ag	Liq. Infil.		72		1				16
W	5	50	Cb-40Ti 10Cr-5Al	Diff. B.	As Fab.	142 / 58 (980°C)	32.7					1
W-3% Re	10	24	Cb-40Ti 9Cr-4Al	Diff. B.	As Fab.	171.0 / 51 (1210°C)		1.6	79			11
W(NF)	10	30	Co	Diff. B.	As Fab.	107.0 / 46 (1100°C)			5.8 (1210°C)			14
W(NF)	10	29	B-1900			137.0 / 45 (1100°C)						14
W(NF)	10	30	Hastelloy X			99.0 / 46 (1100°C)	35.1	1				14

Table 4.2–1 (continued)
DATA FOR METAL MATRIX COMPOSITES

Fiber	Diameter (mils)	Vol % fiber	Matrix	Fabrication process	Condition of the composite	UTS$_{11}$ (10³ psi)	E$_{11}$ (10⁶ psi)	εf$_{11}$ (%)	UTS$_{22}$ (10³ psi)	E$_{22}$ (10⁶ psi)	εf$_{22}$ (%)	Source[a]
W(NF)	10	27	InCO713			145.0 51 (1100°C)		2.5				14
W(NF)	10	23	L-605			93.0 48 (1100°C)						14
W	10	22	Nichrome			73.0 35 (1100°C)		3				14
W(NF)	10	24	Udimet 700			51.0 38 (1100°C)		1				14
W(NF)	10	19	Waspalloy			128.0 27 (1100°C)		4				14
Al₂O₃	1–3	30	Al-2.5Si	Liq. Phase H.P.	As Fab.	55.0 25 (370°C)	20.0 15 (370°C)	0.4 (370°C)				18
Al₂O₃	1–3	18	Al-2.5Si	Liq. Phase H.P.	As Fab.	45.0	16.0	0.9	21.0	12.0	1.0	18
Al₂O₃	2	20	Al	Liq. Infil.	As Fab.	45.0						4
Al₂O₃	2	30	Cu	Liq. Infil.	As Fab.	115.0						4
Al₂O₃	2	10	Fe	Liq. Infil.	As Fab.	120.0						19
α-SiC	1–3	20	Mg Alloy	Liq. Phase H.P.	As Fab.	68.2 41.2 (150°C)						18
β-SiC	1–3	15	2024Al	Liq. Phase H.P.	Ht. Trt.	−270.0	23.0	−2.1	−82.0	14.0	−1.0	18
β-SiC	1–3	25	2024Al	Liq. Phase H.P.	Ht. Trt.	170.8 −200.0	25.2 28.0	0.9 −1.0	56.3 −92.0	17.2 16.0	0.8 −0.7	18

Table 4.2–1 (continued)

DATA FOR METAL MATRIX COMPOSITES

[a]Sources:

1. United Aircraft Research Laboratories
2. General Dynamics
3. TRW, Incorporated
4. General Electric Company
5. McDonnell Douglas
6. North American Rockwell
7. NASA Langley
8. Naval Air Systems Command
9. Whittaker Corporation
10. Allison Division of General Motors
11. Solar Division of International Harvester Corporation
12. Aerospace Corporation
13. NASA
14. Clevite Division of Gould, Incorporated
15. General Technologies Corporation
16. IITRI
17. Massachusetts Institute of Technology
18. Artech Corporation
19. Horizons

REFERENCES

Bilow, G., McDonnell Douglas, St. Louis, Mo.
Brennan, J., United Aircraft Research Laboratories, East Hartford, Conn.
Carlson, R. G., General Electric Company, Cincinnati, O.
Christian, J. L., General Dynamics, Convair Division, San Diego, Cal.
Hahn, H., Artech Corporation, Falls Church, Va.
Hamilton, C. H., North American Rockwell, Los Angeles, Cal.
Machlin, I., Naval Air Systems Command, Washington, D.C.
Metcalfe, A. G., Solar, San Diego, Cal.
Pepper, R. T., Aerospace Corporation, Los Angeles, Cal.
Prewo, K., United Aircraft Research Laboratories, East Hartford, Conn.
Schmidt, D., Naval Air Systems Command, Washington, D.C.
Signorelli, R., NASA Lewis Research Center, Cleveland, O.
Toth, I. J., TRW, Inc., Cleveland, O.

Table compiled by F. S. Galasso.

4.3 EUTECTIC COMPOSITES

F. S. Galasso

United Aircraft Research Laboratories

In Tables 4.3–1 through 4.3–4 are listed data for eutectics which have been studied; attempts have been made to unidirectionally solidify most of them.

The first column in each table lists the two phases, α and β. The next column gives the volume percent or weight percent of one of the phases.

The third column lists the eutectic temperature; the microstructure is given in the fourth column.

The fifth column lists the ultimate tensile strength of the eutectics obtained along the axis of solidification.

Table 4.3–1

BINARY EUTECTIC ALLOYS[a]

System $\alpha + \beta$	Volume %, α	Eutectic temperature, °C	Microstructure	Ultimate tensile strength (10^3 psi)
Ag-Bi	3	262	Broken lamellae of Ag	
Ag-Cu	74	799	Lamellar and rod	
Ag-Ge	78	651	Abnormal	
Ag-Pb	15	304	Rods and broken lamellae	
Ag-Si	90	830	Abnormal	
Al-Ag$_3$Al$_2$	40	566	Lamellar	
Al-AlB$_2$		660		
Al-Al$_4$Ca	69	616	Lamellar and rod	
Al-Al$_4$Ce	88	638	Lamellar and rod	
Al-Al$_9$Co$_2$	98	657	Lamellar	
Al-CuAl$_2$	54	548	Lamellar	75
Al-Al$_3$Fe	97	655	Abnormal	
Al-Ge	66	424	Abnormal	
Al-Al$_3$Ni	89	640	Distorted hexagonal rods and broken lamellae of Al$_3$Ni	51
Al-Al$_3$Pd	67	615	Lamellar	
Al-Al$_3$Pd$_2$	69		Lamellar "Chinese script"	
Al-AlSb	99	657	Abnormal	
Al-Si	88	577	Abnormal and rods	
Al-Sn	1.5	232	Lamellar and rod	
Al-Al$_3$Th	12	634	Spiral lamellar	
Al-Al$_4$U	84	640	Chevron	
Al-Al$_3$Y	26–30	640	Lamellar and rod	
Al-Zn	69	382	Lamellar	
Au-Co	18	997	Rod	
Au-Ge		356	Abnormal interconnected flakes	
Bi-Au$_2$Bi		241	Broken lamellar	
Bi-Cd	57	144	Lamellar and/or abnormal with pyramid L/S interface	
Bi–MnBi	96	262	Rod	
Bi-Pb$_2$Bi	27	125	Abnormal, pyramid L/S interface	
Bi-Sn	40	139	Abnormal, pyramid L/S interface	
Bi-Bi$_2$Tl		198	Abnormal	
Bi-Zn	96	254	Broken lamellae and rod	
Cd-Cd$_3$Cu	19	248	Fiber-ribbon	
Cd-Pb			Lamellar	
Cd-CdSb	81	290	Abnormal	

Table 4.3–1 (continued)
BINARY EUTECTIC ALLOYS

System $\alpha + \beta$	Volume %, α	Eutectic temperature, °C	Microstructure	Ultimate tensile strength (10^3 psi)
Cd-Sn	25	177	Lamellar	
Cd-Zn	83	266	Lamellar	
Co-CoAl	65	1,400	Lamellar and rod	72–85
Co-CoBe	77	1,120	Lamellar	
Co-CoCr		1,470	Dendrites	
Co-CoGe$_2$	33	1,110	Lamellar	
Co-Co$_2$Nb	61	1,235	Lamellar	
Co-CoSb	38	1,095	Lamellar	
Co-Co$_2$Ta	65	1,276	Lamellar and rod	109
Co-Co$_7$W$_6$	77	1,480	Lamellar	
Co-Co$_{17}$Y$_2$	19	1,340	Rod	
Co-Mg$_6$Co$_7$	67	1,340	Lamellar	
Cr-Cr$_{23}$C$_6$	39	1,500	Rod	
Cr-Cr$_2$O$_3$	19	1,660	Rod	
Cu-B	92		Ribbon	
Cu-Cr	98	1,083	Rod	
Cu-Cu$_2$O	90	1,065	Rod	
Cu-Cu$_4$Zn	50		Lamellar	
Cu-Cu$_3$P		714		
Cu-Cu$_2$S		1,065		
Cu-Nb		1,550	Lamellar	
Fe-Fe$_2$B	55	1,149	Square rods	
Fe-Fe$_3$C	41	1,147	Lamellar and rod	
Fe-C	92	1,147	Abnormal	
Fe-Fe$_x$O	10	1,371	Rod	
Fe-FeS	9.5	988	Hexagonal rods of Fe	
Fe-Fe$_x$Sb	18	1,002	Hexagonal rods of Fe	
Fe-FeTi		1,340		
Fe-FeZr			Lamellar	
In-BiIn$_2$	30	72	Lamellar	
β, InSn-γ, SnIn	25	117	Lamellar and rod	
Mg-Mg$_{17}$Al$_{12}$	32	437	Lamellar	
Mg-Mg$_2$Cu	60	485	Lamellar	
Mg-Mg$_2$Ni	72	507	Lamellar and rod	

Table 4.3–1 (continued)
BINARY EUTECTIC ALLOYS

System $\alpha + \beta$	Volume %, α	Eutectic temperature, °C	Microstructure	Ultimate tensile strength (10^3 psi)
Mg-Mg$_2$Si	97	632	Faceted rods	
Mg-Mg$_2$Sn	76	561	Lamellar "Chinese script"	
Mo-Mo$_2$C	63	2,200	Lamellar	
Nb-Nb$_2$C	69	2,335	Rectangular rods	155
Nb-Th	10	1,435	Rod	
Ni-C	90	1,318	Abnormal	
Ni-Ni$_3$B	35	1,080	Lamellar	
Ni-NiBe	60–62	1,157	Lamellar	133
Ni-Ni$_3$Nb	74	1,270	Lamellar	108
Ni,Al-Ni$_3$Nb	78	1,270		164
Ni-Cr	77	1,345	Lamellar and rod	104
Ni-Ni$_{1.5}$Gd$_2$	40	1,290	Rod	
Ni-NiMo	50	1,315	Lamellar	181
Ni-Ni$_3$Sb	40	1,097	Lamellar	
Ni-Ni$_5$Si$_2$	35	1,125	Lamellar	
Ni-Ni$_3$Sn	38	1,130	Lamellar	
Ni-Ni$_3$Ta	87	1,360	Lamellar	
Ni-Ni$_3$Ti	61	942	Lamellar	
Ni-Ni$_7$Th$_2$	38	1,300	Rod and lamellar	90–100
Ni-Pb				
Ni-TaNi$_3$		1,360	Lamellar	
Ni-W	93	1,500	Rod	
Pb-AuPb$_2$	52	215	Lamellar	
Pb-Sb	88	251	Abnormal, pyramid L/S interface	
Pb-Sn	37	183	Lamellar	
Sb-Ag$_3$Sb	28	484	Abnormal, with pyramid L/S interface	
Sb-CdSb	13	445	Abnormal	
Sb-InSb	35	530	Triangular rods	
Sb-MnSb	71	570	Circular rods	
Sb-Sb$_2$Tl$_7$	12	195	Abnormal	
Sn-Ag$_3$Sn	97	221	Abnormal/lamellar	
Sn-Cu$_6$Sn$_5$	98	227	Rods	
Sn-In$_3$Sn	91	117	Lamellar and rods	
Sn-Zn		198	Broken lamellae and rods of Zn	

Table 4.3–1 (continued)
BINARY EUTECTIC ALLOYS

System $\alpha + \beta$	Volume %, α	Eutectic temperature, °C	Microstructure	Ultimate tensile strength (10^3 psi)
Ta-Ta$_2$C	71	2,800	Rectangular rods of Ta$_2$C and lamallae	150
TaFe-TaFe$_2$			Lamellar/abnormal	
Te-Bi$_2$Te$_3$	73	413	Lamellae of Bi$_2$Te$_3$	
Th-Ti	75	1,190	Rod	
Ti-TiB	90	1,670	Abnormal	
Ti-Ti$_5$Si$_3$	75	1,330	Rod	
V-V$_2$C	75	1,650	Rod and lamellar	
V-V$_3$Si		1,840		
Zn-MgZn$_2$	50	367	Spiral lamellar	
Zn-Mg$_2$Zn$_{11}$		367	Lamellar and rod with pyramid L/S interface	
Zn-Zn$_{15}$Ti	96	419	Rod	

[a]For references, see Table 4.3–4.

Table compiled by F. S. Galasso.

Table 4.3–2
QUASI-BINARY EUTECTICS[a]

System $\alpha + \beta$	Volume %, α	Eutectic temperature, °C	Microstructure	Ultimate tensile strength (10^3 psi)
Al-Mg$_2$Si	88		Abnormal	
Co-Cr$_7$C$_3$	70	1300		185
Co-Cr$_{23}$C$_6$	60	1340		180
Co-HfC	85		Rods "arrow feather"	
Co-NbC	88	1365	Rods and lamellar	150
Co-TaC	84		Rods "arrow feather"	150
Co-TiC	84		Rods "arrow feather"	
Co-VC	80		Rods	
Ni-HfC	72–85		Rods "arrow feather"	
Ni-NbC	89		Rods "arrow feather"	
Ni-TiC	95		Rods "arrow feather"	
NiAl-Cr	66–67	1450	Rods	180
NiAl-Mo	89		Hexagonal rod	
Ni$_3$Al-Mo	74	1306		163
Ni$_3$Al-Ni$_3$Cb	56	1280	Lamellar	180
Ni$_3$Al-Ni$_3$Ta	35	1330	Lamellar	135
Ni$_3$Al-Ni$_7$Zr$_2$	58	1192	Lamellar	

[a]For references, see Table 4.3–4.

Table compiled by F. S. Galasso.

Table 4.3–3
SEMICONDUCTOR QUASI-BINARY EUTECTICS[a]

System $\alpha + \beta$	Weight %, β	Eutectic temperature, °C	Microstructure	Ultimate tensile strength (10^3 psi)
GaAs-CrAs	35.4		Rod and lamellar	
GaAs-MoAs	9.4		Rectangular rod and lamellar	
GaAs-VAs	8.4		Rectangular rod and lamellar	
GaSb-CrSb	13.4	690	Rod	
GaSb-CoGa$_{1.3}$	7.9	697	Rod	
GaSb-FeGa$_{1.3}$	7.9	695	Rod	
GaSb-GaV$_3$Sb$_5$	4.9	710	Square and rectangular rods	
GaSb-V$_2$Ga$_5$	4.4	707	Square and rectangular rods	
InAs-CrAs	1.7	937	Rod	
InAs-FeAs	10.5	520	Rod	
InSb-CrSb	0.6		Rod	
InSb-FeSb	0.7	520	Rod	
InSb-Mg$_3$Sb$_2$	2.2	519	Lamellae of Mg$_3$Sb$_2$	
InSb-MnSb	6.5	520	Rod	
InSb-NiSb	1.8	517	Rod	

[a]For references, see Table 4.3–4.

Table compiled by F. S. Galasso.

Table 4.3—4
IONIC SALTS AND QUASI-BINARY EUTECTICS

System $\alpha + \beta$	Weight %, α	Eutectic temperature, °C	Microstructure	Ultimate tensile strength (10^3 psi)
LiF-NaF	48.6–60	652	Lamellar	
NaF-NaCl	22.0–23.1	674	Rectangular rods	
NaBr-NaF	83.4	642	Rectangular rods	
LiF-NaCl	25	680	Rod	
NaF-CaF$_2$		810	Lamellar	
NaF-MgF$_2$	20.0	820	Rod	
NaF-PbF$_2$	20.0	540	Rod	
MnO-MnS	43.5		Lamellar	
FeO-FeS	36.1	940	Rectangular rods	
Zn$_5$B$_4$O$_{11}$-ZnB$_2$O$_{11}$	50.0	1300	Lamellar	
BaFe$_{12}$O$_{19}$-BaFe$_2$O$_4$	28.6	1370	Abnormal	
PbMoO$_4$-PbO		1310	Rods	
Al$_2$O$_3$-Y$_5$Al$_3$O$_{12}$			Rods and platelets of Y$_5$Al$_3$O$_{12}$	
Al$_2$O$_3$-Al$_2$O$_3$·TiO$_2$			Lamellar	
Al$_2$O$_3$·TiO$_2$-TiO$_2$			Rod	

REFERENCES

Galasso, F. S., *High Modulus Fibers and Composites,* Gordon and Breach, New York, 1969.

Hogan, L., Kraft, W., and Lemkey, F., *Advances in Materials Research,* Vol. 5, Wiley Interscience, New York, 1971.

Lemkey, F., United Aircraft Research Laboratories, East Hartford, Conn.

Livingston, J., General Electric Research and Development Center, Schenectady, N.Y.

Sahm, P., Brown Boveri and Company, Baden, Switzerland.

Yue, A., University of California, Los Angeles, Cal.

Table compiled by F. S. Galasso.

4.4 NONMETALLIC COMPOSITES

Generally, the older nonmetallic composite systems have employed shorter length fibers, while in the newer generation of composites the reinforcement runs the entire length of the particular specimen, or along the entire respective axes of angle-ply reinforced composites. In this section a more extended coverage has been given, both to specific matrices, which have been employed with the new continuous filaments such as boron, and the reinforced composite structures.[a] Consideration of angle-ply composites which are becoming increasingly important has also been given. The data have been gathered from industrial sources and from the *Structural Design Guide for Advanced Composite Applications,* Air Force Materials Laboratory, Wright-Patterson Air Force Base, Ohio, November 1968.

Key:

E	=	modulus of elasticity
G	=	modulus of elasticity in shear
ksi	=	1,000 lb per square inch
psi	=	pounds per square inch
RT	=	room temperature
α	=	the principal axis of a lamina, a direction angle (i.e., in the direction of the fibers)
β	=	the direction perpendicular to α in the plane of the lamina, a directional angle
γ	=	shear strain
ϵ	=	linear strain
$\bar{\epsilon}$	=	ultimate linear strain
ϵ_{PL}	=	proportional limit linear strain
ν	=	Poisson's ratio

ρ	=	density
σ	=	stress
$\bar{\sigma}$	=	failing stress
σ_{PL}	=	proportional limit stress
σ_{TP}	=	stress at transition point, the point at which the initial slope of the stress-strain curve changes (for metal matrix composites)
τ	=	shear stress
$\bar{\tau}$	=	failing shear stress
$\nu_{\alpha\beta}$	=	Poisson's ratio for strain in β direction when stressed in α direction
$\nu_{\beta\alpha}$	=	Poisson's ratio for strain in α direction when stressed in β direction
$\bar{\sigma}_{\alpha}$	=	failing stress in the direction of α
$\bar{\sigma}_{\beta}$	=	failing stress in the direction of β

[a]The more conventional reinforced plastics such as the glass-reinforced plastics are listed with the polymeric materials in Section 1, Volume 3 of *CRC Handbook of Materials Science.*

Table 4.4−1
E-787 (EPON® 828/1031) EPOXY RESIN MATRIX

Material Properties

Property	Type of test	Number of specimens	Av	Max	Min	Specimen dimensions, in.	Temp
E (ksi)	Tension	2	528	536	520	w = 0.25, t = 0.125	77°F
$\bar{\sigma}$ (ksi)	Flexure	3	17.2	21.45	14.05	w = 0.50, t = 0.125	77°F
$\bar{\sigma}$ (ksi)	Tension	2	9.27	9.42	9.12	w = 0.25, t = 0.125	77°F
$\bar{\sigma}$ (ksi)	Compressive	3	19.61	20.4	19.16	w = 0.50, t = 0.125	77°F
$\bar{\epsilon}$ (%)	Tension	2	2.04	2.05	2.03	w = 0.25, t = 0.125	77°F
ν	—	—	—	—	—	—	—
G (ksi)	—	—	—	—	—	—	—

Data source: Goodyear Aerospace

Table 4.4—2
EPON® 828/1031 B80 EPOXY RESIN MATRIX

Material Properties

Property	Type of test	Number of specimens	Av	Max	Min	Specimen dimensions, in.^{2}	Temp
E (ksi)	Tension	7	514	533	502	0.133 × 0.246	RT
E (ksi)	Flexure	8	491	500	484	0.266 × 0.498	RT
$\bar{\sigma}$ (ksi)	Tension	4	11.0	12.4	8.2	0.133 × 0.246	RT
$\bar{\sigma}$ (ksi)	Flexure	8	13.1	17.4	9.04	0.266 × 0.498	RT
ν	—	—	—	—	—	—	—
G (ksi)	—	—	—	—	—	—	—

Curing process: 7 hr, 180°F; 1 hr, 250°F; 2 hr, 350°F.

Data source: General Dynamics

Table 4.4—3
BP907/U MODIFIED EPOXY RESIN MATRIX

Material Properties

Property	Type of test	Number of specimens	Av	Max	Min	Specimen dimensions, in.	Temp
E (ksi)	Tension	9	504	520	475	w = 0.46, t = 0.135–0.175	RT
E (ksi)	Compression	5	504	530	485	w = 0.48–0.50, t = 0.122–0.134	RT
$\bar{\sigma}$ (ksi)	Tension	7	11.25	13.3	8.75	w = 0.46, t = 0.135–0.175	RT
$\bar{\sigma}$ (ksi)	Compression	5	16.9	17.95	15.85	w = 0.48–0.50, t = 0.122–0.134	RT
$\bar{\tau}$ (ksi)	Shear	6	8.20	8.34	8.10	w = 0.50, t = 0.120–0.131	RT
ν	Tension	5	0.283	0.291	0.278	w = 0.46, t = 0.135–0.175	RT
$\bar{G}$ (ksi)	Shear	7	189	191	179	w = 0.45–0.47, t = 0.09	RT

Curing process: 0.5 hr at 320°F, 0 psi; 1 hr at 350°F, 100–200 psi.

Data source: Boeing – Vertol Division

Table 4.4–4
NARMCO® 5504 MODIFIED EPOXY RESIN MATRIX

Material Properties

Property	Type of test	Number of specimens	Av	Max	Min	Specimen dimensions, in.³	Temp	Cure process
E (ksi)	Compressive	1		29	–	0.5	RT	a
	Compressive	1		272	–	0.5	RT	b
	Compressive	2		280	250	0.5	RT	c
	Compressive	1		260	–	0.5	RT	d
	Compressive	2		290	280	0.5	RT	e
$\bar{\sigma}$ (ksi)	Compressive	1		34	–	0.5	RT	a
	Compressive	1		43	–	0.5	RT	b
	Compressive	2		47.2	42	0.5	RT	c
	Compressive	1		40	–	0.5	RT	d
	Compressive	2		46.1	45.7	0.5	RT	e

[a]3/4 hr at 350°F.
[b]3/4 hr at 350°F; P.C. 1 1/2 hr at 350°F.
[c]3/4 hr at 350°F; P.C. 3 1/2 hr at 350°F.
[d]3/4 hr at 350°F; P.C. 1 1/2 hr at 350°F.
[e]2 hr at 420°F; 1 hr at 200°F; 15 hr at 300°F.

Data source: Narmco Research and Development

Table 4.4–5
NARMCO® 5504 MATRIX

Material Properties

Property	Type of test	Number of specimens	Av	Max	Min	Specimen dimensions, in.³	Temp
E (ksi)	Compressive	2		54	30	0.5	270°F
	Compressive	1		55	–	0.5	270°F
	Compressive	2		120	110	0.5	270°F
	Compressive	1		120	–	0.5	270°F
	Compressive	1		120	–	0.5	270°F
τ (ksi)	Compressive	2		19.6	17.8	0.5	270°F
	Compressive	1		21.7	–	0.5	270°F
	Compressive	2		25.6	22.0	0.5	270°F
	Compressive	1		22.0	–	0.5	270°F
	Compressive	1		23.8	–	0.5	270°F

Data source: Narmco Research and Development

Table 4.4–6
NARMCO® 5505 MODIFIED EPOXY RESIN MATRIX

Material Properties

Property	Type of test	Number of specimens	Av	Max	Min	Specimen dimensions, in.³	Temp	Cure process
E (ksi)	Compressive	2	–	330	330	0.5	RT	a
	Compressive	2	–	291	255	0.5	RT	b
	Compressive	1	–	284	–	0.5	RT	c
	Compressive	2	–	102	95	0.5	270°F	a
	Compressive	2	–	115	106	0.5	270°F	b
	Compressive	1	–	108	–	0.5	270°F	c
$\bar{\sigma}$ (ksi)	Compressive	2	–	37.6	35.3	0.5	RT	a
	Compressive	2	–	44.3	42.7	0.5	RT	b
	Compressive	1	–	46.9	–	0.5	RT	c
	Compressive	2	–	27.0	19.0	0.5	270°F	a
	Compressive	2	–	27.4	26.4	0.5	270°F	a
	Compressive	1	–	22.5	–	0.5	270°F	a
E (ksi)	Tension	5	529	580	480	–	–	d
$\bar{\sigma}$ (ksi)	Tension	5	6.25	7.05	5.30	–	–	d
ϵ (%)	Tension	5	1.53	3.06	1.05	–	–	d

[a] 20 hr at 250°F.
[b] 15 hr at 250°F; 4 hr at 315°F.
[c] 20 hr at 250°F; 2 hr at 420°F.
[d] Cure not specified.

Data source: Narmco Research and Development

Table 4.4–7
ERLA-2256 EPOXY RESIN MATRIX

Material Properties

Property	Type of test	Number of specimens	Av	Max	Min	Specimen dimensions	Temp	Load rate, %/min
E (ksi)	Tension	–	–	545		–	–	0.15
$\bar{\sigma}$ (ksi)	Compressive	1	–	29.0		1 in. long × 0.5 in. diam	RT	10
$\bar{\sigma}$	Compressive	1	–	49.8		0.5 in. long × 0.5 in. diam	RT	20
$\bar{\sigma}$ (ksi)	Compressive	1	–	34.6		1 in. long × 0.5 in. diam	RT	200
$\bar{\sigma}$ (ksi)	Flexure	Nominal value	22	–		–	–	–
$\bar{\sigma}$ (ksi)	Tension	Nominal value	14	–		0.5 in. × 0.5 in. × 15 in.	RT	–
ν	–	–	–	0.36		–	RT	–
G (ksi)	Torsion	–	–	200		–	RT	–
$\bar{\epsilon}$ (%)	Tension	Nominal value	~4	–		–	RT	–

Data source: Union Carbide Corporation, Carbon Products Division

Table 4.4–8
DEN 438 EPOXY NOVOLAC RESIN MATRIX

Material Properties

Property	Type of test	Number of specimens	Av	Max	Min	Specimen dimensions, in.	Temp
E (ksi)	Flexure	3	553.5	565.2	533.7	w = 0.50, t = 0.125	77°F
E (ksi)	Tension	3	494.4	508.0	480.3	w = 0.25, t = 0.125	77°F
$\bar{\sigma}$ (ksi)	Compressive	3	35.88	43.21	31.88	w = 0.50, t = 0.125	77°F
$\bar{\sigma}$ (ksi)	Tension	3	5.27	6.15	3.85	w = 0.25, t = 0.125	77°F
$\bar{\sigma}$ (ksi)	Flexure	3	15.23	15.63	14.99	w = 0.50, t = 0.125	77°F
$\bar{\epsilon}$ (%)	Tension	3	1.13	1.33	0.80	w = 0.25, t = 0.125	77°F

Curing process: 3 hr at 180°F; 1 hr at 220°F; 16 hr at 300°F.

Data source: Goodyear Aerospace

Table 4.4–9
DEN 438 EPOXY NOVOLAC RESIN MATRIX

Material Properties

Property	Type of test	Number of specimens	Av	Max	Min	Specimen dimensions, in.	Temp	Load rate
E (ksi)	Flexure	6	493	502	482	w = 0.52, t = 0.25	RT	0.05
$\bar{\sigma}$ (ksi)	Flexure	6	16.6	18.0	14.1	w = 0.52, t = 0.25	RT	0.05
$\bar{\epsilon}$ (ksi)	Flexure	6	3.75	4.30	2.97	w = 0.52, t = 0.25	RT	0.05

Curing process: 2 hr at 195°F; 4 hr at 330°F; 16 hr at 395°F.

Data source: General Dynamics

Table 4.4–10
DEN 438 EPOXY NOVOLAC RESIN MATRIX

Material Properties

Property	Type of test	Number of specimens	Av	Max	Min	Specimen dimensions, in.	Temp
E (ksi)	Flexure	7	455	467	441	w = 0.52, t = 0.25	RT
$\bar{\sigma}$ (ksi)	Flexure	6	17.65	19.2	16.8	w = 0.52, t = 0.25	RT
$\bar{\epsilon}$ (%)	Flexure	6	5.94	7.97	4.96	w = 0.52, t = 0.25	RT
σ_{PL} (ksi)	–	–	–	–	–	–	–
ν	–	–	–	–	–	–	–
G (ksi)	–	–	–	–	–	–	–

Curing process: 16 hr at 131°F; 2 hr at 257°F; 2 hr at 347°F.

Data source: General Dynamics

Table 4.4—11
DEN 439-75 PARTS, ERRA 0300-25 PARTS EPOXY NOVOLAC RESIN MATRIX

Material Properties

Property	Type of test	Number of specimens	Av	Max	Min	Specimen dimensions, in.	Temp
E (ksi)	Flexure	7	474	479	466	w = 0.49, t = 0.23	RT
E (ksi)	Tension	6	461	488	417	w = 0.49, t = 0.11−0.13	RT
$\bar{\sigma}$ (ksi)	Flexure	7	17.1	18.3	13.5	w = 0.49, t = 0.23	RT
$\bar{\sigma}$ (ksi)	Tension	7	11.4	13.2	9.6	w = 0.49, t = 0.11−0.13	RT
$\bar{\epsilon}$ (%)	Tension	7	5.16	5.95	3.38	w = 0.49, t = 0.23	RT
$\bar{\epsilon}$ (%)	Tension	6	3.3	4.6	2.6	w = 0.49, t = 0.11−0.13	RT
G (ksi)	−	−	−	−	−	−	−

Curing process: 1 hr at 185°F; 1.5 hr at 275°F; 2 hr at 347°F.

Data source: General Dynamics

Table 4.4—12
HT-424 PHENOLIC-EPOXY RESIN MATRIX

Material Properties

Property	Type of test	Number of specimens	Av	Max	Min	Specimen dimensions, in.	Temp
E (ksi)	Flexure			−	330	1/4 × 1/8 × 4	RT
E (ksi)	Flexure			460	−	1/4 × 1/8 × 4	RT
$\bar{\sigma}$ (ksi)	−			−	−	−	−
σ_{PL} (ksi)	−			−	−	−	−
ν	−			−	−	−	−
G (ksi)	−			−	−	−	−

Curing process: 5 days at 120°F; 3 days at 140°F;
3 days at 160°F; 3 days at 180°F;
1 day at 200°F; 1 day at 220°F;
16 hr at 250°F; 16 hr at 400°F;
postcure: 3 hr at 350°F; 16 hr at
400°F; 4 hr at 450°F.

Data source: General Electric, Missiles and Space Division

Table 4.4–13
EPON® 828/1031-BORON UNIDIRECTIONAL COMPOSITE

Material Properties

Property	Type of test	Number of tests	Av	Max	Min	Width, in.	Temp	Filament volume, %
E_α (psi × 10^6)	Tension	2	31.8	32.2	31.4	0.50	RT	~79
	Flexure	2	38.1	40.9	35.2	0.50	RT	~79
E_β (psi × 10^6)	–	–	–	–	–	–	–	–
$\bar\sigma_\alpha$ (ksi)	Tension	2	123.0	127.0	119.0	0.50	RT	~79
	Compression	1	155.0	–	–	0.50	RT	~79
	Compression	1	232.6	–	–	1.25	RT	~79
$\bar\sigma_\beta$ (ksi)	Tension	4	8.64	9.33	7.95	1.0	RT	~47.5
	Compression	5	27.3	29.9	24.3	0.05	RT	~47.5
$\bar\tau_{\alpha\beta}$ (ksi)	Shear	3	10.39	11.46	9.36	0.252	RT	~65.7
$\nu_{\alpha\beta}$	Tension	2	0.25	0.25	0.24	0.50	RT	~79
$G_{\alpha\beta}$ (psi × 10^6)	Shear	1	1.12	–	–	4.0	RT	~79

Note: Thickness, orientation, number of plies: (a) 0.034 in., 0°, 9; (b) ~0.077 in., 0°, 17; (c) 0.036 in., 90°, 6.

Data sources: General Electric; Boeing – Vertol Division; North American Rockwell

Table 4.4–14
NARMCO® 5504-BORON UNIDIRECTIONAL COMPOSITE

Material Properties

Property	Type of test	Number of tests	Av	Max	Min	Width, in.	Temp	Filament volume, %	Load rate, in./min
E_α (psi × 10^6)	Flexure	4	44.8	49.2	40.5	0.5	RT	74	0.05
	Flexure	4	43.2	44.6	41.9	0.5	270°F	74	0.05
E_β (psi × 10^6)	Flexure	7	1.94	2.37	1.63	0.5	RT	74	0.05
	Flexure	7	2.16	3.07	0.89	0.5	270°F	74	0.05
$\bar\sigma_\alpha$ (ksi)	Flexure	4	375.	410.	337.	0.5	RT	74	0.05
	Flexure	4	337.	372.	263.	0.5	270°F	74	0.05
$\bar\sigma_\beta$ (ksi)	Flexure	7	12.20	14.27	8.34	0.5	RT	74	0.05
	Flexure	7	9.26	11.78	5.47	0.5	270°F	74	0.05
$\bar\tau_{\alpha\beta}$ (ksi)	Flexure	8	13.7	15.3	11.3	0.5	RT	74	0.05
	Flexure	7	8.4	9.3	7.4	0.5	270°F	74	0.05
$\nu_{\alpha\beta}$	–	–	–	–	–	–	–	–	–
$G_{\alpha\beta}$ (psi × 10^6)	–	–	–	–	–	–	–	–	–

Note: Thickness, orientation, and number of plies: 0.057 in., 0 or 90°, 17.

Curing Process:

Time	0.5 hr	0.5 hr	2 hr	4 hr	4 hr
Temp	160°F	200°F	300°F	350°F	400°F
Pressure			← max 50 psi →		

Data source: General Dynamics

Table 4.4—15
N5505 EPOXY/BORON UNIDIRECTIONAL COMPOSITES

Material Properties

Test series no.	Property	Type of test	Number of specimens	Av	Max	Min	Test temp	Density, lb/in.^{3}	Number of plies	Thickness, in.	Fiber volume, %
1	E(psi $\times 10^6$)	Tension	3	33.1	33.7	32.5	RT	0.072	8	0.0405	56.54
2	E(psi $\times 10^6$)	Tension	4	34.5	35.4	32.6	RT	0.071	8	0.039	53.84
3	E(psi $\times 10^6$)	Tension	4	33.5	35.5	32.3	RT	0.0715	16	0.079	56.15
4	E(psi $\times 10^6$)	Tension	4	35.2	43.5	32.5	RT	0.072	16	0.0785	56.19
5	E(psi $\times 10^6$)	Tension	4	33.5	36.2	32.0	RT	0.072	16	0.0805	55.12
1	$\overline{\sigma}$(ksi)	Tension	3	181.7	190.7	164.5	RT	0.072	8	0.0405	56.54
2	$\overline{\sigma}$(ksi)	Tension	4	175.5	202.4	161.1	RT	0.071	8	0.039	53.84
3	$\overline{\sigma}$(ksi)	Tension	4	204.2	213.2	191.2	RT	0.0715	16	0.079	56.15
4	$\overline{\sigma}$(ksi)	Tension	4	192.2	204.6	177.3	RT	0.072	16	0.0785	56.19
5	$\overline{\sigma}$(ksi)	Tension	4	189.1	194.0	183.2	RT	0.072	16	0.0805	55.12
6	$\overline{\tau}$(ksi)	Shear	3	11.4	11.9	10.9	RT	0.072	8	0.0405	56.54
7	$\overline{\tau}$(ksi)	Shear	4	15.3	15.6	15.0	RT	0.0715	16	0.079	56.15
8	$\overline{\tau}$(ksi)	Shear	5	14.2	14.7	13.5	RT	0.072	16	0.0785	56.19
9	$\overline{\tau}$(ksi)	Shear	4	11.4	12.1	10.6	RT	0.072.	16	0.0805	56.12

Precure/Cure Process

Time, min	Temp, °F	Pressure, psig	Heat/pressure type
140	200	100	Steel closed die mold
120	300	100	Steel closed die mold
120	200	75	Steel closed die mold

Postcure Process

Time, min	Temp, °F	Pressure, psig	Heat/pressure type
120	200	75	Steel closed die mold
120	200	50	Steel closed die mold
120	300	50	Steel closed die mold

Date source: Northrup-Norair

Table 4.4–16
N5505 EPOXY-BORON/104 CLOTH UNIDIRECTIONAL COMPOSITE

Material Properties

Test series No.	Property	Type of test	Number of specimens	Av	Max	Min	Standard deviation	Test temp	Fiber volume, %	Number of plies	Thickness, in.	Width, in.	Void content, % by volume
1	$\overline{\sigma}_\alpha$(ksi)	Tension	8	161.1	169.7	148.2	6.8	RT	47	8	0.0432	1.0	5
2	$\overline{\sigma}_\beta$(ksi)	Tension	8	4.5	5.2	3.5	0.88	RT	47	8	0.0432	1.0	5
3	$\overline{\sigma}_\alpha$(ksi)	Tension	9	192.7	211.0	166.5	13.8	RT	55	8 and 15	0.0376/0.0705	0.5	—
4	$\overline{\sigma}_\alpha$(ksi)	Compression	8	162.1	204.5	126.3	24.8	RT	47	8	0.0432	0.5	5
5	$\overline{\sigma}_\beta$(ksi)	Compression	8	21.2	25.6	19.0	2.3	RT	47	8	0.0432	0.5	5
6	$\overline{\sigma}_\alpha$(ksi)	Compression	2	199.8	203.8	115.8	195.8	RT	55	15	0.0705	0.5	—

Curing process: 2 hr at 150°F at 13 psig.
2 hr at 200°F at 13 psig.
2 hr at 250°F at 13 psig.

Postcure process: 2 hr at RT at 113 psig.
2 hr at 200°F at 113 psig.
2 hr at 300°F at 113 psig.

Data source: North American Rockwell

Table 4.4–17

N5505 EPOXY-BORON/SCRIM UNIDIRECTIONAL COMPOSITE 500-hr HEAT AGE STUDY (15 PLIES BORON, 15 PLIES 104 GLASS)

Material Properties[a]

Property	Type of test	Number of specimens	Av	Max	Min	Standard deviation	Coefficient of variation	Test temp	Soak time, hr	Soak temp, °F
E (psi × 10⁶)	Flexure	5	32.6	34.2	31.4	—	—	RT	0	350
E (psi × 10⁶)	Flexure	5	28.1	28.9	27.5	—	—	350°F	0	350
E (psi × 10⁶)	Flexure	5	31.9	32.8	30.5	—	—	RT	6	350
E (psi × 10⁶)	Flexure	5	22.1	25.2	20.1	—	—	350°F	6	350
E (psi × 10⁶)	Flexure	5	32.4	33.1	31.1	—	—	RT	12	350
E (psi × 10⁶)	Flexure	5	28.9	29.9	28.0	—	—	350°F	12	350
E (psi × 10⁶)	Flexure	5	—	—	—	—	—	RT	25	350
E (psi × 10⁶)	Flexure	5	21.0	22.4	19.7	—	—	350°F	25	350
E (psi × 10⁶)	Flexure	5	32.4	33.0	31.5	—	—	RT	50	350
E (psi × 10⁶)	Flexure	5	21.4	24.2	17.1	—	—	350°F	50	350
E (psi × 10⁶)	Flexure	5	33.1	34.5	32.6	—	—	RT	100	350
E (psi × 10⁶)	Flexure	5	29.2	29.4	29.0	—	—	350°F	100	350
E (psi × 10⁶)	Flexure	5	32.8	33.4	32.1	—	—	RT	150	350
E (psi × 10⁶)	Flexure	5	—	—	—	—	—	350°F	150	350
E (psi × 10⁶)	Flexure	5	32.9	33.5	32.5	—	—	RT	200	350
E (psi × 10⁶)	Flexure	5	29.0	29.9	28.2	—	—	350°F	200	350
E (psi × 10⁶)	Flexure	5	32.4	33.3	31.8	—	—	RT	250	350
E (psi × 10⁶)	Flexure	5	28.6	29.3	28.2	—	—	350°F	250	350
E (psi × 10⁶)	Flexure	5	30.9	33.2	29.8	—	—	RT	300	350
E (psi × 10⁶)	Flexure	5	27.9	29.9	25.5	—	—	350°F	300	350
E (psi × 10⁶)	Flexure	5	30.3	32.6	27.4	—	—	RT	350	350
E (psi × 10⁶)	Flexure	5	27.8	29.2	26.1	—	—	350°F	350	350
E (psi × 10⁶)	Flexure	5	32.4	34.4	31.0	—	—	RT	400	350
E (psi × 10⁶)	Flexure	5	26.7	29.2	24.0	—	—	350°F	400	350
E (psi × 10⁶)	Flexure	5	22.2	28.3	17.7	—	—	RT	450	350
E (psi × 10⁶)	Flexure	5	25.5	27.9	23.9	—	—	350°F	—	350
E (psi × 10⁶)	Flexure	5	25.1	33.9	19.4	—	—	RT	500	350
E (psi × 10⁶)	Flexure	5	23.1	26.3	18.3	—	—	350°F	—	350

Table 4.4–17 (continued)
N5505 EPOXY-BORON/SCRIM UNIDIRECTIONAL COMPOSITE 500-hr HEAT AGE STUDY (15 PLIES BORON, 15 PLIES 104 GLASS)

Material Properties (continued)

Property	Type of test	Number of specimens	Av	Max	Min	Standard deviation	Coefficient of variation	Test temp	Soak time, hr	Soak temp, °F
Ultimate strength (ksi)	Flexure	5	283.1	299.4	264.0	15.2	5.4	RT	0	350
Ultimate strength (ksi)	Flexure	5	234.2	242.1	231.9	6.1	2.6	350°F	0	350
Ultimate strength (ksi)	Flexure	5	276.9	290.2	265.2	10.7	3.85	RT	6	350
Ultimate strength (ksi)	Flexure	5	149.3	161.7	141.1	8.82	5.83	350°F	6	350
Ultimate strength (ksi)	Flexure	5	280.0	286.3	275.8	4.52	1.61	RT	12	350
Ultimate strength (ksi)	Flexure	5	242.6	253.8	235.1	8.04	3.31	350°F	12	350
Ultimate strength (ksi)	Flexure	5	278.9	288.5	263.5	10.75	3.85	RT	25	350
Ultimate strength (ksi)	Flexure	5	158.4	166.4	151.4	6.45	4.0	350°F	25	350
Ultimate strength (ksi)	Flexure	5	281.9	293.6	274.2	8.33	2.95	RT	50	350
Ultimate strength (ksi)	Flexure	5	161.8	203.5	139.2	27.6	17.0	350°F	50	350
Ultimate strength (ksi)	Flexure	5	281.0	284.0	278.5	2.35	0.84	RT	100	350
Ultimate strength (ksi)	Flexure	5	243.8	247.5	241.1	2.75	1.13	350°F	100	350
Ultimate strength (ksi)	Flexure	5	277.3	284.6	271.4	5.68	2.05	RT	150	350
Ultimate strength (ksi)	Flexure	5	248.1	260.7	243.3	7.47	3.01	350°F	150	350
Ultimate strength (ksi)	Flexure	5	286.3	297.1	270.8	11.29	3.95	RT	200	350
Ultimate strength (ksi)	Flexure	5	243.1	250.2	234.4	6.79	2.8	350°F	200	350
Ultimate strength (ksi)	Flexure	5	279.9	284.8	277.1	3.5	1.25	RT	250	350
Ultimate strength (ksi)	Flexure	5	232.5	241.5	222.6	8.11	3.48	350°F	250	350
Ultimate strength (ksi)	Flexure	5	264.5	276.6	238.7	16.29	6.16	RT	300	350
Ultimate strength (ksi)	Flexure	5	209.2	219.1	202.7	7.05	3.37	350°F	300	350
Ultimate strength (ksi)	Flexure	5	273.6	282.3	255.4	11.56	4.2	RT	350	350
Ultimate strength (ksi)	Flexure	5	207.4	218.5	195.5	9.89	4.76	350°F	350	350
Ultimate strength (ksi)	Flexure	5	283.2	288.8	274.9	5.98	2.11	RT	400	350
Ultimate strength (ksi)	Flexure	5	200.5	214.3	190.6	10.19	5.1	350°F	400	350
Ultimate strength (ksi)	Flexure	5	282.2	287.3	273.4	5.98	2.1	RT	450	350

Table 4.4–17 (continued)
N5505 EPOXY-BORON/SCRIM UNIDIRECTIONAL COMPOSITE 500-hr HEAT AGE STUDY (15 PLIES BORON, 15 PLIES 104 GLASS)

Materials Properties (continued)

Property	Type of test	Number of specimens	Av	Max	Min	Standard deviation	Coefficient of variation	Test temp	Soak time, hr	Soak temp, °F
Ultimate strength (ksi)	Flexure	5	155.7	166.7	145.4	9.16	5.8	350°F	–	350
Ultimate strength (ksi)	Flexure	5	289.5	294.4	281.2	5.68	1.9	RT	500	350
Ultimate strength (ksi)	Flexure	5	140.3	152.3	126.7	10.9	7.8	350°F	–	350
Ultimate strength (ksi)	Transverse flexure	5	15.3	16.8	14.3	1.075	7.0	RT	0	350
Ultimate strength (ksi)	Transverse flexure	5	8.04	8.76	6.70	0.885	11.0	350°F	0	350
Ultimate strength (ksi)	Transverse flexure	5	14.51	16.12	12.8	1.43	0.984	RT	6	350
Ultimate strength (ksi)	Transverse flexure	5	7.94	8.48	7.25	0.529	6.66	350°F	6	350
Ultimate strength (ksi)	Transverse flexure	5	14.8	15.7	13.2	1.075	7.26	RT	12	350
Ultimate strength (ksi)	Transverse flexure	5	9.82	10.2	9.1	0.559	5.7	350°F	12	350
Ultimate strength (ksi)	Transverse flexure	5	14.96	15.86	13.53	1.00	6.7	RT	25	350
Ultimate strength (ksi)	Transverse flexure	5	7.69	8.28	6.80	0.636	8.28	350°F	25	350
Ultimate strength (ksi)	Transverse flexure	5	14.86	16.54	13.10	1.48	9.95	RT	50	350
Ultimate strength (ksi)	Transverse flexure	5	7.78	8.42	7.35	0.460	5.91	350°F	50	350
Ultimate strength (ksi)	Transverse flexure	5	13.51	15.03	12.03	1.29	9.55	RT	100	350
Ultimate strength (ksi)	Transverse flexure	5	9.58	10.08	9.18	0.387	4.04	350°F	100	350
Ultimate strength (ksi)	Transverse flexure	5	12.8	13.3	11.8	0.645	1.77	RT	200	350
Ultimate strength (ksi)	Transverse flexure	5	9.66	10.4	8.6	0.744	8.01	350°F	200	350
Ultimate strength (ksi)	Transverse flexure	5	14.36	14.60	14.20	0.172	1.2	RT	300	350
Ultimate strength (ksi)	Transverse flexure	5	8.23	8.76	7.51	0.538	6.53	350°F	300	350
Ultimate strength (ksi)	Transverse flexure	5	14.78	16.18	14.02	0.929	6.28	RT	400	350
Ultimate strength (ksi)	Transverse flexure	5	7.62	7.87	7.49	0.163	2.14	350°F	400	350
Ultimate strength (ksi)	Transverse flexure	5	14.84	16.58	12.44	1.78	12.0	RT	500	350
Ultimate strength (ksi)	Transverse flexure	5	6.07	6.67	5.61	0.456	7.51	350°F	500	350
Ultimate strength (ksi)	Short beam shear	5	16.0	16.4	15.6	0.344	2.15	RT	0	350
Ultimate strength (ksi)	Short beam shear	5	7.53	7.84	7.29	0.24	3.1	350°F	0	350
Ultimate strength (ksi)	Short beam shear	5	14.97	15.16	14.68	0.245	1.64	RT	6	350
Ultimate strength (ksi)	Short beam shear	5	6.62	6.77	6.46	0.133	2.01	350°F	6	350
Ultimate strength (ksi)	Short beam shear	5	16.4	16.6	16.2	0.172	1.04	RT	12	350

Table 4.4–17 (continued)
N5505 EPOXY-BORON/SCRIM UNIDIRECTIONAL COMPOSITE 500-hr HEAT AGE STUDY (15 PLIES BORON, 15 PLIES 104 GLASS)

Material Properties (continued)

Property	Type of test	Number of specimens	Av	Max	Min	Standard deviation	Coefficient of variation	Test temp	Soak time, hr	Soak temp, °F
Ultimate strength (ksi)	Short beam shear	5	6.6	6.8	6.5	0.129	1.95	350°F	12	350
Ultimate strength (ksi)	Short beam shear	5	15.15	15.56	14.96	0.258	1.70	RT	25	350
Ultimate strength (ksi)	Short beam shear	5	7.16	7.50	6.85	0.280	3.9	350°F	25	350
Ultimate strength (ksi)	Short beam shear	5	15.00	15.16	14.84	0.138	0.917	RT	50	350
Ultimate strength (ksi)	Short beam shear	5	7.01	7.34	6.66	0.292	4.17	350°F	50	350
Ultimate strength (ksi)	Short beam shear	5	16.4	16.6	16.3	0.129	0.78	RT	100	350
Ultimate strength (ksi)	Short bean shear	5	6.3	6.5	6.1	0.172	2.73	350°F	100	350
Ultimate strength (ksi)	Short beam shear	5	16.5	16.7	16.0	0.301	1.82	RT	150	350
Ultimate strength (ksi)	Short beam shear	5	6.6	6.8	6.4	0.129	1.95	350°F	150	350
Ultimate strength (ksi)	Short beam shear	5	16.7	17.0	16.5	0.215	1.29	RT	200	350
Ultimate strength (ksi)	Short beam shear	5	6.6	7.2	6.3	0.387	5.86	350°F	200	350
Ultimate strength (ksi)	Short beam shear	5	16.4	16.6	16.2	0.172	1.0	RT	250	350
Ultimate strength (ksi)	Short beam shear	5	6.7	6.8	6.5	0.129	1.92	350°F	250	350
Ultimate strength (ksi)	Short beam shear	5	16.7	17.0	16.6	0.172	1.03	RT	300	350
Ultimate strength (ksi)	Short beam shear	5	7.07	7.20	6.83	0.159	2.25	350°F	300	350
Ultimate strength (ksi)	Short beam shear	5	16.8	17.3	16.6	0.301	1.80	RT	350	350
Ultimate strength (ksi)	Short beam shear	5	7.42	7.62	7.22	0.172	2.31	350°F	350	350
Ultimate strength (ksi)	Short beam shear	5	16.6	16.8	16.3	0.215	1.30	RT	400	350
Ultimate strength (ksi)	Short beam shear	5	6.43	6.50	6.33	0.073	1.14	350°F	400	350
Ultimate strength (ksi)	Short beam shear	5	15.45	15.65	15.34	0.133	0.862	RT	450	350
Ultimate strength (ksi)	Short beam shear	5	6.49	6.62	6.31	0.133	2.05	350°F	450	350
Ultimate strength (ksi)	Short beam shear	5	15.37	15.56	15.21	0.150	0.979	RT	500	350
Ultimate strength (ksi)	Short beam shear	5	5.72	5.97	5.48	0.211	3.68	350°F	500	350

a 55 vol % fiber.

Curing process: 2 hr at 200°F at 20 psig.
2 hr at 300°F at 50 psig.
Cool under pressure.

Postcure process: 2 hr at 350°F at 0 psig.

Data source: Narmco Research and Development

Table 4.4–18
N5505-BORON/BETA GLASS OVERWRAP UNIDIRECTIONAL COMPOSITE

Material Properties[a]

Property	Type of test	Number of specimens	Av	Max	Min	Standard deviation	Temp	Load rate
E_α(psi $\times$ 10^6)	Tension	2	30.0	30.0	30.0	0.0	RT	0.025
E_β(psi $\times$ 10^6)	Tension	2	2.6	2.6	2.6	0.0	RT	0.025
$\bar{\sigma}_\beta$(ksi)	Compression	3	25.1	25.6	24.8	0.438	RT	0.025
$\bar{\sigma}_\beta$(ksi)	Tension	7	5.9	6.8	4.9	0.63	RT	0.025
$\bar{\sigma}_\alpha$(ksi)	Compression	4	180.6	199.2	171.2	12.7	RT	0.025
$\bar{\sigma}_\alpha$(ksi)	Tension	8	169.8	198.6	145.2	19.1	RT	0.025
$\bar{\sigma}_\alpha$(ksi)	Compression	2	189.3	192.3	186.4	4.18	RT	0.025

[a]53 vol % fiber.

Cure process: 2 hr at 200°F at 13 psig.
            2 hr at 300°F at 13 psig.

Data source: North American Rockwell

Table 4.4–19
SP272-BORON UNIDIRECTIONAL COMPOSITES, LOAD PARALLEL TO FIBERS

Material Properties[a]

Property	Type of test	Number of specimens	Av	Max	Min	Temp	Load rate, in./min
E_α(psi $\times$ 10^6)	Tension	2	30.5	31.4	29.6	75	0.00184
E_α(psi $\times$ 10^6)	Tension	2	28.6	28.8	28.4	350	0.0014
$\bar{\sigma}_\alpha$(ksi)	Tension	2	186	197	174	75	0.00184
$\bar{\sigma}_\alpha$(ksi)	Tension	2	164	166	163	350	0.0014
$\bar{\epsilon}_\alpha$(in./in. $\times$ 10^{-3})	Tension	2	6.30	6.52	6.07	75	0.00184
$\bar{\epsilon}_\alpha$(in./in. $\times$ 10^{-3})	Tension	2	5.90	6.00	5.81	350	0.0014
$\bar{\sigma}_{PL\alpha}$(ksi)	Tension	2	106	109	103	75	0.00184
$\bar{\sigma}_{PL\alpha}$(ksi)	Tension	2	97.0	102	92.1	350	0.0014
$\bar{\epsilon}_\alpha$(in./in. $\times$ 10^{-3})	Tension	2	3.475	3.48	3.47	75	0.00184
$\bar{\epsilon}_\alpha$(in./in. $\times$ 10^{-3})	Tension	2	3.40	3.60	3.210	350	0.0014

[a]55 vol % fiber.

Cure process: ½ hr at 250°F at 100 psig.
                1 hr at 350°F at 100 psig.

Postcure process: 4 hr at 350°F at 0 psig.

Data source: General Dynamics

Table 4.4–20
SP272-BORON UNIDIRECTIONAL COMPOSITES, LOAD PERPENDICULAR TO FIBERS

Material Properties

Property	Type of test	Number of specimens	Av	Max	Min	Test temp	Fiber volume, %	Load rate, in./min.	Number plies/ thickness, in.
E_β(psi × 10⁶)	Tension	2	3.92	4.03	3.82	75°F	52.0	0.00112	8/0.041
E_β(psi × 10⁶)	Tension	2	1.96	2.03	1.90	350°F	54.0	0.0021	8/0.040
$\bar\sigma_\beta$(ksi)	Tension	2	17.8	17.9	17.7	75°F	52.0	0.00112	8/0.041
$\bar\sigma_\beta$(ksi)	Tension	2	10.7	10.9	10.5	350°F	54.0	0.0021	8/0.040
$\bar\epsilon_\beta$(in./in. × 10⁻³)	Tension	2	5.22	5.37	5.07	75°F	52.0	0.00112	8/0.041
$\bar\epsilon_\beta$(in./in. × 10⁻³)	Tension	2	12.4	12.4	12.3	350°F	54.0	0.0021	8/0.040
$\bar\sigma_{PL\beta}$(ksi)	Tension	2	7.29	7.82	6.76	75°F	52.0	0.00112	8/0.041
$\bar\sigma_{PL\beta}$(ksi)	Tension	2	1.30	1.32	1.28	350°F	54.0	0.0021	8/0.040
$\bar\epsilon_{PL\beta}$(in./in. × 10⁻³)	Tension	2	1.84	1.94	1.75	75°F	52.0	0.00112	8/0.041
$\bar\sigma_{PL\beta}$(in./in. × 10⁻³)	Tension	2	0.655	0.690	0.620	350°F	54.0	0.0021	8/0.040
E_β(psi × 10⁶)	Compression	1	4.17	—	—	75°F	54.0	0.0022	8/0.040
E_β(psi × 10⁶)	Compression	1	2.08	—	—	350°F	52.0	0.0032	8/0.041
$\bar\sigma_\beta$(ksi)	Compression	1	49.0	—	—	75°F	54.0	0.0022	8/0.040
$\bar\sigma_\beta$(ksi)	Compression	1	19.8	—	—	350°F	52.0	0.0032	8/0.041
$\bar\epsilon_\beta$(in./in. × 10⁻³)	Compression	1	16.4	—	—	75°F	54.0	0.0022	8/0.040
$\bar\epsilon_\beta$(in./in. × 10⁻³)	Compression	1	15.7	—	—	350°F	52.0	0.0032	8/0.041
$\bar\sigma_{PL\beta}$(ksi)	Compression	1	16.8	—	—	75°F	54.0	0.0022	8/0.040
$\bar\sigma_{PL\beta}$(ksi)	Compression	1	4.0	—	—	350°F	52.0	0.0032	8/0.041
$\bar\epsilon_{PL\beta}$(in./in. × 10⁻³)	Compression	1	4.08	—	—	75°F	54.0	0.0022	8/0.040
$\bar\epsilon_{PL\beta}$(in./in. × 10⁻³)	Compression	1	1.93	—	—	350°F	52.0	0.0032	8/0.041

Note: Cure and postcure process same as for Table 4.4–19.

Data source: General Dynamics

Table 4.4–21
BORON-104 GLASS CLOTH-PR279 EPOXY UNIDIRECTIONAL COMPOSITES

Material Properties

Property	Number of Specimens	Type of test	Av	Max	Min	Standard deviation	Test temp	Fiber volume, %
E_α(psi $\times 10^6$)	1	Tension	32.0	–	–	–	RT	51
E_β(psi $\times 10^6$)	2	Tension	3.2	3.3	3.1	0.14	RT	51
$\bar{\sigma}_\alpha$(ksi)	2	Tension	232.8	233.5	232.2	–	RT	51
$\bar{\sigma}_\beta$(ksi)	2	Tension	9.8	10.2	9.4	0.57	RT	51
$\bar{\sigma}_\alpha$(ksi)	5	Compression	188.1	207.2	178.1	11.3	RT	51
$\nu_{\alpha\beta}$	1	Tension	0.2	–	–	–	RT	51

Data source: North American Rockwell

Table 4.4–22
BP 907-BORON UNIDIRECTIONAL COMPOSITE

Material Properties

Property	Type of test	Number of tests	Av	Max	Min	Width, in.	Temp	Filament volume, %	Load rate, in./min
E_β(psi $\times 10^6$)	Flexure	3	24.4	27.9	22.1	0.498	RT	62.0	0.05
E_α(psi $\times 10^6$)	Compression	6	106.4	171.0	77.2	0.497	RT	79.4	0.05
	Flexure	3	176.9	184.5	168.4	0.498	RT	62.0	0.05
$\bar{\sigma}_\alpha$(ksi)	–	–	–	–	–	–	–	–	–
$\bar{\sigma}_\beta$(ksi)	–	–	–	–	–	–	–	–	–
$\bar{\tau}_{\alpha\beta}$(ksi)	–	–	–	–	–	–	–	–	–
$\bar{\nu}_{\alpha\beta}$	–	–	–	–	–	–	–	–	–
$G_{\alpha\beta}$(psi $\times 10^6$)	–	–	–	–	–	–	–	–	–

Note: Thickness, orientation, number of plies: varies, $0°$, varies.

Curing process:

Time	25 min	30 min	30 min
Temp	175–185°F	280–290°F	330–350°F
Pressure	50 psig	50 psig	50 psig
Vacuum	15 in.	15 in.	15 in.

Data source: Boeing – Vertol Division

Table 4.4–23
HT-424/BORON UNIDIRECTIONAL COMPOSITE

Material Properties

Property	Type of test	Number of tests	Av	Max	Min	Width, in.	Temp	Filament volume, %	Cure process
E_α(psi × 10^6)	Tension	2	35.5	36.6	34.4	1.0	RT	59	II[b]
E_β(psi × 10^6)	–	–	–	–	–	–	–	–	–
$\bar{\sigma}_\alpha$(ksi)	Tension	2	144.8	149.0	140.5	1.0	RT	59	I[a]
	Tension	2	115.0	118.0	112.0	1.0	300°F	59	I
	Tension	2	127.5	130.0	125.0	1.0	400°F	59	I
$\bar{\sigma}_\beta$(ksi)	–	–	–	–	–	–	–	–	–
$\bar{\tau}_{\alpha\beta}$(ksi)	Shear	2	8.90	9.35	8.45	–	RT	55	I
	Shear	2	5.11	6.16	4.05	–	300°F	55	I
	Shear	2	3.26	3.28	3.24	–	400°F	55	I
$\bar{\nu}_{\alpha\beta}$	–	–	–	–	–	–	–	–	–
$G_{\alpha\beta}$(psi × 10^6)	–	–	–	–	–	–	–	–	–

Note: Thickness, orientation, number of plies: varies, 0°, varies.

[a]Curing process I

Time	2 hr	4 hr	4 hr	6 hr
Temp	180°F	200°F	250°F	325°F
Pressure	150 psi	150 psi	150 psi	150 psi

Postcuring process: time, 16 hr; temp, 400°F

[b]Curing process II

Time	2 hr	2 hr	4 hr	6 hr
Temp	180°F	200°F	250°F	225°F
Pressure	150 psi	150 psi	150 psi	150 psi

Postcuring process: time, 16 hr; temp, 400°F

Data source: General Electric

Table 4.4–24
ERLA-2256-GRAPHITE UNIDIRECTIONAL COMPOSITE

Material Properties

Property	Type of test	Number of tests	Av	Max	Min	Width, in.	Temp	Filament volume, %
E_α(psi × 10⁶)	Tension	6	12.8	13.4	11.5	0.25	RT	51.4
	Flexure	6	10.0	10.9	9.5	0.25	RT	51.4
	Compression	2	14.5	14.8	14.1	0.25	RT	52.0
E_β(psi × 10⁶)	–	–	–	–	–	–	–	–
$\bar{\sigma}_\alpha$(ksi)	Flexure	6	93.0	97.5	90.3	0.25	RT	51.4
	Compression	2	67.2	73.4	61.0	0.25	RT	52.0
$\bar{\sigma}_\beta$(ksi)	–	–	–	–	–	–	–	–
$\bar{\tau}_{\alpha\beta}$(ksi)	–	–	–	–	–	–	–	–
$\nu_{\alpha\beta}$	–	–	–	–	–	–	–	–
$G_{\alpha\beta}$(psi × 10⁶)	–	–	–	–	–	–	–	–

Note: Thickness, orientation, number of plies: ?, 0°, ?

Data source: Air Force Materials Laboratory

Table 4.4–25
ERLA-2256-GRAPHITE (THORNEL 25) UNIDIRECTIONAL COMPOSITE

Material Properties

Property	Type of test	Number of tests	Av	Width, in.	Temp	Filament volume, %
E_α(psi × 10⁶)	Tensile	1	11.07	0.250	RT	50
	Compression	1	10.4	0.375	RT	50
E_β(psi × 10⁶)	Tensile	1	1.05	0.250	RT	50
	Compression	1	1.03	0.375	RT	50
$\bar{\sigma}_\alpha$(ksi)	Tensile	1	92.0	0.250	RT	50
	Compression	1	65.7	0.375	RT	50
$\bar{\sigma}_\beta$(ksi)	Tensile	–	2.0[4]	–	RT	50
	Compression	1	17.7	0.375	RT	50
$\bar{\tau}_{\alpha\beta}$(ksi)	Edgewise Shear	–	4.0[4]	–	RT	50
$\nu_{\alpha\beta}$	Tensile	1	0.29	0.250	RT	50
$G_{\alpha\beta}$(psi × 10⁶)	Tensile	1	0.604	0.250	RT	50
$\bar{\tau}_{IS}$(ksi)	Interlaminar shear	4	5.49	0.12–0.4	RT	50

Note: Thickness, orientation, number of plies: 0.120 in., 0°, 90°.

Curing process

Time	1 hr	1 hr	1 hr	1 hr
Temp	167°F	194°F	230°F	320°F
Pressure		90 psi or 180 psi		

Postcuring process: time, 2 to 4 hr; temp, 320°F; pressure, C-clamps.

Data source: Union Carbide Corporation, Carbon Products Division

Table 4.4–26
ERLA-2256-GRAPHITE (THORNEL 40) UNIDIRECTIONAL COMPOSITE

Material Properties

Property	Type of test	Number of tests	Av	Temp	Filament volume
E_α(psi $\times 10^6$)	Tensile		22.0	RT	No data
	Compression		23.6	RT	No data
	Flexure		20.1	RT	No data
E_β(psi $\times 10^6$)	–		–	–	
$\bar{\sigma}_\alpha$(ksi)	Tensile		112.5	RT	No data
	Compression		94.0	RT	No data
	Flexure		84.3	RT	No data
$\bar{\sigma}_\beta$(ksi)	–		–	–	
$\bar{\tau}_{\alpha\beta}$(ksi)	–		–	–	
$\nu_{\alpha\beta}$	–		–	–	
$G_{\alpha\beta}$(psi $\times 10^6$)	–		–	–	
$\bar{\tau}_{IS}$(ksi)	Interlaminar shear		3.4	RT	No data

Note: Thickness, orientation, number of plies: ?, 0°, ?

Data source: Union Carbide Corporation, Carbon Products Division

Table 4.4–27
ERLA-2256-GRAPHITE (THORNEL 50) UNIDIRECTIONAL COMPOSITE

Material Properties

Property	Type of test	Number of tests	Av	Temp	Filament volume	Test specifications
E_α(psi $\times 10^6$)	NOL Ring (tension)		22.0	No data	No data	ASTM
	NOL Ring (compression)		23.6	No data	No data	ASTM
	NOL Ring (flexure)		20.1	No data	No data	ASTM
$\bar{\sigma}_\alpha$(ksi)	NOL Ring (tension)		112.5	No data	No data	ASTM
	NOL Ring (compression)		94.0	No data	No data	ASTM
	NOL Ring (flexure)		84.3	No data	No data	ASTM
$\bar{\tau}_{IS}$(ksi)	NOL Ring (horizontal shear)		3.4	No data	No data	ASTM

Note: Thickness, orientation, number of plies: ?, 0°, ?

Data source: Union Carbide Corporation, Carbon Products Division

Table 4.4–28
EPON 828/1031-BORON ANGLE-PLY COMPOSITE

Material Properties

Property	Type of test	Number of tests	Av	Max	Min	Width, in.	Temp	Filament volume, %
E_x(psi × 10^6)	Tension	7	2.5	2.8	2.3	0.45	RT	60
	Flexure	7	3.4	3.8	2.8	0.45	RT	60
E_y(psi × 10^6)	–	–	–	–	–	–	–	–
$\bar{\sigma}_x$(ksi)	Tension	4	32.7	40.4	25.5	0.47	RT	60
	Compression	2	67.8	75.6	60.0	0.47	RT	60
$\bar{\sigma}_y$(ksi)	–	–	–	–	–	–	–	–
τ_{xy}(ksi)	Shear	3	4.81	5.10	4.29	0.242	RT	49.6
ν_{xy}	Tension	7	0.91	1.10	0.72	0.45	RT	60
	Flexure	7	0.71	0.84	0.60	0.45	RT	60
G_{xy}(psi × 10^6)	Shear	2	11.3	13.3	9.2	4.00	RT	60

Note: Thickness, orientation, number of plies: (a) ~0.044 in., ±45°, 8 or 9; (b) ~0.101 in., ±45°, 17.

Curing process: ¾ hr at 250°F at 50 psig.
1½ hr at 330 to 350°F at 50 psig.

Data source: General Electric

Table 4.4–29
BP 907-BORON ANGLE-PLY COMPOSITE

Material Properties

Property	Type of test	Number of tests	Av	Max	Min	Width in.	Temp	Filament volume, %	Load rate, in./min
E_x(psi × 10^6)	Tension	9	2.822	3.282	2.641	0.532	RT	59.4	0.05
E_y(psi × 10^6)	–	–	–	–	–	–	–	–	–
$\bar{\sigma}_x$(ksi)	Tension	9	22.72	30.12	16.37	0.532	RT	59.4	0.05
	Compression	7	27.98	31.30	24.21	0.494	RT	60.4	0.05
$\bar{\sigma}_y$(ksi)	–	–	–	–	–	–	–	–	–
$\bar{\tau}_{xy}$(ksi)	Shear	5	6.07	7.14	5.58	0.245	RT	62.2	0.05
ν_{xy}	–	–	–	–	–	–	–	–	–
G_{xy}(psi × 10^6)	–	–	–	–	–	–	–	–	–

Note: Thickness, orientation, number of plies: varies, ±45°, varies.

Curing process

Time	25 min	30 min	30 min
Temp	175–185°F	280–290°F	330–350°F
Pressure	50 psig	50 psig	50 psig
Vacuum	15 in.	15 in.	15 in.

Data source: Boeing – Vertol Division

Table 4.4–30
N5505 EPOXY-BORON ANGLE-PLY COMPOSITE

Material Properties

Property	Type of test	Number of specimens	Av	Max	Min	Test temp	Fiber volume, %	Load rate, in./min	Number of plies/orientation/thickness, in.
E (psi × 10⁶)	Tension	1	21.2	–	–	75	51.0	0.0020	8/±20°/0.041
	Tension	1	10.7	–	–	RT	51.0	0.0018	8/±30°/0.041
	Tension	5	2.80	2.89	2.68	RT	54.2	0.02	8/±45°/0.0415
	Tension	–	11.9	–	–	–	51	0.001	12/0±60°/0.062
	Tension	1	11.7	–	–	75	51.0	0.0016	8/0±45±90°/0.041
	Tension	5	16.40	17.30	14.10	RT	54.2	0.02	8/0−90°/0.041
E_α → Tension	Tension	3	14.63	14.80	14.35	RT	51.0	0.05	10/0−90±45°/0.051
	Tension	2	5.82	6.00	5.64	75	51.0	0.0012	10/90±45°/0.052
E_β → Tension	Tension	2	9.56	9.97	9.16	RT	51.4	0.05	10/0−90±45°/0.0515
	Compression	1	24.6	–	–	75	51.0	0.004	8/±20°/0.041
	Compression	1	11.1	–	–	75	51.0	0.0018	8/±30°/0.041
	Compression	1	6.67	–	–	75	51.0	0.0011	10/90±45°/0.052
$\bar{\sigma}$ (ksi)	Tension	1	135.0	–	–	75	51.0	0.002	8/±20°/0.041
	Tension	1	101.0	–	–	RT	51.0	0.0018	8/±30°/0.041
	Tension	5	22.4	22.8	21.9	RT	54.2	0.02	8/±45°/0.0415
	Tension	–	71.0	–	–	–	51	0.001	12/0±60°/0.062
	Tension	1	74.9	–	–	75	51.0	0.0016	8/0±45±90°/0.041
	Tension	5	88.2	92.7	85.3	RT	54.2	0.02	8/0−90°/0.041
	Tension	3	94.24	95.13	93.34	RT	51.0	0.05	10/0−90±45°/0.051
	Tension	2	49.52	51.28	47.76	RT	51.4	0.05	10/0−90±45°/0.0515
	Tension	2	20.5	21.4	19.6	75	51.0	0.0012	10/90±45°/0.052
	Compression	1	120.0	–	–	75	51.0	0.0040	8/±20°/0.041
	Compression	1	45.0	–	–	75	51.0	0.0018	8/±30°/0.041
	Compression	1	68.5	–	–	75	51.0	0.0011	10/90±45°/0.052
$\bar{\tau}$ (ksi)	Shear	5	4.30	4.84	3.92	RT	54.2	0.05	8/±45°/0.0415
	Short beam	5	7.2	7.7	6.6	RT	54.2	0.05	8/0−90°0.041
$\bar{\epsilon}$ (in./in. × 10⁻³)	Tension	1	7.02	–	–	75	51.0	0.0020	8/±20°/0.041
	Tension	1	14.10	–	–	RT	51.0	0.0018	8/±30°/0.041
	Tension	–	6.52	–	–	–	51	0.001	12/0±60°/0.062
	Tension	1	7.35	–	–	75	51.0	0.0016	8/0±45°/0.052
	Tension	2	3.93	4.00	3.85	75	51.0	0.0012	10/90±45°/0.052

Table 4.4–30 (continued)
N5505 EXPOXY-BORON ANGLE-PLY COMPOSITE

Material Properties (continued)

Property: $\overline{\epsilon}$ (in./in. $\times 10^{-3}$) (continued)

Type of test	Number of specimens	Av	Max	Min	Test temp	Fiber volume, %	Load rate, in./min	Number of plies/orientation/ thickness, in.
Compression	1	5.57	—	—	75	51.0	0.0040	8/±20°/0.041
Compression	1	4.76	—	—	75	51.0	0.0018	8/±30°/0.41
Compression	1	12.4	—	—	75	51.0	0.0011	10/90±45°/0.052

Data source: General Dynamics

Table 4.4-31
N5505 EPOXY-BORON ANGLE-PLY COMPOSITES, 0 ± 45°

Material Properties, 0 ± 45°[a]

Property	Type of test	Number of specimens	Av	Max	Min	Test temp, °F	Load rate, in./min
E (psi × 10⁶)	Tension	2	20.5	20.6	20.4	75	0.0018
E (psi × 10⁶)	Tension	2	19.6	20.0	19.2	350	0.0013
$\bar{\sigma}$ (ksi)	Tension	2	127.0	136.0	117.0	75	0.0018
$\bar{\sigma}$ (ksi)	Tension	2	112.0	115.0	110.0	350	0.0013
$\bar{\epsilon}$ (in./in. × 10⁻³)	Tension	2	6.64	7.21	16.06	75	0.0018
$\bar{\epsilon}$ (in./in. × 10⁻³)	Tension	2	6.37	6.55	6.18	350	0.0013
E (psi × 10⁶)	Compression	1	22.4	–	–	75	0.0013
E (psi × 10⁶)	Compression	1	19.2	–	–	350	0.0013
$\bar{\sigma}$ (ksi)	Compression	1	260	–	–	75	0.0013
$\bar{\sigma}$ (ksi)	Compression	1	114	–	–	350	0.0013
$\bar{\epsilon}$ (in./in. × 10⁻³)	Compression	1	12.0	–	–	75	0.0013
$\bar{\epsilon}$ (in./in. × 10⁻³)	Compression	1	5.94	–	–	350	0.0013

[a]51 vol % fiber.

Data source: General Dynamics

Table 4.4−32
N5505 EPOXY-BORON ANGLE-PLY COMPOSITES − PERMANENT SET TEST

Material Properties[a]

Property	Type of test	Number of tests	Number of specimens	Value	Number of plies/orientation/thickness
E_α (psi $\times 10^6$)	Tension	1[b]	1	15.38	10/0−90 ±45°/0.051
E_β (psi $\times 10^6$)	Tension	2[c]	1	9.41	10/0−90 ±45°/0.051
$\bar{\sigma}_\alpha$ (ksi)	Tension	1	1	78.47[d]	10/0−90 ±45°/0.051
$\bar{\sigma}_\beta$ (ksi)	Tension	2	1	37.70[e]	10/0−90 ±45°/0.051
$\nu_{\alpha\beta}$	Tension	1	1	0.314	10/0−90 ±45°/0.051
$\nu_{\beta\alpha}$	Tension	2	1	0.206	10/0−90 ±45°/0.051

[a]51.4 vol % fiber.
[b]Specimen cycled 16 times from 50-lb preload in 200-lb increments to 3000 lb (maximum).
[c]Specimen cycled 16 times from 50-lb preload in 200-lb increments to 1500 lb (maximum).
[d]Maximum stress at 3000-lb load.
[e]Maximum stress at 1500-lb load.

Data source: Northrup-Norair

Table 4.4−33
N5505 EPOXY-BORON/BETA GLASS OVERWRAP ANGLE-PLY COMPOSITE, ± 45°

Material Properties[a]

Property	Type of test	Number of tests	Av	Max	Min	Standard deviation	Test temp	Load rate, in./min
E (psi $\times 10^6$)	Tension	2	2.8	3.2	2.4	5.6	RT	0.025
$\bar{\sigma}$ (ksi)	Tension	3	18.1	18.8	17.5	0.63	RT	0.025
$\bar{\sigma}$ (ksi)	Compression	4	23.4	24.3	22.1	1.0	RT	0.025
ν	Tension	1	0.45	−	−	−	RT	0.025

[a]45 vol % fiber.

Data source: North American Rockwell

Table 4.4–34
SP272-BORON ANGLE-PLY COMPOSITES, 0–90°

Material Properties[a]

Property	Type of test	Number of specimens	Av	Max	Min	Test temp, °F	Load rate, in./min
E (psi $\times 10^6$)	Tension	2	17.2	17.2	17.2	75	0.001
E (psi $\times 10^6$)	Tension	2	15.6	16.0	15.1	350	0.001
$\bar{\sigma}$ (ksi)	Tension	2	108	111	105	75	0.001
$\bar{\sigma}$ (ksi)	Tension	2	89.8	90.6	89.0	350	0.001
$\bar{\epsilon}$ (in./in. $\times 10^{-3}$)	Tension	2	7.03	7.20	6.86	75	0.001
$\bar{\epsilon}$ (in./in. $\times 10^{-3}$)	Tension	2	6.01	6.10	5.92	350	0.001
$\bar{\sigma}_{PL}$ (ksi)	Tension	2	33.6	35.1	31.5	75	0.001
$\bar{\sigma}_{PL}$ (ksi)	Tension	2	27.6	29.6	25.7	350	0.001
$\bar{\epsilon}_{PL}$ (in./in. $\times 10^{-3}$)	Tension	2	1.96	2.08	1.83	75	0.001
$\bar{\epsilon}_{PL}$ (in./in. $\times 10^{-3}$)	Tension	2	1.68	1.74	1.62	350	0.001
E (psi $\times 10^6$)	Compression	1	17.8	–	–	75	0.001
E (psi $\times 10^6$)	Compression	1	15.1	–	–	350	0.001
$\bar{\sigma}$ (ksi)	Compression	1	225	–	–	75	0.001
$\bar{\sigma}$ (ksi)	Compression	1	104	–	–	350	0.001
$\bar{\epsilon}$ (in./in. $\times 10^{-3}$)	Compression	1	13.2	–	–	75	0.001
$\bar{\epsilon}$ (in./in. $\times 10^{-3}$)	Compression	1	6.9	–	–	350	0.001
$\bar{\sigma}_{PL}$ (ksi)	Compression	1	147	–	–	75	0.001
$\bar{\sigma}_{PL}$ (ksi)	Compression	1	–	–	–	350	0.001
$\bar{\epsilon}_{PL}$ (in./in. $\times 10^{-3}$)	Compression	1	8.25	–	–	75	0.001
$\bar{\epsilon}_{PL}$ (in./in. $\times 10^{-3}$)	Compression	1	–	–	–	350	0.001
G (psi $\times 10^6$)	Shear	1	1.08	–	–	75	0.0068
$\bar{\tau}$ (ksi)	Shear	1	29.0	–	–	75	0.0068
γ (in./in. $\times 10^{-3}$)	Shear	1	116.0	–	–	75	0.0068

[a]52 vol % fiber.

Data source: General Dynamics

Table 4.4–35

SP272-BORON ANGLE-PLY COMPOSITES, ± 45°, TEMPERATURE TESTS

Material Properties

Property	Type of test	Number of specimens	Av	Max	Min	Test temp, °F	Fiber volume, %	Load rate, in./min	Number of plies/thickness, in.
E (psi × 10^6)	Tension	1	4.92	–	–	-65	52.0	0.0044	8/0.041
E (psi × 10^6)	Tension	2	3.85	3.89	3.81	75	54.0	0.0044	8/0.040
E (psi × 10^6)	Tension	1	1.54	–	–	350	54.0	0.0113	8/0.040
$\bar{\sigma}$ (ksi)	Tension	1	38.5	–	–	-65	52.0	0.0044	8/0.041
$\bar{\sigma}$ (ksi)	Tension	2	35.5	36.2	34.8	75	54.0	0.0044	8/0.040
$\bar{\sigma}$ (ksi)	Tension	1	20.7	–	–	350	54.0	0.0113	8/0.040
$\bar{\epsilon}$ (in./in. × 10^{-3})	Tension	1	14.5	–	–	-65	52.0	0.0044	8/0.041
$\bar{\epsilon}$ (in./in. × 10^{-3})	Tension	2	22.2	23.6	20.7	75	54.0	0.0044	8/0.040
$\bar{\epsilon}$ (in./in. × 10^{-3})	Tension	1	56.7	–	–	350	54.0	0.0113	8/0.040
E (psi × 10^6)	Compression	2	3.98	4.02	3.93	75	52.0	0.0054	8/0.041
E (psi × 10^6)	Compression	1	1.6	–	–	350	52.0	0.012	8/0.041
$\bar{\sigma}$ (ksi)	Compression	2	43.0	43.2	42.7	75	52.0	0.0054	8/0.041
$\bar{\sigma}$ (ksi)	Compression	1	17.3	–	–	350	52.0	0.012	8/0.041
$\bar{\epsilon}$ (in./in. × 10^{-3})	Compression	2	33.8	34.0	33.6	75	52.0	0.0054	8/0.041
$\bar{\epsilon}$ (in./in. × 10^{-3})	Compression	1	34.5	–	–	350	52.0	0.012	8/0.041

Data source: General Dynamics

Table 4.4–36
HT-424/BORON MATERIAL SYSTEM – ANGLE-PLY COMPOSITE

Material Properties

Property	Type of test	Number of tests	Av	Max	Min	Width in.	Temp	Filament volume, %
E_x (psi × 10⁶)	Tension	2	18.9	18.9	18.8	1.0	RT	55
	Tension	2	13.4	14.5	12.2	1.0	300°F	55
	Flexure	2	19.0	19.4	18.6	1.0	RT	55
	Flexure	2	18.2	19.5	16.9	1.0	300°F	55
	Flexure	2	19.9	20.4	19.4	1.0	400°F	55
E_y (psi × 10⁶)	–	–	–	–	–	–	–	–
$\bar{\sigma}_x$ (ksi)	Tension	2	75.8	76.0	75.5	1.0	RT	55
	Tension	2	78.9	80.5	77.2	1.0	300°F	55
	Tension	2	74.4	79.8	69.0	1.0	400°F	55
	Flexure	2	142.3	146.0	138.5	1.0	RT	55
	Flexure	2	137.2	138.0	136.5	1.0	300°F	55
	Flexure	2	143.5	148.0	139.0	1.0	400°F	55
$\bar{\sigma}_y$ (ksi)	–	–	–	–	–	–	–	–
$\bar{\tau}_{xy}$ (ksi)	–	–	–	–	–	–	–	–
ν_{xy}	–	–	–	–	–	–	–	–
G_{xy} (psi × 10⁶)	–	–	–	–	–	–	–	–

Note: Thickness, orientation, number of plies: ~0.045 in., 0 ± 90°, 9.

Data source: General Electric

Table 4.4–37
ERLA-2256-GRAPHITE – COMPOSITE

Material Properties

Property	Type of test	Number of tests	Av	Max	Min	Temp	Filament volume, %
E_x (psi × 10⁶)	Tension	2	3.22	3.39	3.05	RT	41.1
	Flexure	3	7.52	8.05	6.99	RT	49.5
E_y (psi × 10⁶)	–	–	–	–	–	–	–
$\bar{\sigma}_x$ (ksi)	Tension	2	30.1	31.3	28.8	RT	41.1
	Flexure	3	74.2	76.4	71.6	RT	45.9
$\bar{\sigma}_y$ (ksi)	–	–	–	–	–	–	–
$\bar{\tau}_{xy}$ (ksi)	–	–	–	–	–	–	–
ν_{xy}	–	–	–	–	–	–	–
G_{xy} (psi × 10⁶)	Shear	1	5.30	–	–	RT	41.6
	Shear	–	5.79	–	–	RT	48.9
	Shear	–	7.58	–	–	RT	61.6

Note: Thickness, orientation, number of plies: ?, 0 to 90°, 8.

Data source: Air Force Materials Laboratory

Index

T

W

Z